D0467704

'As MPLS has grown in popularity, most books on the subject have concentrated on explaining the basics, or illustrating how the core function can be successfully deployed. Minei and Lucek, however, have taken the subject on to the next level. They explain the new and significant developments in MPLS showing how the technology has come of age and how it will be used to build advanced services in complex, interconnected networks. This book provides an important tutorial on the recent advances in MPLS, and brings together many of the threads from the IETF to provide a comprehensive overview of the next generation of MPLS networking.'

Adrian Farrel, Old Dog Consulting Ltd., IETF CCAMP and PCE working groups co-chair

'While MPLS is in itself simple, its apparent complexity lies in the proliferation of applications, which shows no signs of ceasing. To make things worse, catching up involves reading a large number of documents written by various authors at various times in various styles. Here at last is a single, all-encompassing resource where the myriad applications sharpen into a comprehensible text that first explains the whys and whats of each application before going on to the technical detail of the hows.'

Kireeti Kompella, Juniper Fellow, Juniper Networks.

'*MPLS-Enabled Applications* thoroughly covers the MPLS base technology and applications on MPLS-enabled IP networks. It guides you to a comprehensive understanding of standards, problems, and solutions in networking with MPLS. Before it had been necessary to go through material from many different sources, here we have everything in one place. All the MPLS protocols are covered, as are the applications of these protocols. This should be the textbook for MPLS courses, both for training of experienced networking professionals and for universities.'

Loa Andersson, Acreo AB, IAB-member and IETF MPLS working group co-chair

'This is the MPLS text that the industry has been waiting for. On one hand, the text presents MPLS technology clearly enough that the reader can absorb its content in a few easy sittings. On the other hand, the text provides a sufficiently in depth treatment that even an MPLS expert can learn from it. The authors offer a clear and complete description of MPLS, its inner workings and its applications, in a manner that could only be achieved by persons who have been significant contributors to the MPLS development effort. Every network operator who has deployed or is considering the deployment of MPLS technology should read this book. It is appropriate reading for everyone from the CTO to the tier 1 NOC engineer.'

Ron Bonica, Engineer, Juniper Networks, IETF L3 VPN working group co-chair

'My first impression after reading a few selected chapters of *MPLS-Enabled Applications* has been very good. It attempts to describe a number of modern techniques to address today's customer requirements. It should be a good reference for network operators who face significant challenges to keep up to date with latest developments of new and emerging networking services.'

Robert Raszuk, Technical Leader, Routing Protocols Deployment and Architecture, Cisco Systems

'*MPLS-Enabled Applications* provides an excellent review of the fundamentals and key applications of the MPLS suite of protocols. Its balanced mix of both the technical and business motivations for using these applications make it a must-read for anyone involved in enterprise or service-provider networks.'

Dave Cooper, Sr. Manager IP Engineering, Global Crossing, Ltd.

'This is a highly recommended book for those wanting to update themselves with the latest MPLS developments, or those just wanting to learn this technology thoroughly. In addition to the impressive technology coverage and depth, the book is also a delightful reading!'

Lei Wang, IP Network Architect, Telenor

'*MPLS-Enabled Applications* is an excellent read for network engineers involved in the design of MPLS networks and services. It can serve as an introduction to MPLS networking or as a reference book for the advanced engineer. It discusses practical issues that must be considered in the design of MPLS networks and services, including MPLS-TE, MPLS-IPVPNs and MPLS L2 VPNs. It also discusses current topics that are still evolving in the industry such as inter-AS/area MPLS-TE, point-to-multipoint LSPs and IPVPN multicast, providing a good overview of the issues being addressed and the current industry direction.'

Nabil N. Bitar, Principal member of Technical Staff and lead network architect, Verizon

'*MPLS Enabled Applications* presents the current state of the art in the specification, development, and application of MPLS and its related technologies. I believe, the readers will find the book to be a very valuable resource.'

Bijan Jabbari, PhD, Founder of Isocore, and Professor of Electrical Engineering, George Mason University

'Too few people have the scarce skill for clearly explaining things. Fewer still have a clear understanding of the MPLS protocol and architectures. Minei and Lucek seem to have them both.'

Dr. Eyal Felstaine, senior staff member, Ben Gurion University, Israel

'This book provides an excellent overview of MPLS mechanisms and applications. It allows understanding what we can do today and what we will be able to do tomorrow with MPLS, as it covers not only stable and deployed technologies such as MPLS Fast Reroute, L3 VPNs, L2 VPN, but also new technologies under development within the IETF, such as IP/LDP Fast Reroute, inter-domain Traffic Engineering and Point-To-Multipoint MPLS. Hence this book will be highly useful for network designers, operators, students as well as anyone interested with the MPLS architecture.'

Jean-Louis Le Roux, Senior MPLS architect, France Telecom

MPLS-Enabled Applications

MPLS-Enabled Applications

Emerging Developments and New Technologies

Ina Minei
Juniper Networks

Julian Lucek
Juniper Networks

John Wiley & Sons, Ltd

Telephone (+44) 1243 779777

Email (for orders and customer service enquiries): cs-books@wiley.co.uk
Visit our Home Page on www.wiley.com

Other Wiley Editorial Offices

John Wiley & Sons Inc., 111 River Street, Hoboken, NJ 07030, USA

Jossey-Bass, 989 Market Street, San Francisco, CA 94103-1741, USA

Wiley-VCH Verlag GmbH, Boschstr. 12, D-69469 Weinheim, Germany

John Wiley & Sons Australia Ltd, 42 McDougall Street, Milton, Queensland 4064, Australia

John Wiley & Sons (Asia) Pte Ltd, 2 Clementi Loop #02-01, Jin Xing Distripark, Singapore 129809

John Wiley & Sons Canada Ltd, 22 Worcester Road, Etobicoke, Ontario, Canada M9W 1L1

Library of Congress Cataloging in Publication Data

Minei, Ina.
MPLS-enabled applications : emerging developments and new technologies / Ina Minei, Julian Lucek.
p. cm.
Includes bibliographical references.
ISBN 0-470-01453-9 (alk. paper)
1. MPLS standard. 2. Extranets (Computer networks) I. Lucek, Julian. II. Title.
TK5105.573.M56 2005
621.382'16—dc22

2005022611

British Library Cataloguing in Publication Data

A catalogue record for this book is available from the British Library

ISBN 0-470-01453-9

Typeset in 11/13pt Palatino by Integra Software Services Pvt. Ltd, Pondicherry, India
Printed and bound in Great Britain by Antony Rowe Ltd, Chippenham, Wiltshire
This book is printed on acid-free paper responsibly manufactured from sustainable forestry in which at least two trees are planted for each one used for paper production.

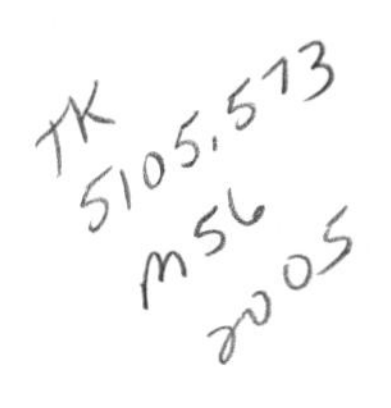

Contents

MPLS-Enabled Applications: Emerging Developments and New Technologies Ina Minei and Julian Lucek

About the Authors

Ina Minei has been a network protocols engineer at Juniper Networks for the past five years working on next-generation network technologies for core routers, where she focuses on MPLS protocols and applications. She previously worked at Cisco for two years in various software development projects for routers and switches. Ms Minei is an active participant in industry fora and conferences and has co-filed several patents in the area of IP and MPLS. She holds a Master's degree in computer science from the Technion, Israel.

Julian Lucek joined Juniper Networks in 1999 and is currently the Technical Leader for Junos for the Europe, Middle East and Africa region, where he has been working with many service providers on the design and evolution of their networks. He previously worked at BT for several years, at first in the Photonics Research Department and later in the data transport and routing area. During this time he gained a PhD in ultrahigh-speed data transmission and processing. He has a Master's degree in Physics from Cambridge University and holds Juniper Networks Certified Internet Expert (JNCIE) certification number 21.

Foreword

Yakov Rekhter, Juniper Fellow, Juniper Networks, June 2005

Multi-Protocol Label Switching (MPLS) began in the mid-1990s with just two modest design objectives. The first was a better integration of ATM with IP by providing a single IP-based control plane that could span both ATM switches and IP routers. The second objective was to augment the IP control plane with some additional functionality, namely traffic engineering using constraint-based routing that was already present in the ATM control plane.

Not long after it started, MPLS usage was extended to applications such as Circuit Cross Connect (CCC), ATM and Frame Relay service over an IP/MPLS infrastructure (draft-martini), and BGP/MPLS VPNs (2547 VPNs). The original constraint-based routing functionality evolved beyond traffic engineering to applications such as fast reroute and Differentiated Services Traffic Engineering (DiffServ-TE).

The idea of a single control plane for both ATM switches and IP routers evolved into Generalized Multi-Protocol Label Switching (GMPLS), which provides a single control plane that could span not only routers and ATM switches but SONET/SDH and optical cross connects as well.

More recently, MPLS usage has been extended to provide Virtual Private LAN Services (VPLS). There are two recent MPLS

MPLS-Enabled Applications: Emerging Developments and New Technologies Ina Minei and Julian Lucek

developments deserving of mention here, the first being the work on point-to-multipoint Label Switched Paths (P2MP LSPs), which is expected to be used for applications such as video distribution, IP multicast with BGP/MPLS VPNs and IP multicast with VPLS. The second development is the work on extending constraint-based routing, including applications such as traffic engineering, fast reroute and Differentiated Services Traffic Engineering to multiple OSPF/IS-IS areas and even multiple autonomous systems.

It is important to keep in mind that in all of the applications mentioned above MPLS is just one of the components of such applications, albeit a critical one. If we look back at the time when MPLS was created, and compare its design objectives with what MPLS is used for today, we notice several things. First of all, most of the applications of MPLS that we have today were not conceived of during the original design of MPLS. Furthermore, while the original design goal of a better integration of ATM and IP routers is still with us, the way MPLS today supports this is completely different from the way it was originally conceived. Instead of having a single control plane that could span both ATM switches and routers, 'better integration' today has a dual meaning of being able to offer the ATM service over an IP/MPLS infrastructure that has no ATM switches at all, as well as the ability to interconnect ATM switches over such an infrastructure. While originally MPLS was conceived as a technology solely for the Service Providers, we see today how MPLS is gradually penetrating the enterprise environment. Additionally, over time the whole MPLS concept evolved from Multi-*Protocol* Label Switching to Multi-*Purpose* Label Switching.

A new technology quite often generates opposition, and MPLS was by no means an exception. You may all remember how MPLS was branded by its opponents in negative terms as 'bad', 'evil', 'a social disease' or 'a nightmare of unprecedented proportions'. To put this in a proper perspective, we need to keep in mind that technologies exist not for their own sake but for the purpose of solving business problems. Therefore, talking about 'good' technologies versus 'bad/evil' technologies has little practical relevance; yet what is of great relevance is how well a particular technology meets business needs.

One might wonder how they could judge how well a particular technology, like MPLS, meets business needs. To answer this question I would like to remind you of the expression that 'the proof of the pudding is in the eating' (and not in the debate about the pudding). That being said, the ultimate judge of how well a particular technology meets business needs is the marketplace. It is the judgment of the marketplace that determines whether a particular technology deserves to live or to die; and with respect to MPLS the market made its judgment loud and clear – MPLS is here to stay.

Preface

Our aim in writing this book was to describe the latest developments in MPLS. The field is moving so fast that some new applications of MPLS have already been deployed in production networks, yet are not described anywhere in book form. In many cases, the only available resources are the IETF drafts which list the extensions that vendors must implement in order to produce interoperable implementations. These documents often assume familiarity with the problem at hand and omit the details on why a particular solution has been chosen or what are the pros and cons of using it. This book attempts to fill this gap and provide the reader with an understanding of both the problem and why the solution looks the way it does.

Therefore, when we describe the protocol mechanisms underpinning an MPLS application, the emphasis is on giving an overview of the protocol machinery without delving into the bits and bytes of each protocol message. This allows us to convey the concepts without making it difficult to see the wood for the trees. Also, some of the mechanisms that we write about are currently being defined, so some of the details of the protocol messages may change, but the concepts are less likely to. References at the end of each chapter point to the documents describing the message formats and processing rules.

Although we both happen to work for the same router vendor, the book is not vendor-specific. Occasionally we point out some vendor-specific quirks, if they are relevant to the discussion at hand or aid in understanding any particular topic. Many of the topics discussed are still under discussion and debate in the IETF, and naturally sometimes our personal views on one topic or another may be stated more strongly than the opposing view.

WHO SHOULD READ THIS BOOK?

The intended audience of this book includes employees of service providers and network equipment vendors, customers of service providers who are interested in the mechanisms underpinning the services that they buy, network professionals who want to keep up to date with the latest advances in MPLS and students of network technology. We assume that the reader has some degree of familiarity with network technology and routing protocols, in particular BGP and the link-state IGPs, but these are not a requirement to benefit from the book. Although the main aim of the book is to cover the cutting-edge developments of MPLS, the Foundation chapter allows the reader unfamiliar with MPLS to get up to speed in order to benefit from the remainder of the book. Even when discussing basic topics such as TE or fast reroute, an attempt is made to look at the more interesting and advanced aspects of the technology.

HOW THIS BOOK IS ORGANIZED

The book is divided into two parts, each containing several chapters. Part One describes the MPLS infrastructure tools used as the foundation to build services and Part Two of the book covers the MPLS-based services themselves.

The structure of Part One

Chapter 1, the Foundations chapter, reviews the control plane and forwarding plane mechanisms associated with MPLS. In that chapter, we give an overview of the LDP and RSVP signaling protocols and compare the two.

Chapter 2 discusses MPLS Traffic Engineering, which gives service providers control over the path taken by traffic through their network and the ability to give bandwidth guarantees. In this context, we look at the impact of TE on network scalability, as well as at solutions for TE in LDP networks.

Chapter 3 explores the topic of Protection and Restoration in MPLS networks, essential to allowing MPLS networks to carry mission-critical traffic. We cover link and node protection, their respective scaling properties, the cost of bandwidth protection, as well as more advanced topics such as fate sharing and the new developments for providing fast restoration in IP and LDP networks.

Chapter 4 presents Differentiated Services (DiffServ) Aware Traffic Engineering, which allows traffic engineering to be applied with per-class granularity, bringing QoS to the network.

Chapter 5 introduces Interdomain Traffic Engineering. This is likely to be of increasing importance in the future as MPLS-based services extend across multiple IGP areas and AS boundaries. Both the signaling and computation aspects are discussed, and path-computation elements are also reviewed.

Chapter 6 is devoted to Point-to-Multipoint MPLS functionality, both RSVP and LDP. This is currently of great interest as it allows MPLS to be used in broadcast TV applications and because it is an essential part of the new L3VPN multicast solutions that are currently under discussion within the IETF.

The structure of Part Two

Chapters 7, 8 and 9 are devoted to Layer 3 VPNs, which is to date the most widespread application of MPLS. Chapter 7 provides a tutorial on L3VPN and explains the basic concepts, while the next chapters discuss more advanced topics such as route target filtering, scalability analysis and hierarchical VPNs.

Chapter 10 describes the rapidly growing area of Layer 2 transport over MPLS, including pseudowires and Layer 2 VPNs. These allow service providers to migrate ATM and Frame Relay services to an IP/MPLS network and to offer Ethernet-based alternatives to those services.

Chapter 11 describes the Virtual Private LAN Service (VPLS). This allows a service provider to offer a very simple-to-use service to enterprise customers, in which the customer's sites appear to be attached to the same LAN.

Chapter 12 covers some aspects of the management and troubleshooting of MPLS networks. The subject of management of MPLS networks could fill an entire book by itself and a single chapter does not do it justice. However, we attempt to show some of the challenges (such as ICMP tunneling) and some of the new tools, such as LSPing.

The final chapter takes a look at the achievements of MPLS to date and how MPLS may in future extend beyond the service provider core to access networks and to enterprise networks.

REFERENCES

At the end of each chapter, there is a list of references. In the body of the text, these references appear in brackets, like this [REF1]. Many of the references are IETF documents. As these documents progress in the IETF process, their revision number and document name may change. Therefore, when looking up a reference online, search by the author and title rather than by the document name.

In some chapters, we have included a section with further reading. These are documents that the authors thought would be useful for those wanting to broaden their knowledge on a particular topic.

Ina Minei, Sunnyvale, California
Julian Lucek, Ipswick, UK
June 2005

Acknowledgements

This book would not have existed if it were not for the following three people: Yakov Rekhter, Aviva Garrett and Patrick Ames, and to them we extend our most heartfelt thanks.

Yakov Rekhter encouraged us to pursue this project and provided valuable insight, starting with the book proposal and ending with a detailed technical review of numerous chapters. Most importantly, his faith in our ability to do this work was one of the main factors that determined us to go ahead with this project.

Aviva Garrett was the first person to hear about this idea, encouraged it and arranged all the required support within Juniper Networks. In addition to this, Aviva provided an editorial review of the entire book.

Patrick Ames guided us through the intricate process of bringing a book from proposal to the printing press and provided moral support and appropriate scolding as necessary. We would not have been able to pull this off without him. Patrick also did all the hard work of preparing the manuscript (and in particular the art manuscript) for editing.

This book benefited from the contribution of many people. We would like to thank our following colleagues:

Pedro Marques, for his thorough review of almost all chapters, for many technical discussions and for contributing the analysis of VPN scaling and RR scaling.

Arthi Ayyangar, for her insight on all topics RSVP and TE-related, for many technical discussions throughout our years of working together and for the careful review of numerous chapters.

Steven Lin, for reading and commenting on the entire manuscript, on a very tight schedule.

Der-Hwa Gan, for his mentoring role on TE, RSVP and MPLS, and for his very thorough technical review and comments.

Chaitanya Kodeboyina (CK), for very detailed reviews and discussions on several chapters.

Josef Buchsteiner, for always bringing up tough customer problems and for the timely reviews of selected chapters.

Serpil Bayraktar, for never leaving open ends and for very careful reading of the VPN and interdomain TE chapters.

Amir Tabdili, for always asking the hard questions and for reviewing selected chapters.

Quaizar Vohra, for his insight into Layer 2 circuits and the IGP, and for his technical review of these topics.

Margarida Correia, for always questioning proposed solutions to customer problems, and for technical review of selected chapters.

Hector Avalos, for valuable technical discussions and technical review of selected chapters.

Nischal Sheth, for being a mentor on LDP and for numerous discussions on all topics MPLS-related.

Kireeti Kompella, for many technical discussions and for his insight into all aspects of the MPLS technology.

As always, any errors and omissions are the responsibility of the authors.

We would also like to acknowledge the support and assistance of Juniper Networks and in particular our managers **Andy Heffernan, Nischal Sheth** and **Hector Avalos** for encouraging us to work on this project and providing the necessary resources.

Last but not least we would like to thank the wonderful team at Wiley: **Joanna Tootill, Birgit Gruber, Richard Davies, Julie Ward** and **Pat Bateson** for their support and guidance.

Finally, the authors would like to express their personal thanks to family and friends:

Ina Minei – First of all, I would like to thank my husband Pedro Marques for being my strongest supporter and harshest critic, for having infinite patience throughout the entire project and for the many sacrificed weekends. I would not have been able to do this without his support. Second, I would like to thank my father for the long hours he spent with me in my high school years, teaching me English and writing.

Julian Lucek – I would like to thank my partner Rachel and our daughters Emma and Hannah for their great patience and support during the writing of this book. Also I would like to thank my parents for their encouragement and for looking after my daughters during my writing sessions.

Part One

1

Foundations

1.1 HISTORICAL PERSPECTIVE

In only a few years, Multi-Protocol Label Switching (MPLS) has evolved from an exotic technology to a mainstream tool used by service providers to create revenue-generating services. There is rapid deployment of MPLS-enabled services and active development of new mechanisms and applications for MPLS in the standards bodies. This book aims to describe the fundamental mechanisms used by MPLS and the main service types that MPLS enables, such as Virtual Private Networks (VPNs). We include descriptions of new applications of MPLS that are currently under development.

The history of MPLS and its precursors is described in [Davie Rekhter] and [Doyle Kolon]. The first Internet Engineering Task Force (IETF) MPLS Working Group Meeting took place in April 1997. That working group still exists, and MPLS has grown to the extent that it underpins much of the activity of several other working groups in the IETF, such as Layer 3 VPN (l3vpn), Layer 2 VPN (l2vpn), Pseudo Wire Emulation Edge to Edge (pwe3) and Common Control and Measurement Plane (ccamp). Part of the original MPLS problem statement [MPLS97] from the first MPLS working group meeting is shown below. It contains four items that the group aimed to address through the development of

MPLS-Enabled Applications: Emerging Developments and New Technologies Ina Minei and Julian Lucek

MPLS. It is interesting to examine these to see which items are still relevant today:

1. Scalability of network layer routing. Using labels as a means to aggregate forwarding information, while working in the presence of routing hierarchies.

Layer 3 VPNs have proved to be a good example of aggregation of forwarding information. As described in Chapter 7 of this book, edge routers need to contain routing information pertaining to each VPN that they service, but the core routers do not. Thus, assuming that any edge router services only a subset of the VPNs pertaining to the network, no router in the network needs to hold the entire set of routes present in the network.

2. Greater flexibility in delivering routing services. Using labels to identify particular traffic which are to receive special services, e.g. QoS. Using labels to provide forwarding along an explicit path different from the one constructed by destination-based forwarding.

MPLS has the ability to identify particular traffic flows which must receive special services such as Quality-of-Service (QoS). It also has traffic engineering properties that allow it to provide forwarding along a particular explicit path. These two properties are combined in DiffServ Aware Traffic Engineering, which is described in more detail in Chapter 4 of this book.

3. Increased performance. Using the label-swapping paradigm to optimize network performance.

Because modern routers perform packet forwarding in hardware, the forwarding rates for IP and MPLS packets are similar. However, 'optimizing network performance' implies a wider context than simply the performance of individual nodes. Certainly MPLS has helped in this wider context, e.g. through the use of traffic engineering to avoid congestion and the use of fast reroute to reduce the interruption to traffic when a link in the network fails.

4. Simplify integration of routers with cell switching based technologies: a) making cell switches behave as peers to routers (thus reducing the number of routing peers that a router has to maintain), b) by making information about physical topology available to Network Layer routing procedures, and

c) by employing common addressing, routing, and management procedures.

When this item in the problem statement was written, many networks had a core of asynchronous transfer mode (ATM) switches surrounded by routers. The routers were typically fully meshed with ATM connections. This overlay model was proving difficult to scale because the number of routing adjacencies required grew as the square of the number of routers involved; hence there was a requirement to make the ATM switches act as peers to the routers. It is interesting to note that the situation has now been turned inside out: now many networks have an MPLS-based core, and service providers are migrating ATM services to this core network by interconnecting ATM switches with Layer 2 connections over the MPLS core! This has the problem that the number of adjacencies between ATM switches grows as the square of the number of ATM switches involved. Hence, currently there is work on making ATM switches behave as peers to routers [MPLS ALLI]. This is to avoid having a full mesh of adjacencies between ATM switches rather than to avoid having a full mesh of adjacencies between routers, as stated in the problem statement. The concept expressed in the problem statement of using MPLS as a control plane for multiple technologies has manifested itself in Generalized MPLS (GMPLS). In GMPLS, a common control plane covers a wide range of network devices, such as routers, ATM switches, SONET/SDH equipment and optical cross-connects [RFC3945].

In summary, much of the original problem statement is still relevant today. Many of the mechanisms of MPLS described in Part 1 of this book were developed to address the items listed above, to the benefit of the MPLS applications discussed in Part 2 of this book.

1.2 CURRENT TRENDS

At the time of writing this book, the most widely deployed customer-visible MPLS service is the Layer 3 VPN (also known as an IP VPN or 2547bis VPN, after the IETF document describing them). MPLS is also used in some networks as an infrastructure tool to provide traffic engineering and fast-reroute capabilities. Another rapidly growing application is point-to-point Layer 2 transport, either as means of carrying a customer's Ethernet traffic across the wide area or as a component of ATM or Frame Relay Service emulation.

Finally, Virtual Private LAN Service (VPLS) offerings, in which the service provider gives the impression to the customer that their sites are attached to the same Local Area Network (LAN), are also becoming available.

Many service providers are investigating the possibility of using an MPLS-based network to provide a common platform for a wide range of services that are currently typically delivered over multiple distinct networks. Such a multiservice network might carry Public Switched Telephone Network (PSTN) traffic, public Internet and private IP data services, Layer 2 ATM and Frame Relay services, Broadcast TV and TDM traffic. This offers capital and operational cost savings to the network operators by allowing them to operate a single network rather than a separate network for each service type. A key aim of this book is to show how MPLS can provide the necessary mechanisms for this network convergence, e.g. through the use of DiffServ Aware Traffic Engineering (TE), which allows the MPLS network to provide connection-orientated characteristics to particular traffic flows.

1.3 MPLS MECHANISMS

This section gives an overview of the mechanisms underpinning MPLS. Readers who are familiar with these may wish to skip this section.

A fundamental property of an MPLS network is that it can be used to tunnel multiple traffic types through the core of the network. Tunneling is a powerful tool because only the routers at the ingress and the egress of the tunnel need to understand the 'context' of the underlying traffic carried over the tunnel (e.g. the protocol that the traffic belongs to and the reachability information required to route and forward it in its native form). This detail is hidden from routers in the core of the network. As a consequence, core devices only need to carry sufficient state to enable them to switch MPLS-encapsulated packets without regard to their underlying content. Besides these aggregation properties, which apply to tunnels in general, MPLS tunnels have the following particular properties:

1. Traffic can be explicitly routed, depending on which signaling protocol is used.
2. Recursion is provided for; hence tunnels can exist within tunnels.

3. There is protection against data spoofing, as the only place where data can be injected into an MPLS tunnel is at the head end of that tunnel. In contrast, data can be injected into an IP tunnel from any source that has connectivity to the network that carries the tunnel.
4. The encapsulation overhead is relatively low (4 bytes per MPLS header).

An MPLS network consists of edge devices known as Label Edge Routers (LERs) or Provider Edge (PE) routers and core routers known as Label Switching Routers (LSRs) or Provider (P) routers. A mesh of unidirectional tunnels, known as Label Switched Paths (LSPs) is built between the LERs in order that a packet entering the network at the ingress LER can be transported to the appropriate egress LER. When packets enter a network, the ingress router determines which Forwarding Equivalence Class (FEC) the packets belong to. Packets that are to be forwarded to the same egress point in the network along the same path and with the same forwarding treatment along that path are said to belong to the same FEC. Packets belonging to the same FEC are forwarded with the same MPLS label. In a simple case, packets whose destination addresses correspond to the same Border Gateway Protocol (BGP) next-hop are regarded by the ingress router as belonging to the same FEC. In other cases, there may be a more granular assignment of packets to FECs. For example, in DiffServ Aware TE, each egress point in the network may have multiple FECs, each belonging to a different traffic class.

It is the role of the ingress LER to determine the appropriate egress LER and LSP to that egress LER associated with the FEC. MPLS has the property that multiple traffic types can be multiplexed on to a single LSP. Therefore, if desired by the network operator, a single LSP can be used to carry all the traffic (e.g. L3 VPN, public IP and Layer 2) between a particular ingress LER and a particular egress LER. Transit routers along the path of the LSP make their forwarding decision on the basis of a fixed-format MPLS header, and hence do not need to store 'routes' (L3 VPN routes, external IP routes, Layer 2 forwarding information) pertaining to the underlying tunneled packets. This is an important scaling property, as otherwise each of the core routers would have to carry routing information equivalent to the sum of the routing information carried by all the edge routers in the network.

The following sections describe the fundamental forwarding plane and control plane mechanisms underpinning MPLS.

1.3.1 Forwarding plane mechanisms

Data carried over an MPLS-capable network has one or more MPLS headers applied in order to transport it across the network. The MPLS header structure is shown in Figure 1.1. It contains the following fields:

1. A 20-bit label value. MPLS packets are forwarded on the basis of this field. This value is used as an index into the MPLS forwarding table.
2. EXP field (3 bits). These bits are known as the experimental bits. In practice, they are used to convey the Class of Service to be applied to the packet. For example, LSRs and LERs can use these bits to determine the queue into which the packet should be placed. Note that in some cases, as described later in this chapter, the MPLS label value also determines the queuing behaviour applied to the packet.
3. Bottom of stack bit (S-bit). As described later in this chapter, MPLS headers can be stacked. The S-bit is set on the header of the MPLS packet at the bottom of the stack.
4. Time-to-live (TTL) field. This is used to avoid forwarding loops and can also be used for path-tracing. The value is decremented at each hop and the packet is discarded should the value reach zero.

Packets arriving into the network have one or more MPLS headers applied by the ingress LER. The ingress LER identifies the egress LER to which the packet must be sent and the corresponding LSP. The label value used corresponds to the LSP on to which the packet is placed. The next router performs a lookup of that label and determines the output label that must be used for the next leg of the LSP. The lookup operation on a P router involves reading

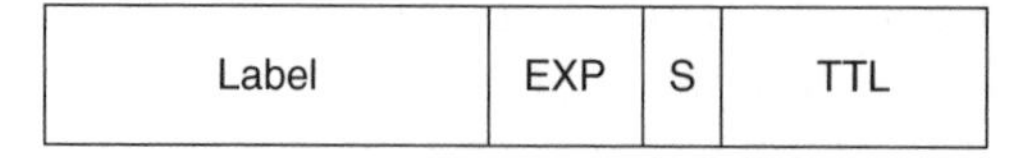

Figure 1.1 MPLS header structure

the incoming label; this yields a new label value to use and the output interface(s) on which the packet should be forwarded. In this way, through this label-swapping paradigm, the packet is conveyed along the LSP from the ingress to the egress LER.

In some simple cases, the use of a single MPLS label is sufficient, e.g. when transporting public IP traffic across a network. In this case, once the packet arrives at the egress LER, the LER performs a normal IP lookup in order to determine which egress link to use. Usually a scheme called Penultimate Hop Popping (PHP) is used. In this scheme, the LSR before the egress LER (i.e. the penultimate router along the LSP) pops the MPLS label and forwards it to the egress LER as an IP packet. This simplifies the processing required at the egress node, as otherwise it would be necessary to pop the label and perform an IP lookup at the egress node.

In other cases, a single MPLS header is insufficient. This is because the LERs in a particular network may be involved in multiple services – Layer 3 VPN, Layer 2 VPN, VPLS – rather than just the public IP. In this case, the egress LER needs to know which service and which instance of that service (i.e. which customer) the packet belongs to. This is achieved by having an additional MPLS header, which is applied by the ingress LER, corresponding to the service and service instance that the packet must be directed to by the egress LER once the packet has crossed the network. This is illustrated in Figure 1.2.

Let us see how an MPLS packet with two headers is transported between the ingress and egress LERs. The inner header with label Y denotes the service and service instance, and the outer header, often called the 'transport' header, is the one required to transport the packet from the ingress LER, PE1, to the correct egress LER, PE2. For example, a particular LER may be running several Layer 3 VPN, VPLS and Layer 2 VPN instances. Label Y tells the egress LER that the packet in question corresponds to the Layer 3 VPN service being provided to Company A, rather than any of the other

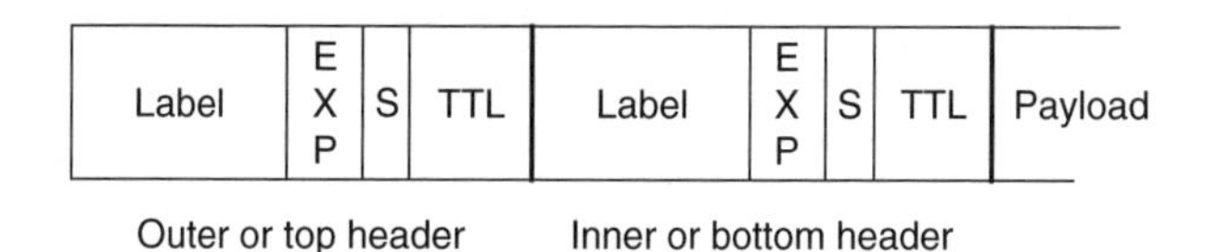

Figure 1.2 MPLS header stack

Layer 3 VPN instances or the VPLS or Layer 2 VPN instances. The ability to stack headers in this way gives MPLS key multiplexing and hierarchical properties, allowing a single LSP between a particular ingress and egress point to carry all traffic between those points. As Figure 1.3 shows, the packet leaves the ingress LER, PE1, with an inner label value of Y and an outer label value of X. Routers P1 and P2 perform a lookup based on the outer transport label and do not need to read or take any action based on the inner label. P1 swaps outer label X with outer label W. If PHP is in use, which is typically the case, router P2 pops the outer header, and sends the remainder of the packet to PE2. Thus, when the packet arrives at PE2, the outermost (and only) label is the original inner label, Y, which PE2 uses to identify the packet as belonging to the Layer 3 VPN instance pertaining to Company A.

How does the ingress LER know the label value(s) to use? The transport label is learnt through either the RSVP or LDP signaling protocols, which are described in more detail later on in this chapter. The inner label in the case of most services is learnt via BGP (e.g. Layer 3 VPNs, BGP-signaled Layer 2 VPNs). However, there are also cases where LDP is used, e.g. LDP-signaled Layer 2 transport circuits.

1.3.1.1 MPLS support of DiffServ

DiffServ was developed as a solution to provide Quality-of-Service (QoS). It does so by dividing traffic into a small number of classes and allocating network resources on a per-class basis. To avoid the need for a signaling protocol, the class is marked directly within

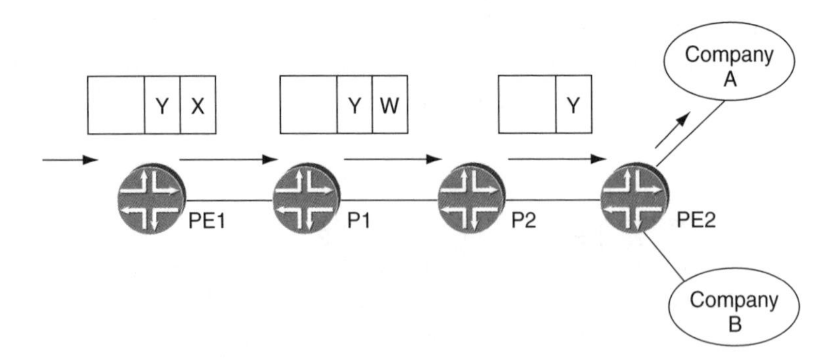

Figure 1.3 Forwarding a packet having two MPLS headers

the packet header. The DiffServ solution was targeted at IP networks so the marking is in the 6-bit DiffServ Code Point (DSCP) field in the IP header. The DSCP determines the QoS behavior of a packet at a particular node in the network. This is called the per-hop behavior (PHB) and is expressed in terms of the scheduling and drop preference that a packet experiences. From an implementation point of view, the PHB translates to the packet queue used for forwarding, the drop probability in case the queue exceeds a certain limit, the resources (buffers and bandwidth) allocated to each queue and the frequency at which a queue is serviced.

The first challenge with supporting DiffServ in an MPLS network is that LSRs make their forwarding decisions based on the MPLS header alone, so the per-hop behavior (PHB) needs to be inferred from it. The IETF solved this problem by assigning the three experimental (EXP) bits in the MPLS header to carry DiffServ information in MPLS.

This solution solves the initial problem of conveying the desired PHB in the MPLS header, while introducing a new one: how does one map DSCP values expressed in a 6-bit field that can encode up to 64 values into a 3-bit EXP field that can carry at most eight distinct values? There are two solutions to this problem, discussed separately below.

The first solution applies to networks that support less than eight PHBs. Here, the mapping is straightforward: a particular DSCP is equivalent to a particular EXP combination and maps to a particular PHB (scheduling and drop priority). During forwarding, the label determines where to forward the packet and the EXP bits determine the PHB. The EXP bits are not a property that is signaled when the label-switched path (LSP) is established; the mapping of EXP to PHB is configured on each node in the network. The EXP bits can be set according to the DSCP bits of the IP packets carried in the LSP, or they can be set by the network operator. LSPs for which the PHB is inferred from the EXP bits are called E-LSPs (where E stands for 'EXP-inferred'). E-LSPs can carry packets with up to eight distinct per-hop behaviors in a single LSP.

The second solution applies to networks that support more than eight PHBs. Here, the EXP bits alone cannot carry all the necessary information to distinguish between PHBs. The only other field in the MPLS header that can be used for this purpose is the label itself. During forwarding, the label determines where to forward the packet and what scheduling behavior to grant it, and the EXP bits convey information regarding the drop priority assigned to a packet.

Thus, the PHB is determined from both the label and the EXP bits. Because the label is implicitly tied to a per-hop behavior, this information needs to be conveyed when the LSP is signaled. LSPs that use the label to convey information about the desired PHB are called L-LSPs (where L stands for 'label-inferred'). L-LSPs can carry packets from a single PHB or from several PHBs that have the same scheduling regimen but differ in their drop priorities (e.g. the set of classes AFxy where x is constant are treated the same from the scheduling point of view but differ in their drop priority according to the value of y). Table 1.1 summarizes the differences between E-LSPs and L-LSPs.

1.3.2 Control plane mechanisms

So far we have seen how MPLS uses labels for forwarding, but how are the bindings between labels and FECs distributed throughout the network? Since manual configuration is not an option, there clearly is a need for a protocol to disseminate this information. From a practical point of view, there are two options: (a) invent a new protocol for distributing label bindings or (b) extend

Table 1.1 Comparison of E-LSPs and L-LSPs

E-LSP	L-LSP
PHB is determined by the EXP bits	PHB is determined by the label or by the label and EXP bits together
Can carry traffic with up to 8 distinct PHBs in a single LSP	A single PHB per LSP or several PHBs with the same scheduling regimen and different drop priorities
Conservative label usage and state maintenance, because the label is used only for conveying path information	Uses more labels and keeps more state, because the label conveys information about both the path and the scheduling behavior
No signaling is required to convey the PHB information	The PHB information needs to be signalled when the LSP is established
Up to 8 PHBs can be supported in the network when only E-LSPs are used. E-LSPs can be used in conjunction with L-LSPs when more PHBs are required	Any number of PHBs can be supported in the network

an existing protocol to carry labels in addition to routing information. The question of whether to invent a new protocol or extend an existing one is a popular one in the MPLS world, and we will discuss it in detail in later chapters. At this point, suffice it to say that when the question arises, the result is usually that both approaches are followed.

Regarding the distribution of label bindings, the engineering community invented a new protocol (LDP, or Label Distribution Protocol) and extended two existing protocols (RSVP, or Resource Reservation Protocol, and BGP, or Border Gateway Protocol). The packet formats and basic operation of these protocols are explained in detail in many introductory texts [Doyle Kolon, Osborne Simha]. Instead of repeating this information here, let us instead examine the properties of the different protocols, and see the benefits and limitations of each of them.

1.3.2.1 LDP

LDP [RFC3036] is the result of the MPLS Working Group [MPLS WG] in the IETF. Unlike RSVP or BGP, which existed well before MPLS and were extended to do label distribution, LDP was specifically designed to distribute labels in the network. Since the goal of LDP is label distribution, LDP does not attempt to perform any routing functions and relies on an Interior Gateway Protocol (IGP) for all routing-related decisions. The original LDP specification was defined for setting up LSPs for FECs representing an IPv4 or IPv6 address. This is the functionality described in this section. The extensions of LDP used for pseudowire and VPLS signaling will be discussed in the appropriate chapters.

LDP was designed with extensibility in mind. All the information exchanged in LDP is encoded as TLVs (type–length–value triplets). The type and length are at the start of the encoding, and their length is known in advance. The type identifies which information is exchanged and determines how the rest of the encoding is to be understood. The value is the actual information exchanged and the length is the length of the value field. TLVs make it easy to: (a) add new capabilities by adding a new type and (b) skip unknown objects by ignoring the amount of data specified in the length field. Over the years, many new capabilities were added to the protocol thanks to this built-in extensibility.

LDP operation is driven by message exchanges between peers. Potential peers, also known as neighbors, are automatically discovered via hello messages multicast to a well-known UDP port. The protocol also allows for discovery of remote peers using targeted hello messages. Once a potential peer is discovered, a TCP connection is established to it and an LDP session is set up. At session initialization time, the peers exchange information regarding the features and mode of operation they support. After session setup, the peers exchange information regarding the binding between labels and FECs over the TCP connection. The use of TCP ensures reliable delivery of the information and allows for incremental updates, rather than periodic refreshes. LDP uses the regular receipt of protocol messages to monitor the health of the session. In the absence of any new information that needs to be communicated between the peers, keepalive messages are sent.

The association between an FEC and a label is advertised via label messages: label mapping messages for advertising new labels, label withdraw messages for withdrawing previously advertised labels, etc. The fundamental LDP rule states that LSR A that receives a mapping for label L for FEC F from its LDP peer LSR B will use label L for forwarding if and only if B is on the IGP shortest path for destination F from A's point of view. This means that LSPs set up via LDP always follow the IGP shortest path and that LDP uses the IGP to avoid loops.

Relationship between LDP and the IGP

The fact that LDP relies on the IGP for the routing function has several implications:

1. LDP-established LSPs always follow the IGP shortest path. The LSP path shifts in the network when the IGP path changes, rather than being nailed down to a pre-defined path.
2. The scope of LDP-established LSPs is limited to the scope of the IGP. Thus, LDP LSPs cannot traverse autonomous system (AS) boundaries. The need for Inter-AS LSPs, as well as the solution proposed by the IETF for establishing them, is explained in the Interdomain Traffic Engineering chapter of this book (Chapter 5).
3. During reconvergence, traffic may be blackholed or looped. The existence of loops and the possibility of blackhole traffic is a fact

of life for the IGPs during reconvergence. The same properties are inherited by LDP, by virtue of it relying on the IGP for routing decisions. We will discuss how such loops are created and what their impact is in the Protection and Restoration chapter of this book (Chapter 3).

4. The IGP convergence time poses a lower bound on the LDP convergence time. Assuming that the IGP implements smart fast-convergence mechanisms the traffic loss is in the range of 1–2 seconds, orders of magnitude larger than RSVP's fast-reroute time. The IETF is currently working on adding fast-reroute capabilities to LDP. This is discussed in more detail in the Protection and Restoration chapter of this book (Chapter 3).
5. Loss of synchronization between the IGP and LDP can result in traffic loss. As always, for situations where two protocols must operate in tandem, there is a potential for race conditions.

Let us take a closer look at a race condition caused by the loss of synchronization between LDP and the IGP. In the diamond-shaped topology in Figure 1.4, LSR A is advertising a binding for its loopback FEC A. To start with, all links have the same metric, and the link C–D does not exist in the topology. From D's point of view, the LSP for FEC A follows the path D–B–A. At a later time the link C–D is added to the topology with a metric that is better than the metric of link B–D, causing the IGP shortest path from D's

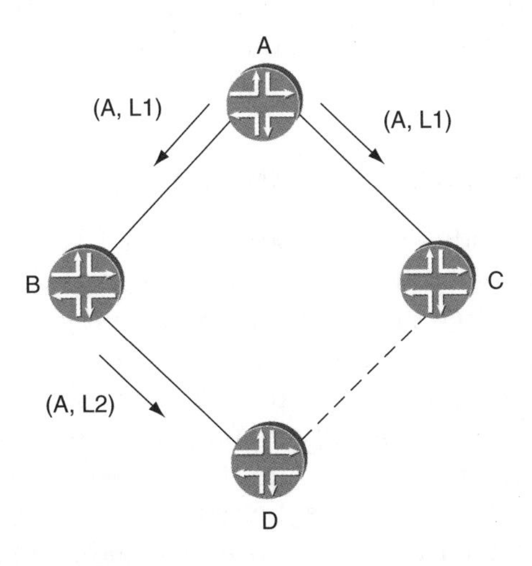

Figure 1.4 Race condition between the IGP and LDP

point of view to be D–C–A. Assume that the IGP reacts faster than LDP. As soon as D finds out about the routing change, it stops using the binding it received from B, thus breaking the LSP. The LSP stays down until a binding for FEC A is received on the LDP session C–D. This may take a while, depending on how fast the session establishment takes place. The situation described here is particularly unattractive, since an alternate path exists in the topology and could have been used until the LDP session comes up on the link C–D.

The above example shows a loss of synchronization caused by the fact that the LDP session on the new link comes up after the IGP session. This is not the only way in which loss of synchronization can occur: forgetting to enable LDP on the new interface, mis-configuring the LDP session authentication, setting up firewall filters that block LDP traffic, or any other event that would cause the IGP to take into account a link but would cause LDP not to use the link, has the same effect.

One solution to this problem is to tie (through configuration) the IGP metric for a particular link to the existence of an LDP session on the link [LDP-IGP-SYNC]. When the LDP session is down, the IGP metric advertised for the link is very high. Therefore, if an alternate path is available, the LDP labels on that path can be used. This is discussed in more detail in the MPLS Management chapter of this book (Chapter 12).

So far we have seen the implications of having LDP rely on the IGP for the routing function. Next, let us take a look at the choice of label distribution and retention modes made by common LDP implementations.

Label retention and label distribution modes

Label retention mode – which labels to keep? The LDP specification allows the use of both liberal and conservative label retention modes. Conservative retention means keeping only those labels which are used for forwarding, and discarding the rest. This policy makes sense for devices where the label space is a precious resource that must be carefully managed (such as ATM switches). The savings in the label usage come at a cost. Since the 'uninteresting' labels are discarded, they must be requested again if they become 'interesting' at a later point (e.g. due to a change in routing). Until the requested label arrives, traffic is lost. This undesirable property, coupled with the fact that label space is not a concern in modern routers means that most implementations today use liberal retention.

Label distribution mode – who assigns the labels? The key function of LDP is to distribute bindings between labels and FECs. The goal is to build a forwarding table containing a mapping between an incoming label and an outgoing label. Traffic arriving at the LSR labeled with the incoming label is forwarded labeled with the outgoing label. When building the forwarding table, the question is whether to use the locally picked label as the incoming or the outgoing label. The MPLS architecture [RFC3031] uses downstream label assignment, which means that the router expects to receive the traffic with the label that it picked locally. For example, if LSR A receives label L1 for FEC F and advertises label L2 for it, then it expects traffic destined for FEC F to come labeled with label L2. When forwarding traffic for FEC F, LSR A labels the traffic with label L1. The traffic flows in the opposite direction from the distribution of labels. The method is called downstream because the label that is assigned to the traffic at point P in the network was actually picked by a router who is one hop further down in the direction of the traffic flow (downstream) from P.

The next question is: should labels be advertised only to those asking for them (on-demand label distribution) or to everyone (unsolicited label distribution)? We have already seen that on-demand label distribution has the undesirable property that traffic is blackholed until the request for the label is satisfied. For this reason, most implementations use the unsolicited label distribution mode. Since LDP uses downstream label allocation, the label distribution mode is usually referred to as downstream unsolicited.

Liberal retention, coupled with unsolicited label advertisements, ensures that labels received from peers are readily available. This is important for handling routing changes in a seamless fashion. To better understand this, let us look at LSR A, which receives two unsolicited label advertisements for FEC F: one with label L1 from peer B and one with label L2 from peer C. LSR A keeps both labels, since it is doing liberal retention. Assuming that the IGP route for FEC F points to peer B, LSR A installs label L1 in its forwarding table. If at some later point the IGP route changes and starts pointing at peer C, all that LSR A has to do is change its forwarding table to use label L2.

Control over the LSP setup

The sole purpose of distributing bindings between labels and FECs is to establish label-switched paths in the network. So far we have discussed a lot of interesting properties of LDP but have not yet answered two key questions: (a) which FEC to advertise a binding for and (b) when to advertise this binding.

The choice of FECs is derived from the LSPs that must be set up in the network. It is independent of the LDP protocol and therefore the LDP specification is silent on this topic. All vendors allow control over the choice of FECs through configuration, but the behavior in the absence of a user-defined configuration is different for different vendors. Some advertise a binding for every prefix in their routing table, while others only advertise a binding for the FEC corresponding to the LSR's loopback address. The outcome in terms of the numbers of LSPs that are set up and of the destinations reachable via these LSPs is quite different. There is no right or wrong decision here, as different implementations may have different constraints. However, from a network operations point of view, it is a bad idea to allow LDP to advertise bindings for FECs that will not be used for forwarding. The extra binding and LSP information uses up resources in the network and makes troubleshooting extremely difficult.

The choice of FEC determines which LSPs are set up. The decision when to advertise the label binding determines who has control over the LSP setup. The LDP specification allows two modes of operation: ordered control and independent control. Since not all vendors implement the same mode, let us take a closer look at the two options and their properties, by reference to Figure 1.5. For the purposes of this discussion, assume that link if5 does not exist. This link will be used for a later discussion in this section.

Ordered control. Under ordered control, egress LSR PE1 initiates the LSP setup by assigning label L1 to the FEC corresponding to its loopback address PE1 and advertising this mapping to its peer A. Upon receipt of the label mapping, A evaluates whether PE1 is on the IGP shortest path for that FEC. Since the check is successful, A assigns label L2 for FEC PE1, installs forwarding state swapping labels L2 and L1 and advertises a binding for label L2 and FEC PE1 to its peer B, who will do similar processing. If the check is not successful, A would not advertise the FEC any further. In this fashion, the LSP setup proceeds in an orderly way from egress to

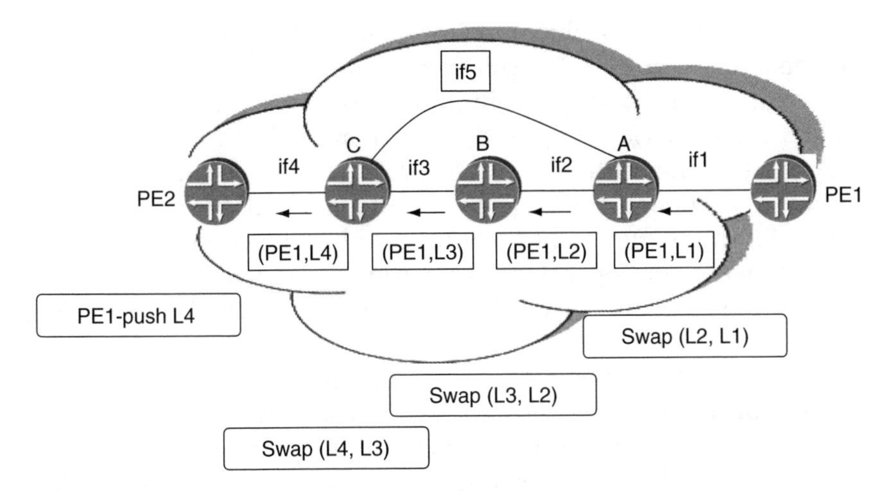

Figure 1.5 Different behavior for the ordered and independent label distribution modes

ingress. Each LSR consults the IGP for two decisions: (a) whether to advertise a mapping for an FEC and (b) whether to use a label for forwarding.

Independent control. With independent control, each LSR assigns a label for FEC PE1 and advertises this binding independently of the peers. Each LSR uses the locally assigned label as its incoming label in the forwarding table. The outgoing label in the forwarding table is filled in when the LSR receives a label for PE1 from a peer lying directly on the IGP shortest path for prefix PE1. The LSRs use the IGP for just one decision: whether to use a label for forwarding or not. The success of the LSP establishment depends on all LSR advertising labels for the same set of FECs. If LSR A were configured not to advertise a label for FEC PE1, the LSP to PE1 would never be established.

At this point, it is probably already clear that the default behavior regarding the choice of FECs that are advertised, which we discussed earlier in this section, is not an arbitrary one. With ordered control, the router who is the egress of the LSP decides which FECs to initiate LSPs for. Thus, a reasonable default behavior for an implementation performing ordered control is to advertise a mapping for the loopback address of the egress. With independent control, all routers in the network must advertise the same set of

FECs. Thus, the reasonable thing for an implementation performing independent control is to advertise a mapping for all prefixes in the routing table. Another point to note is that when changing the default behavior via configuration, with ordered control the change is applied to one router only (the egress), while with independent control the change must be uniformly applied throughout the network. The requirement for a uniformly applied change is due to the independent operation of the routers in the network: unless they agree on the same set of FECs to advertise, LSPs will not establish end-to-end throughout the network, causing traffic blackholing. This situation is made worse by the fact that the protocol has no built-in mechanisms for detecting such misconfigurations.

The different behavior with regards to the propagation of labels has important implications regarding the setup of LSPs. With ordered control, the bindings must propagate from the egress to the ingress before the LSP is established and traffic can be forwarded on to it. If an application (such as a Layer 3 VPN) relies on the existence of the LSP, then it cannot forward traffic. This behavior is not limited to the initial setup of LSPs. The same dynamics apply when routing changes. With ordered control labels must propagate to the routers in the new IGP path, while with independent control the labels are already available on these routers. This, however, is not as bad as it looks: when routing changes, the IGP messages themselves must propagate and new routes computed, so the propagation of LDP labels is no worse than the propagation of IGP messages.

A more interesting scenario is a failure case where LDP cannot follow the IGP. Let us go back to the example in Figure 1.5. Assume that the interface if5 does not yet exist in the network. The LSP for FEC PE1 (the loopback of router PE1) establishes along the routers PE2–C–B–A–PE1. At this point, the operator decides to add the interface if5 and includes it in the IGP, but forgets to enable LDP on it. As a result, the IGP best path from router C for FEC PE1 is C–A–PE1.

With ordered control, LSR C notices that the label advertisement that it received for FEC PE1 from LSR B does not match the IGP best path, withdraws its advertisement for FEC PE1 and removes its forwarding state. When LSR PE2 receives the withdrawal, it removes the forwarding state for FEC PE1. PE2 knows that the LSP is not operational and will not attempt to forward labeled traffic on it. With independent control, LSR C notices that the routing changed and that the outgoing label it installed in the forwarding table for FEC PE1 is no longer valid and removes the forwarding state for FEC PE1. PE2 does not change its forwarding state, since from its point of view the best path to PE1 is still through C. The net effect is that the LSP for PE1 is broken at point C, but PE2 is unaware of the failure. It will continue to send labeled traffic on this LSP and the traffic will be dropped at C. This type of silent failure is very problematic in a VPN environment, as we will see in later chapters. A solution to this issue is the scheme described in [LDP-IGP-SYNC], in which the IGP metric for a link is given a high value if LDP is not fully operational over the link. As described earlier, this scheme is also a solution to race conditions between LDP and the IGP.

Implementations supporting each of the two modes of operation can be and are deployed together in the same network [LDP-OP]. The key to interoperability is the fact that LSRs do not assume anything regarding the behavior of their peers, except consistent installation of the forwarding state following the IGP path.

Now that we have discussed the way LDP labels are distributed, let us look at an example of an LDP LSP. Figure 1.6 illustrates the fact that LDP forms an 'inverted tree' rooted at each egress point in the network through the mechanisms already described. The figure shows the IGP metric on each link and the LDP-signaled

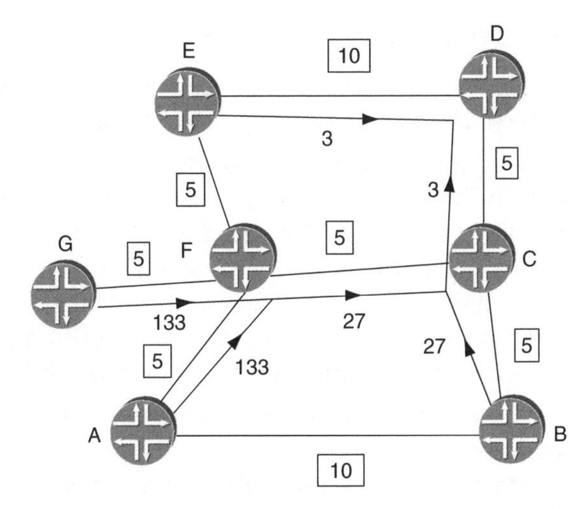

Figure 1.6 Inverted tree formed by LDP rooted at D

LSP rooted at D. The arrows show the direction of the data flow. It can be seen that the LSP path follows the best route as determined by the IGP. On any particular link, the label used to reach a particular destination router is the same, regardless of the origin of the packet. Thus, for example, on link F–C all packets whose destination is D have a label value of 27, regardless of whether they originated at G or A or F. Also, if per-platform label space is used, router C (for example) announces the same label value in order to reach D to all its neighbors, so all traffic passing via C to reach D has the same label value on all links into C. Hence traffic from B to D also has a label value of 27 on the B–C link. Note that in the example, penultimate hop popping is used, so D announces a label value of 3 to its neighbors. The diagram only shows the tree rooted at D. In reality, there would be multiple overlapping trees, each rooted at a different router in the network. As a result, on any particular link various labels may be in use if multiple routers are reachable over that link.

As with the IGPs, typically LDP implementations install multiple forwarding table entries in Equal Cost Multi-Path (ECMP) situations. For example, in Figure 1.6, if the metric between E and D were 5 rather than 10, there would be two equal cost paths from F to D, F–E–D and F–C–D. Hence F installs two forwarding entries for D, one corresponding to each path. Traffic arriving at F for D is load-balanced over the two paths.

LDP key properties

Here is a summary of the key properties of LDP:

- Automatic discovery of peers. LDP uses discovery messages to find peer LSRs. This yields two important benefits:
  - Ease of configuration. The operator does not need to configure each peer individually. Adding a new LSR in the network requires configuration of the new LSR, but not of any of the other LSRs in the network (in contrast to RSVP). The automatic discovery built into the LDP protocol is one of the most compelling reasons for picking LDP as the label distribution protocol in networks where traffic engineering is not required.
  - Session maintenance. The amount of session state an LSR must maintain is proportional to the number of neighbors. In the absence of targeted peers, this number is constant, regardless of the size of the network.

- Reliable transport. LDP uses TCP as the transport protocol for all except the discovery messages. Once advertised, information does not need to be refreshed. Keepalive messages are sent periodically for session maintenance, but their number is proportional to the number of sessions, not to the amount of information that was exchanged over the session.
- Extensible design. LDP uses TLVs for passing information around. This has proven itself over and over as the protocol was extended over the years.
- Reliance on the IGP.[1] LDP relies on the IGP for the routing-related decisions. LDP-established LSPs follow the IGP shortest path and are influenced by changes in routing. During periods of network convergence, LDP LSPs are affected, and traffic may be looped or blackholed.
- Liberal label retention and downstream unsolicited label distribution. The labels are advertised to all peers and kept by the peers even if they are not actively used for forwarding. Thus LDP reacts quickly to changes in the IGP routing.

1.3.2.2 RSVP

Another scheme for distributing labels for transport LSPs is based on the Resource Reservation Protocol (RSVP). RSVP was invented before MPLS came into being, and was originally devised as a scheme to create bandwidth reservations for individual traffic flows in networks (e.g. a video telephony session between a particular pair of hosts) as part of the so-called 'int-serv' model. RSVP includes mechanisms for reserving bandwidth along each hop of a network for an end-to-end session. However, the original int-serv application of RSVP has fallen out of favor because of concerns about its scalability: the number of end-to-end host sessions passing across a service provider network would be extremely large, and it would not be desirable for the routers within the network to have to create, maintain and tear down state as sessions come and go.

In the context of MPLS, however, RSVP has been extended to allow it to be used for the creation and maintenance of LSPs and to create associated bandwidth reservations [RFC 3209]. When used in this context, the number of RSVP sessions in the network is

[1] Recall that the discussion in this section is for FECs that are IP addresses.

much smaller than in the case of the int-serv model because of the way in which traffic is aggregated into an LSP. A single LSP requires only one RSVP session, yet can carry all the traffic between a particular ingress and egress router pair, containing many end-to-end flows.

An RSVP-signaled LSP has the property that its path does not necessarily follow the path that would be dictated by the IGP. RSVP, in its extended form, has explicit routing properties in that the ingress router can specify the entire end-to-end path that the LSP must follow, or can specify that the LSP must pass through particular transit nodes. Here are a few consequences of the explicit routing properties of RSVP:

1. The path does not necessarily follow the IGP. The path can be computed to comply with different constraints that may not be taken into account when the IGP paths are computed. As such, RSVP-signaled LSPs are a key component of MPLS-based traffic engineering, enabling the network administration to control the path taken by traffic between a particular pair of end-points by placing the LSP accordingly.
2. The path may be computed online by the router or offline using a path computation tool. In the case of online computation, typically only the ingress router needs to be aware of any constraints to be applied to the LSP. Moreover, use of the explicit routes eliminates the need for all the routers along the path to have a consistent routing information database and a consistent route calculation algorithm.
3. The path is not restricted to a single IGP domain. As long as a path was specified in some way, RSVP is not restricted to a single IGP domain (unlike LDP). The restriction of a single IGP domain does come into play during online path computation, as will be seen in the Traffic Engineering chapter of this book (Chapter 2).
4. An LSP can be signaled in such a way that its path can only be changed by the head end. This is in contrast to LDP, where each LSR updates its forwarding state independently of all other LSRs as it tracks the IGP state. This property is very important in the context of traffic protection schemes such as Fast Reroute, discussed in detail in the Protection and Restoration chapter of this book (Chapter 3). Fast Reroute schemes involve each router along the path of an LSP computing a local repair path that bypasses

a failure in the downstream link or downstream neighbor node. Traffic sent on the LSP is guaranteed to reach the router where the local repair path has been set up, since the routers do not change their forwarding state after a failure (this again is in contrast to the looping that may happen with LDP following a failure).

The creation of an RSVP-signaled LSP is initiated by the ingress LER. The ingress LER sends an RSVP Path message. The destination address of the Path message is the egress LER. However, the Path message has the Router Alert option set so that transit routers can inspect the contents of the message and make any necessary modifications.

Here are some of the objects contained in a Path message:

1. *Label Request Object*. Requests an MPLS label for the path. As a consequence, the egress and transit routers allocate a label for their section of the LSP.
2. *Explicit Route Object (ERO)*. The ERO contains the addresses of nodes through which the LSP must pass. If required, the ERO can contain the entire path that the LSP must follow from the ingress to the egress.
3. *Record Route Object (RRO)*. RRO requests that the path followed by the Path message (and hence by the LSP itself once it is created) be recorded. Each router through which the Path message passes adds its address to the list within the RRO. A router can detect routing loops if it sees its own address in the RRO.
4. *Sender TSpec*. TSpec enables the ingress router to request a bandwidth reservation for the LSP in question.

In response to the Path message, the egress router sends an Resv message. Note that the egress router addresses the message to the adjacent router upstream, rather than addressing it directly to the source. This triggers the upstream router to send a Resv message to its upstream neighbor and so on. As far as each router in the path is concerned, the upstream neighbor is the router from which it received the Path message. This scheme ensures that the Resv message follows the exact reverse path of the Path message. Figure 1.7 illustrates the Path and Resv message exchange along the path of an LSP.

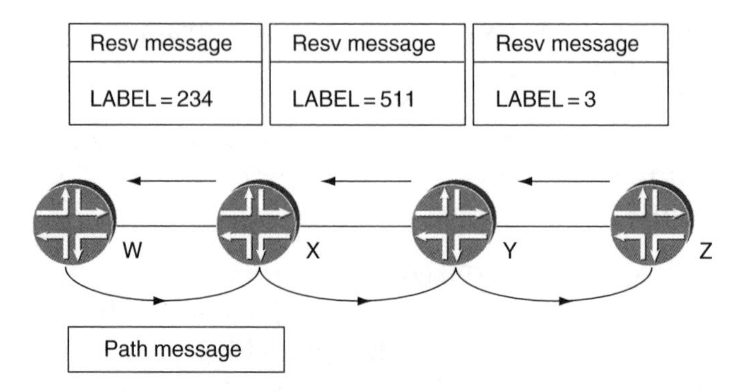

Figure 1.7 Illustration of the RSVP Path and Resv message exchange

Here are some of the objects contained in an Resv message:

1. *Label Object*. Contains the label to be used for that section of the LSP. For example, in Figure 1.7 when the Resv message is sent from the egress router Z to the upstream neighbor Y, it contains the label value that Y must use when forwarding traffic on the LSP to Z. In turn, when Y sends the Resv message to X, it overwrites the Label Object with the label value that X must use when forwarding traffic on the LSP to Y. In this way, for the LSP in question, Y knows the label with which traffic arrives at Y and the label and outgoing interface that it must use to forward traffic to Z. It can therefore install a corresponding label swap entry in its forwarding table.
2. *Record Route Object*. Records the path taken by the Resv message, in a similar way to the RRO carried by the Path message. Again, a router can detect routing loops if it sees its own address in the Record Route Object.

As can be seen, RSVP Path and Resv messages need to travel hop-by-hop because they need to establish the state at each node they cross, e.g. bandwidth reservations and label setup.

As a consequence of the scheme described above, an RSVP-signaled LSP only requires configuration at the ingress router. In typical implementations, properties of the LSP and the underlying RSVP session, such as the ERO and RRO and requested bandwidth, can be viewed on any router along the path of the LSP since that information is known to all routers along the path.

RSVP requires a periodic exchange of messages once an LSP is established in order to maintain ('refresh') its state. This can be achieved by periodically sending Path and Resv messages for each active LSP. If a router does not receive a certain number of consecutive Path or Resv messages for a particular LSP, it regards the LSP as no longer required and removes all states (such as forwarding entries and bandwidth reservations) pertaining to that LSP. The processing overhead of such a scheme can become a scaling concern for a router maintaining a very large number of LSPs. In order to address this, the 'Refresh Reduction Extensions' to RSVP were devised to reduce this overhead. These include a Summary Refresh Extension that allows multiple RSVP sessions (and hence multiple LSPs) to have their state refreshed by a single message sent between RSVP neighbors for refresh interval [RFC2961].

RSVP has an optional node failure detection mechanism, in which hello messages are sent periodically between RSVP neighbors. Without this mechanism, a node might only become aware of the failure of a neighbor through the timeout of RSVP sessions, which can take a relatively long time.

Note that there is no concept of ECMP in RSVP as there is in LDP. A particular LSP follows a single path from ingress to egress. If, in performing the path computation, the ingress router finds that there are multiple potential paths for an LSP that have equal merit, it chooses one of those paths for the LSP and signals for its creation via RSVP. Hence, once traffic has entered an RSVP-signaled LSP, there is no splitting and merging of traffic as sometimes occurs in the LDP case.

In some cases, a network may only have a handful of RSVP-signaled LSPs, as a tactical way of controlling traffic flows around particular hot-stops in the network. In those situations, RSVP-signaled LSPs would be created between certain pairs of end-points to achieve this aim. In other networks, the reason for deploying RSVP-signaled LSPs might be in order to make use of fast reroute, in which case the administrator may choose to fully mesh the PEs in the network with RSVP-signaled LSPs.

By way of summary, here are the key properties of RSVP:

- Explicit routing. The ingress LER has control over the path taken by the LSP, either by specifying the entire path or by specifying particular nodes that the LSP must pass through. As a consequence,

RSVP lends itself to traffic engineering and traffic protection schemes that operate independently of, and faster than, the IGP.

- Periodic message exchange is required to renew the state of an LSP, although the RSVP Refresh Reductions reduce this overhead.
- The amount of session state on a node is proportional to the number of LSPs traversing the node. This tends to grow as the network grows (assuming a high degree of meshing of RSVP-signaled LSPs).

1.3.2.3 RSVP and LDP comparison

A frequently asked question is whether LDP or RSVP is the better protocol to use in a deployment. Let us compare the two protocols with regard to the factors that affect the choice of which to use:

1. Ease of configuration:
    (a) Initial configuration. LDP has the advantage that it is easy to configure, only requiring one line of configuration in some implementations, to allow the protocol to run on a particular interface. RSVP, on the other hand, requires explicit configuration of the LSPs on the ingress router. Each router must know all other routers to which it must establish LSPs.
    (b) Incremental configuration when new edge devices are added. For LDP, only the new device must be configured. For RSVP, adding a new router to the edge means configuring LSPs to it from all the existing routers, potentially requiring configuration changes on all other edge routers in the network.

There are currently moves to reduce the configuration effort when using RSVP. One scheme is an automatic meshing capability, where each edge router in the network automatically creates an RSVP-signaled LSP to the other edge routers in the network. Another is an autobandwidth capability, where the bandwidth reservation for an LSP changes in accordance with the volume of traffic using that LSP. Used in combination, the configuration effort would not be very different to that associated with LDP. Such schemes may not help in all cases, however, e.g. when each LSP has particular constraints associated with it or requires a fixed bandwidth reservation rather than one that dynamically varies.

2. Scalability:
   (a) Control plane sessions. For LDP, each router must maintain a number of sessions equal to the number of LDP neighbors. For RSVP, the number of sessions is equal to the total number of LSPs that the router is involved with (whether in the role of ingress, transit or egress router). For a fully meshed topology, the total number of LSPs in the network is of order *N*-squared in the RSVP case, where *N* is the number of edge routers, but is proportional to *N* in the LDP case.
   (b) State maintenance. LDP sends periodic keepalive and hello messages, but only for a limited and constant number of neighbors/sessions. RSVP must refresh all sessions for the LSPs traversing a router, a number over which it has no control. RSVP refresh reduction reduces the number of RSVP messages that have to be created and sent in order to refresh the sessions; however, the router still needs to track the state of each session.
   (c) Forwarding state. LDP maintains the forwarding state for all FECs in the network. By nature of the protocol each FEC is reachable from every point in the network. The ability of LDP to support ECMP means that often more than one path is maintained. RSVP, on the other hand, only keeps the state for the LSPs traversing it, and potentially their protection paths.

For practical purposes, the above considerations may not be of practical importance unless one has a very large number of routers that need to be fully meshed with RSVP-signaled LSPs, resulting in an unsustainably large number of LSPs to be maintained by routers in the core of the network. In those cases, either the LDP over RSVP or the LSP hierarchy schemes described later in this section can be used.

3. Features supported. Currently, only RSVP supports traffic engineering and fast reroute.

From the above analysis it should come as no surprise that if the traffic engineering or fast-reroute properties offered by RSVP are not required, LDP is almost always chosen. Let us take a closer look at the choice of protocol in the context of the application for which the MPLS connectivity is required:

1. L3 VPN. These services often do not have stringent SLAs in terms of outage time in the event of a link failure and although

they may offer several Diff-Serv traffic classes, none of the traffic classes have associated bandwidth reservations through the core. The main considerations in this case are ease of management and provisioning. Therefore, to date, LDP has received wider deployment than RSVP in such networks.

2. Migration of Layer 2 services to MPLS networks. Emulation of services such as ATM and Frame Relay over MPLS networks often requires tangible bandwidth guarantees. For example, if a service provider offers a particular access rate at a particular class of service between two access points in the network, it is necessary to ensure that the bandwidth between those points is reserved and uncontended. In addition to the bandwidth guarantees, Layer 2 services require fast restoration following a link failure. Due to its fast reroute and traffic engineering capabilities (and in particular DiffServ Aware Traffic Engineering), RSVP is better suited than LDP in such deployments.
3. Services requiring fast restoration, such as voice services. In some cases, there may be no TE requirement, because link utilization is low and bandwidth plentiful. However, fast-reroute capabilities may still be required, due to the nature of the service (e.g. voice). RSVP is the only protocol that supports fast restoration today. To cater for service providers (SPs) that require faster restoration times but do not require traffic engineering, there are moves to improving the convergence time of traffic traveling down LDP-signaled LSPs. In some cases, it is advantageous to use a combination of RSVP and LDP-signaled LSPs.

In many deployments, each Point-of-Presence (PoP) consists of several access routers and one or two core facing routers. The SP may wish to use RSVP for its traffic engineering properties in the core, but has no need for traffic engineering within the PoP. Similarly, there may be a need for fast reroute in the core but not within the PoP infrastructure, on the premise that intra-PoP link failure is relatively rare.

In these cases, the SP can use LDP within the PoPs and RSVP-signaled LSPs in the core. Targeted LDP sessions are created between the ingress and egress routers of each RSVP-signaled LSP so that LDP labels are exchanged without involving the transit routers of the RSVP-signaled LSPs. If the number of core-facing routers in the network is X and the number of edge routers in the network is Y, then the number of RSVP-signaled LSPs is reduced from $Y\,(Y-1)$

to X $(X-1)$. This could be a large reduction if the ratio Y to X is large. For example, consider a network that has 30 PoPs, each containing two core-facing routers and five edge routers. In the case where the edge routers are fully meshed with RSVP-signaled LSPs, there would be 22 350 (i.e.150×149) RSVP-signaled LSPs in the network. In the case where only the two core-facing routers in each PoP are fully meshed, there would be a total of 3480 (i.e. 60×58) RSVP-signaled LSPs in the network. This is almost an order of magnitude smaller than the full mesh case. The smaller number of LSPs means a lighter load on the protocols and the routers. This, in itself, is only of practical consequence if the load in the fully meshed edge router case is unsustainably high. More importantly, fewer LSPs means easier provisioning and management from the operator's point of view.

The LDP over RSVP process is illustrated in Figure 1.8, which shows a cross-section through the edge and core of a network. Routers A, B and C are within the same PoP. Routers F, G and H are within another PoP. D and E are core routers. LDP is used within the PoPs. In the network, the core-facing routers in the PoPs are fully meshed with RSVP-signaled LSPs. Hence there is a pair of RSVP-signaled LSPs between C and F (one in each direction). Also, there are targeted LDP sessions between the core-facing routers in each PoP, i.e. between C and F in the diagram. The targeted LDP session allows C and F to directly exchange labels for the FECs associated with the edge routers in their respective PoPs. For example, C learns the label from F to use when forwarding traffic to H. Routers D and E are not involved in the LDP signaling process and do not store LDP labels.

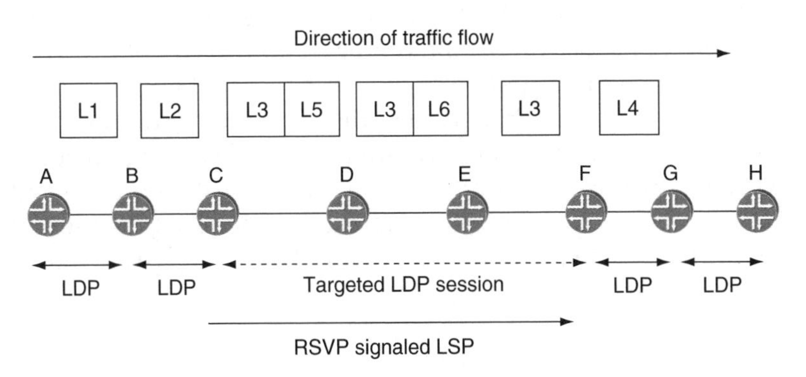

Figure 1.8 LDP over RSVP forwarding

Let us consider the transport of packets arriving into the network at router A and leaving the network at router H. The forwarding plane operation is as follows: ingress router A pushes a label which is learnt via LDP. In the example, the label value is L1, and is the label associated with H, the egress point of the packet. Router B swaps the label for one having the value L2. Router C is the ingress router for the RSVP-signaled LSP across the core. C swaps the existing label L2 for a label value L3 that it learnt via the targeted LDP session with F. Also, it pushes on to the packet a label of value L5 learnt via RSVP. Hence, at this point, the label stack consists of an outer label of value L5 and an inner label of value L3. The core routers D and E are only aware of the RSVP-signaled LSP and hence only carry out operations on the outer label. D swaps the outermost label of value L5 for a label having value L6. Note that the underlying label having value L3 is left untouched. If PHP is in use, router E pops the label learnt via RSVP, thus exposing the label, L3, learnt via LDP. Router F swaps the LDP label for one having value L4. If PHP is in use, router G pops the label, exposing the header of the underlying packet. This could be an IP header or could be another MPLS header, e.g. a VPN label.

In cases where the properties brought by RSVP are required from edge to edge, the above LDP over RSVP scheme is not suitable. However, in the case of very large networks, it may not be feasible either to fully mesh all the edge routers with RSVP-signaled LSPs because of the resulting amount of the RSVP state in the core of the network. The concept of LSP hierarchy [LSP HIER] was introduced to solve this problem. In this scheme, a layer of routers are fully meshed with RSVP-signaled LSPs. The layer is chosen such that the number of routers involved in the mesh is less than the number of edge routers. For example, the routers chosen might be the core-facing routers within each PoP. The edge routers are also fully meshed with RSVP-signaled LSPs which are nested within the LSPs between the core-facing routers.[2] The LSPs in the core of the network are called forwarding adjacency (FA) LSPs. In this way, routers in the heart of the network only have to deal with the session state corresponding to the core LSPs, and are unaware of the fact that LSPs from edge to edge are nested within them.

[2] Note that, as a consequence, the use of the LSP hierarchy does not solve the issue of the overhead of configuring a full mesh of RSVP-signaled LSPs.

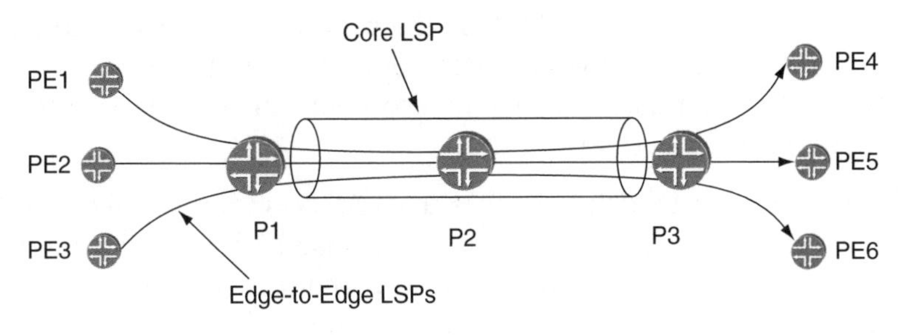

Figure 1.9 LSP hierarchy

The LSP hierarchy concept is illustrated in Figure 1.9. The diagram shows six PEs, three in each of two PoPs. P1 is a core-facing router in one PoP and P3 is a core-facing router in the other PoP. The diagram shows an RSVP-signaled LSP between P1 and P3. Using LSP hierarchy, edge-to-edge LSPs between the PE routers in the two PoPs can be nested within the core LSP between P1 and P3. For example, there is an LSP between PE1 and PE4, another between PE2 and PE5 and so on. However, P2 in the core of the network is unaware of the existence of these LSPs and is only involved in the maintenance of the core LSP. This is because the RSVP messages associated with the edge-to-edge LSPs pass directly between P1 and P3 without being processed by the control plane of P2.

1.3.2.4 BGP label distribution

The third type of label distribution also relies on a preexisting protocol, BGP. BGP has support for multiple address families, which make it straightforward to define and carry new types of reachability information and associated attributes. Thus, by adding a new address family to BGP, it is possible to advertise not just a prefix but also one or more labels associated with the prefix. In the Hierarchical and Inter-AS VPNs chapter of this book (Chapter 9), we will see that this capability is essential in the context of inter-AS MPLS/VPNs. The chapter describes several solutions in which BGP is used to:

(a) distribute the 'inner' labels (VPN labels) required by the egress PE to identify the service and service instance that the packet belongs to and/or

(b) distribute the outer label required to transport a packet to the appropriate egress PE.

The reasons for picking BGP as the protocol for the solution are discussed in detail in the Hierarchical and Inter-As VPNs chapter (Chapter 9). At this point, let us see some of added benefits of using BGP for label distribution:

- The ability to establish LSPs that cross AS boundaries. An example of where this is required is an MPLS-based VPN service having attachment points within multiple providers. In this case, it is necessary to distribute labels pertaining to PE reachability, so that the transport label required to reach a PE in another AS is known. BGP is protocol that is used today to convey reachability information across AS boundaries; therefore it can easily convey label information across AS boundaries.
- Reduction in the number of different protocols running in the network. Rather than deploying an entirely new protocol, reuse one of the existing protocols to provide one more function.
- Reuse of existing protocol capabilities. BGP supports a rich set of attributes that allow it to filter routing information, control the selection of exit points, prevent loops, etc. All these capabilities are readily available when label information is distributed along with a prefix.

1.4 CONCLUSION

We have started this chapter by looking at the original goals of the MPLS Working Group back in 1997. As is often the case for successful technologies, MPLS has become a key component in the development of new applications that were not envisioned at the time MPLS started out. The following chapters take a closer look at many of the innovations made possible by MPLS.

1.5 REFERENCES

[Davie Rekhter] B. Davie and Y. Rekhter, *MPLS: Technology and Applications*, Morgan Kaufmann, 2000

[Doyle Kolon] J. Doyle and M. Kolon (eds), *Juniper Networks Routers: The Complete Reference*, McGraw-Hill, 2002

[LDP-IGP-SYNC] M. Jork, A. Atlas and L. Fang, 'LDP IGP synchronization', draft-jork-ldp-igp-sync-01.txt (work in progress)

[LDP-OP] L. Andersson, I. Minei and B. Thomas, 'Experience with the LDP protocol', draft-minei-ldp-operational-experience-00.txt (work in progress)

[LSP HIER] K. Kompella and Y. Rekhter, 'LSP Hierarchy with Generalized MPLS TE', draft-ietf-mpls-lsp-hierarchy-08.txt (work in progress)

[MPLS97] Original problem statement for the IETF MPLS Working Group, http://www.ietf.org/proceedings/97apr/97apr-final/xrtftr90.htm

[MPLS ALLI] T. Walsh and R. Cherukuri, 'Two reference models for MPLS control plane interworking', MPLS/FR Alliance Technical Committee document mpls2005.050.00, March 2005

[MPLS WG] IETF MPLS Working Group, http://ietf.org/html.charters/mpls-charter.html

[Osborne Simha] E. Osborne and A. Simha, *Traffic Engineering with MPLS*, Cisco Press, 2002

[RFC2961] L. Berger, D. Gan, G. Swallow, P. Pan, F. Tommasi and S. Molendini, *RSVP Refresh Overhead Reduction Extensions*, RFC2961, April 2001

[RFC3031] E. Rosen, A. Viswanathan and R. Callon, *Multiprotocol Label Switching Architecture*, RFC 3031, January 2001

[RFC3036] L. Andersson, P. Doolan, N. Feldman, A. Fredette and R. Thomas, *LDP Specification*, RFC 3036, January 2001

[RFC3209] D. Awduche, L. Berger, D. Gan, T. Li, V. Srinivasan and G. Swallow, *RSVP-TE: Extensions to RSVP for LSP Tunnels*, RFC 3209, December 2001

[RFC3945] E. Mannie, 'Generalized multi-protocol label switching (GMPLS) architecture', October 2004

1.6 FURTHER READING

[LDP-MTU] B. Black and K. Kompella, *Maximum Transmission Unit Signaling Extensions for the Label Distribution Protocol*, RFC3988, January 2005

[RFC3478] M. Leelanivas, Y. Rekhter and R. Aggrawal, *Graceful Restart Mechanism for Label Distribution Protocol*, RFC3478, February 2003

2

Traffic Engineering with MPLS (MPLS-TE)

2.1 INTRODUCTION

Controlling the path taken by traffic through a network is called traffic engineering (TE). There are many reasons why network operators want to influence the path traffic is taking in their networks. The most popular reason is improving utilization of network resources. The goal is simple: avoid a situation where parts of the network are congested while others are underutilized. Other reasons for using traffic engineering include ensuring that the path has certain characteristics (e.g. it does not use high-latency links), ensuring that transmission resources are available along a particular path, and determining which traffic gets priority at a time of resource crunch (e.g. following a link cut).

This chapter describes why MPLS is a useful technology for implementing traffic engineering and how it accomplishes the goal of steering traffic around the network.

MPLS-Enabled Applications: Emerging Developments and New Technologies Ina Minei and Julian Lucek

2.2 THE BUSINESS DRIVERS

Influencing the path that traffic takes in the network can increase revenues in two ways:

1. Offering new services with extra guarantees.
2. Lowering the investment in new network resources (primarily bandwidth) by improving the utilization of existing resources.

'Offering new services' means any guarantee that the operator can charge extra money for. One example is the 'guaranteed bandwidth service', which simply means that a certain amount of bandwidth is available for a particular customer's traffic, both in the steady state and under failure conditions.

'Improving resource utilization' means avoiding a situation where part of the network is congested while other parts are underutilized. For example, if some part of the traffic is routed around a congested link on to a path where enough bandwidth is available, the upgrade of the congested link can be delayed. Avoiding congestion also means better quality for the customer traffic: less loss, less delay and better throughput.

Another important cost-saving measure achieved through traffic engineering is increasing the maximum percentage of link utilization. Operators constantly monitor link utilization to determine at which point it is necessary to schedule a link upgrade. Each network has its own upgrade rules, expressed in terms of the percentage of link utilization that triggers an upgrade. A typical rule is to upgrade at 50% utilization, to be able to accommodate traffic shifting from a failed link. The benefit of traffic engineering in this context is that it can allow a higher utilization of the links, because there is more control over the path that traffic takes in the network, both under normal operation and in the failure case. By increasing the average percentage of link utilization, the upgrade of links can be delayed.

From this discussion, it should be clear that traffic engineering is not always required. If bandwidth resources are plentiful or utilization is low, there will be no congestion, not even following a link failure. If there are no high-latency links in the network, there is no need to worry about traffic crossing high-latency links. Indeed, as seen in the Foundations chapter of this book (Chapter 1), not all MPLS deployments are used for traffic engineering, and depending on the label distribution protocol used, not all MPLS networks can indeed provide traffic engineering. Thus, although traffic engineering is often equated with MPLS, an MPLS network does not necessarily mean a traffic engineered network.

An important thing to bear in mind when discussing the benefits of traffic engineering is that the means through which traffic engineering is achieved must be simple enough to deploy and maintain. In financial terms, the added cost of operating a more complex network must be justified by the new revenue brought in by traffic engineering. MPLS provides the required operational simplicity, along with the flexibility for implementing complex traffic engineering policies.

2.3 APPLICATION SCENARIOS

[RFC2702] lays out the requirements for traffic engineering with MPLS by listing the desirable properties of a TE solution. Rather than discussing the requirements in abstract terms, the following section illustrates them by looking at three application scenarios. At the end of this section we discuss why MPLS is a powerful tool for satisfying these requirements.

The network topology used in all scenarios is shown in Figure 2.1. Two sources, A and B, send traffic to destination D through node C. The cost of a path is measured in terms of the link metrics. When all metrics are equal, the cost simply translates to the number of

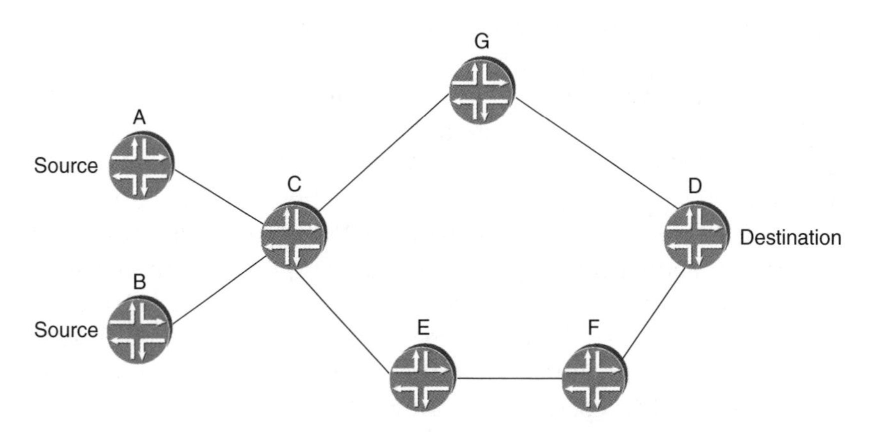

Figure 2.1 A network with two sources, A and B, and two unequal cost paths to the destination

hops in the path. In the sample network, two unequal cost paths exist between C and D. All links are of the same capacity and speed, unless specified otherwise.

The first application scenario highlights the need to forward traffic along a predefined path. In the example network in Figure 2.1, assume that A and B are two customers and that customer A buys a service that guarantees low latency. Assume that the link G–D is a satellite link. From the operator's point of view, the requirement is that traffic originating at A should avoid the high-latency link G–D. Thus, a path must be found from A to D that avoids the link G–D. In the simple network from Figure 2.1, the path from A to D can be easily identified as A–C–E–F–D. Regardless of how this path is computed, the problem becomes one of how to forward traffic originating at A along the path A–C–E–F–D.

It is difficult to satisfy this requirement in an IP network. The challenge with IP lies in the fact that forwarding is done independently at every hop, based on the destination of the traffic. Therefore, enforcing a forwarding policy that takes into account the source of the traffic is not always possible and when possible it is not always straightforward. In this example, it is required not only that both paths to D be of equal cost but also that node C must know that it handles a packet from A in order to forward the traffic towards E rather than towards G. In this simple network, the source can be identified based on the incoming interface, but in more complex topologies this may not be the case (e.g. in a similar network where A and B are connected to router C1, which is directly connected to C). The requirement for traffic engineering is the ability to forward traffic along a path specified by the source – in other words, explicit routing.

The second application scenario shows the requirement for improving resource utilization. In the network shown in Figure 2.1, assume that the capacity of all links is 150 Mbps and that source A sends 120 Mbps and source B sends 40 Mbps towards destination D. If traffic follows the shortest path, 160 Mbps of traffic cross path C–G–D, exceeding link capacity and causing congestion and loss. By splitting the traffic between both the shortest path (C–G–D) and the longer path (C–E–F–D), congestion is avoided and 80 Mbps of traffic traverse both links. This can be achieved, for example, by manipulating the costs of the links to make them behave as equal cost. Once two equal paths exist, traffic can be split (load balanced) between them.

However, this approach is not foolproof. Imagine now that under the same conditions, the capacity of all links is 150 Mbps, except for link E–F, which has a capacity of only 50 Mbps. In this case, the link E–F will be congested under the 80 Mbps load provided by the previous solution. The requirement is to specify the bandwidth requirements between each source/destination pair, find a path that satisfies these requirements and forward the traffic along this path.

The final scenario shows the need for control over the resources at a time of resource contention. The same network as the previous example is used and all links are 150 Mbps except E–F, which is 50 Mbps. Assume A sends 100 Mbps of traffic and B 40 Mbps. Also assume that customer B buys a service with strict service guarantees, while A does not. Under normal conditions, the 140 Mbps can be placed on the shortest path. However, when the link G–D fails, there are not enough resources on the alternate path to carry both A and B's traffic, so congestion and loss occur. To protect B's traffic, one could take a DiffServ-based approach and map B's traffic to a more preferred class. However, such an approach is not always feasible, for two reasons:

1. Under normal conditions, it may well be the case that both A and B's traffic should receive the same treatment.
2. Operators typically strive to minimize the number of behavioral aggregates they support in the network and implementing a priority scheme between traffic originated by two different sources increases the number of DiffServ code points.

In this case, the requirements are to find a path from the source to the destination that complies with the bandwidth constraints and to enforce the priority of the path sourced at B over the path sourced at A. Thus, after the link G–D fails, only the path sourced at B can set up on the alternate links. The path sourced at A will not find enough resources and so traffic from A will not interfere with traffic from B.

The three application scenarios described in this section boil down to two requirements: computing a path between source and destination that complies with a set of constraints and forwarding traffic along this path. As discussed in the Foundations chapter (Chapter 1), MPLS can easily forward traffic along an arbitrary path. The explicit routing capabilities of MPLS, implemented in

RSVP [RFC3209] with the Explicit Route Object (ERO), allow the originator of the LSP to establish the MPLS forwarding state along a path defined at the source. Once a packet is mapped on to an LSP, forwarding is done based on the label, and none of the intermediate hops makes any independent forwarding decisions based on the packet's IP destination. In the following sections, we will see how the constrained-path computation is accomplished.

2.4 SETTING UP TRAFFIC-ENGINEERED PATHS USING MPLS-TE

As seen in the application scenarios in the previous section, traffic engineering is accomplished in two steps: computing a path that satisfies a set of constraints and forwarding traffic along this path. These steps are discussed in detail in the following sections. However, it is first necessary to introduce the concept of LSP priorities.

2.4.1 LSP priorities and preemption

MPLS-TE uses LSP priorities to mark some LSPs as more important than others and to allow them to confiscate resources from less important LSPs (preempt the less important LSPs). Doing this guarantees that:

1. In the absence of important LSPs, resources can be reserved by less important LSPs.
2. An important LSP is always established along the most optimal (shortest) path that fits the constraints, regardless of existing reservations.
3. When LSPs need to reroute (e.g. after a link failure), important LSPs have a better chance of finding an alternate path.

MPLS-TE defines eight priority levels, with 0 as the best and 7 as the worst priority. An LSP has two priorities associated with it: a setup priority and a hold priority. The setup priority controls access to the resources when the LSP is established and the hold priority controls access to the resources for an LSP that is already established. When an LSP is set up, if not enough resources are available, the setup priority of the new LSP is compared to the hold priority of the LSPs using the resources in order to determine whether the new LSP can preempt any of the existing LSPs and take over their resources. If so, the other LSP(s) are torn down. By using different LSP priorities in the third application scenario from Section 2.2, the requirement to give traffic from B priority after a failure can be easily satisfied by simply giving the LSP B–D better priority than the LSP A–D.

So far so good, but is it ever necessary to assign distinct setup and hold priorities to an LSP? The answer is 'yes', and doing so is the default for many implementations. Assigning an important hold priority (say 0) and a less important setup priority (say 7) to an LSP creates a stable network environment. Using these priorities, a new LSP can never preempt an existing LSP and in turn can never be preempted. Conversely, assigning an unimportant hold priority (say 7) and an important setup priority (say 0) is a recipe for disaster, because it guarantees constant churn if two LSPs compete for the same resource. Imagine that LSP1 has been established over a particular path and that LSP2 wants to use the same links. LSP2's setup priority is better than LSP1's hold priority; thus LSP2 can preempt LSP1. When LSP1 attempts to reestablish, it notices that it can preempt LSP2, and so the cycle of preemption continues indefinitely. For this reason, most implementations disallow the configuration of a hold priority that is worse than the setup priority.

Priorities determine the treatment of an LSP in cases of resource contention in the network. They are essential for ensuring that 'important' traffic obtains the necessary resources at a time of shortage (e.g. after a link failure). However, this is not their only application. In a network where large LSPs and small LSPs exist, large LSPs are usually given better priorities to prevent setup failures. The reasoning is that smaller LSPs have a better chance of finding the necessary resources over an alternate path.

Having introduced the concept of priorities, we are now ready to start the discussion of path computation.

2.4.2 Information distribution – IGP extensions

As seen in the example scenarios, the requirement is to find a path in the network that meets a series of constraints. Therefore, the constraints must be taken into account when calculating feasible paths to a destination. Some of the constraints are:

1. The bandwidth requested for a particular LSP (such as 10 Mbps from source x to destination y).
2. The administrative attributes ('colors') of the links that the traffic is allowed to cross. An example of a constraint expressed in terms of link colors is to avoid high-latency links, where these links are marked with a particular administrative attribute. Link coloring is discussed in more detail in Section 2.4.3.
3. The metric that is assigned to a link for the purpose of traffic engineering.
4. The number of hops that the traffic is allowed to transit.
5. The setup priority of the LSP.

Other constraints are also possible, such as the inclusion or exclusion of a particular hop in the path or the requirement to place two related LSPs on different links, to ensure that failure of a single link does not affect both LSPs. Note that the constraints fall into two categories: (a) link properties such as available bandwidth, link color and traffic engineering metric and (b) LSP properties such as number of hops or priority.

Calculating a path that satisfies a set of constraints requires that the information about whether the constraints can be met is available for each link and that this information is distributed to all the nodes that perform path computation. Therefore, the relevant link properties have to be advertised throughout the network. This is achieved by adding TE-specific extensions to the link-state protocols IS-IS (Intermediate System-to-Intermediate Sytem) and OSPF (Open Shortest Path First) [RFC3784, RFC3630], which allow them to advertise not just the state (up/down) of the links but also the link's administrative attributes and the bandwidth that is available for use by LSPs at each of the eight priority levels. In this way, each node has knowledge of the current properties of all the links in the network. This information is stored in the traffic engineering database (TED) on each router and used in the path computation.

The question, however, is not just what to distribute but also when to distribute it. Link-state advertisements are sent periodically at typically large intervals (on the order of 30 minutes). New advertisements must be sent whenever the link information changes (e.g. when the available bandwidth changes) and they must be propagated throughout the network. To protect the network from being overwhelmed by link-state advertisements, new advertisements are not sent on every change, only on changes that are deemed significant (e.g. a change in the available bandwidth by more than a certain percentage). This necessary throttling creates a tradeoff between the accuracy of the information stored in the TED and the number of link advertisements that the network elements must process.

To summarize, the IGP extensions for traffic engineering ensure that the TE-related link attributes are available at all the nodes in the network. Next we will see how they are used in the computation of the constrained path.

2.4.3 Path calculation - CSPF

Like conventional SPF, constrained SPF (CSPF) computes a shortest path with regard to some administrative metric. CSPF takes into account only paths that satisfy one or more user-defined constraints (such as available bandwidth) by pruning out of the network topology links that do not satisfy the constraints. For example, if the constraint is bandwidth, CSPF prunes from the topology links that do not have enough bandwidth. In the second application scenario in Section 2.2, once the LSP A–D is set up for 120 Mbps, only 30 Mbps are available along the path A–C–G–D. Thus, when computing the path for LSP B–D, with a requirement of 40 Mbps, the links C–G and G–D are removed from the topology and CSPF picks the alternate path as the best available.

Another frequently used constraint is link coloring (also called administrative attributes). The concept of link colors is very intuitive. Links are marked with different colors through configuration and a link can be marked with multiple colors if desired, or no colors at all. Up to 32 different colors are available.[1] Figure 2.2 shows an example network where links E–F and F–D are colored 'red', link C–D is colored 'blue', link C–G is not colored at all while link C–E is colored both 'red' and 'green'. There is no restriction on how link colors are assigned, but they typically correspond to link properties such as latency, loss, operational cost or geographic location.

[1] The limitation is because of the way link colors are encoded in the IGP advertisements.

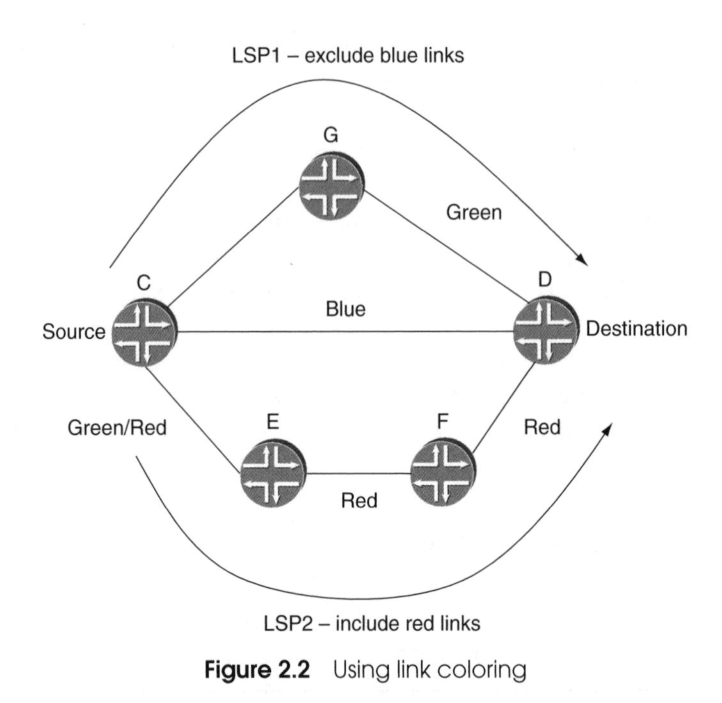

Figure 2.2 Using link coloring

The colors are used to express the desire to include or exclude a link or set of links from a particular path. For example, if the operator marks all high latency links with the color 'blue', he or she can then compute a path that does not cross high-latency links by excluding links marked with the color 'blue' from the path. For example, in Figure 2.2, assume link C–D is a high-latency link. LSP1 is set up between C and D with the constraint 'exclude blue links'. This means that none of the links in the path can be marked 'blue'. Thus, although the shortest path is through link C–D, this link is excluded from the computation due to its coloring and the LSP must establish the best path in a topology that does not include link C–D, yielding path C–G–D. Similarly, LSP2 is set up between C and D with a constraint of 'include red links'. Thus, all links in the path must be marked red. Note that for this purpose link C–E, which is marked with two colors (red and green), is acceptable. Although the example shown includes and excludes constraints separately, they can be used together for the same LSP. In effect, link coloring creates several TE topologies in the network, where some of the links belong to several topologies.

The reason administrative attributes (colors) are sometimes perceived as intimidating is because of how this feature is implemented by some vendors. Administrative attributes are encoded in a bit field in the IGP link advertisement. From an implementation point of view, the inclusion or exclusion of a link from the computation is accomplished by encoding the user-defined constraints in a similar bit-field format and then performing the necessary bit arithmetic between the link bit field and the constraint bit field. Some implementations force the user to express the constraints in bit format, something nonintuitive for most users. Other implementations offer friendlier interfaces, where instead of bits the user deals with attribute names expressed as strings (words). In both cases, however, the concept is equally simple: tag a link to be able to reference it in the computation. From the CSPF point of view, the links whose colors do not match the constraints are pruned from the topology.

CSPF bases its computation on the information stored in the TED, which is built from information disseminated by the IGP. Therefore:

1. The computation is restricted to a single IGP area. The solutions currently under evaluation in the IETF for extending LSP across area and AS boundaries are explored in the chapter discussing Interdomain TE (Chapter 5).
2. The result of the computation is only as good as the data on which it is based. Recall from Section 2.4.2 that the generation of link advertisements on link attributes changes is throttled. Therefore, the TED is not 100% up to date and the result of the computation is not always accurate. In Section 2.4.4 a scenario is presented where a path is found when in fact there are no available resources, leading to a failure during the path setup.

The LSP path may not have been optimal when it was computed, or may have been optimal at the time but became nonoptimal later on, due to changes in the network. Recall that link-state advertisements are sent both periodically and whenever significant events occur. The temptation is to recompute the LSPs based on the updated information and move the traffic to the new paths if they are deemed more optimal. This process is called reoptimization. Reoptimization ensures optimality of the paths, at the cost of stability, shifting traffic patterns in the network and possibly preempting LSPs in networks where multiple priority levels are enforced between the different LSPs. Stable and known traffic patterns are desirable both for debugging and for capacity planning. For this reason, reoptimization is off by default in most vendor implementations and can be turned on with different levels of granularity (periodic, event-driven or manual). The requirements for signaling paths following a reoptimization are discussed in more detail in section 2.4.4.

Just like SPF, the result of CSPF is a single path. Even if several equally good paths are available, only one is chosen. The tie-breaking rule is one of the following: random, least-fill (causing links to be least full) or most-fill (causing links to be most full). Without a global view of all reservations, both present and future, it is always possible to find a case where any of the algorithms is not optimal.

Let us see an example for least-fill. In Figure 2.1, assume all links are 150 Mbps and the metrics are such that the two paths C–G–D and C–E–F–D are considered of equal cost. The following three LSPs must be set up between C and D: LSP1 and LSP2, with 75 Mbps each, and LSP3, with 150 Mbps. There is enough bandwidth in the network to accommodate all three LSPs. However, depending on the order of the computation and on the tie-breaking algorithm used, the LSPs may not all be created. Under the least-fill algorithm, LSP1 is placed on C–G–D, LSP2 is placed on C–E–F–D and LSP3 cannot be placed. What is needed to make the correct decision in this case is knowledge of all the reservations ahead of time. This is possible when doing offline (rather than dynamic online) path computation. Offline path computation is discussed further in Section 2.8.

Regardless of how the path is actually computed, a label-switching forwarding state must be set up along the path in order to ensure that traffic does not stray from the desired course. The mechanisms for setting up the label-switching path are described in the next section.

2.4.4 Path setup – RSVP extensions and admission control

After a path has been successfully calculated, it is set up using RSVP-TE.[2] As discussed in previous sections, the path is specified at the LSP head end in the Explicit Route Object (ERO). However,

[2]Although CR-LDP [RFC3212] also supports explicit routing, it never gained much traction. In the context of MPLS, RSVP has become synonymous with TE. The IETF decided in [RFC3468] to abandon new development for CR-LDP.

the ERO is not the only TE-related information that must be carried in the RSVP messages. RSVP must also carry:

(a) the TE information that intermediate nodes must keep track of, such as the bandwidth requested by the LSP, and
(b) the information that is relevant in the path setup, such as the setup and hold priorities of the LSP.

As the RESV messages travel from the LSP tail end towards the LSP head end, admission control is performed at each node. Admission control during RSVP signaling is required for the following reasons:

1. The LSP may not have necessarily been computed with CSPF.
2. Even if it was computed with CSPF, the state of the available resources between the time the computation was performed and the path was signaled may have changed (e.g. because another LSP was set up, sourced at a different node).
3. The result of CSPF is only as accurate as the information in the TED (which may not always be up to date because of link advertisement throttling).

If enough resources are available at a particular node, admission control is successful, the path is set up through the node and the available resources are updated. This information is fed back into the IGP so that other nodes in the network become aware of the new state of the available resources. The information may not be immediately distributed, owing to the throttling of link-state advertisements, discussed in Section 2.4.2.

It is important to understand that the bandwidth reservations are in the control plane only and that there is no enforcement of the reservations in the data plane. This means that the data plane usage may be higher than the control plane reservation. When it is important to keep the two equal, policing must to be enforced at the ingress of the LSP to ensure that traffic stays within the bounds of the reservation.[3]

[3] It is not always required to keep the control plane and data plane usage equal. For example, overbooking can be implemented by reporting higher available resources in the control plane than in the data plane.

If not enough resources are available, it may be necessary to preempt other LSPs passing through the node. This is where the setup and hold priorities of the LSPs come into play, as explained in Section 2.4.1. If preemption cannot solve the resource problem, the reservation fails and an error message is sent to the head end.

On receipt of the admission control error message, the head end of the LSP recomputes the path. However, if the TED at the head end was not updated in the meantime, it is very likely that the same path is recomputed and the path setup fails again. The IETF standards do not specify a method to avoid this problem. In practice, two things can be done:

1. Exclude the link where the admission control failure was encountered from the CSPF computation for a period of time. Thus, the new path is guaranteed not to use the problematic link. The advantage of this approach is that it is localized to the head end and does not require any extra actions on the node where the failure occurs. The drawback is that the TED database does not get updated, so LSPs sourced by other nodes will encounter a similar failure.
2. Force the generation of a link-state advertisement on an admission control failure, regardless of the throttling mechanism. This ensures that the TED is up to date and the new path does not use the problematic link. The advantage of this approach is that the link information is updated in the TED on all the nodes in the network. The drawbacks are: (a) it requires the computation to happen after a delay, to make sure that the TED was updated, (b) it relies on help from a downstream node which may not implement the same behavior because it is not standardized in any document and (c) it generates extra link-state advertisements that need to be flooded through the network and processed by all the nodes in the network.

Note that the two approaches described above are not mutually exclusive. In fact, they complement each other and are often implemented together.

Another interesting admission control problem arises in the context of reoptimization. Recall from Section 2.4.3 that reoptimization finds a more optimal path in the network, based on new information in the TED. Switching the traffic from the old path to the new must happen without any traffic loss. Therefore, the new path must be set up before the old one is torn down. This method is known as make-before-break. After the new path is set up, traffic is switched to it and the old path is torn down. This means that for a short period of time, the forwarding state is maintained for both the old path and the new path throughout the network, causing the LSP to consume twice the forwarding resources it would normally use.

Another challenge with make-before-break arises because the new path may use some of the same links as the old path. To avoid double-counting of the resources used by the LSP, which can lead to admission control failures, it is necessary for the old path and the new path to share the bandwidth resources that they reserve. To accomplish this, two pieces of information must be conveyed to all nodes along the path: (a) the desire for reservation sharing and (b) the fact that the two paths belong to the same reservation. The shared explicit (SE) reservation style in RSVP provides support for this behavior.

Once an LSP is set up, traffic can be forwarded along it from the source to the destination. But how does traffic actually get mapped to the LSP?

2.5 USING THE TRAFFIC-ENGINEERED PATHS

The simplest, most basic way to map traffic to LSPs is through static routing. The LSR can be configured to send traffic to a destination by sending it over the LSP. However, the fact that the route must be manually configured to use the LSP is both restrictive and unscalable from an operational point of view, thus limiting widespread use.

To reap the benefits of the traffic-engineered paths, it is necessary for the routing protocols to become aware of the LSPs. From the routing protocol's point of view, an LSP is treated as an interface (a tunnel) and has a metric associated with it. The metric can be the same as that of the underlying IP path or it can be configured to a different value to influence the routing decision. Different routing protocols have different properties and therefore their use of the LSP is different.

The rule for LSP usage in BGP is that when an LSP is available to the BGP next-hop of a route, the LSP can be used to forward traffic to that destination. This property is crucial for the implementation of Layer 3 BGP/MPLS VPNs, as will be seen in the chapter discussing the basics of VPNs (Chapter 7). In a plain IP/MPLS network (non-VPN), this means that if an LSP is set up between the AS border routers (ASBRs), all traffic transiting the AS uses the LSP, with the following consequences:

1. Forwarding for transit traffic is done based on MPLS labels. Thus, none of the routers except the ASBRs need to have knowledge of the destinations outside the AS, and the routers in the core of the network are not required to run BGP. By using an LSP to carry traffic inside the domain it is thus possible to achieve a 'BGP-free core'.
2. The use of an LSP allows tight control over the path that transit traffic takes inside the domain. For example, it is possible to ensure that transit traffic is forwarded over dedicated links, making it easier to enforce service-level agreements (SLAs) between providers.

The use of LSPs by the IGPs makes it possible to mix paths determined by constraint-based routing with paths determined by IP routing. Therefore, even when traffic engineering is applied to only a portion of the network, label-switched paths are taken into account when computing paths across the entire network. This is a very important property from a scalability point of view, as will be seen in Section 2.6.1.

In the context of IGPs, there are two distinct behaviors:

1. Allow the IGP on the LSP head end to use the LSP in the SPF computation.
2. Advertise the LSP in the link-state advertisements so that other routers can also take it into account in their SPF (shortest path first).

There is often a lot of confusion about why two different behaviors are needed and how they differ. This confusion is not helped by the fact that the two behaviors are individually configurable and that vendors use nonintuitive names for the two features. To illustrate the difference between the two, refer to Figure 2.3, which shows a simple network topology, with a single LSP set up between E and D, along the path E–F–D, with a metric of 15. Note that the LSP metric in this case is smaller and therefore better than the IGP metric of the path E–F–D, which is 50.

Traffic is forwarded towards destination W from two sources, E and A. The goal is to forward the traffic along the shortest path.

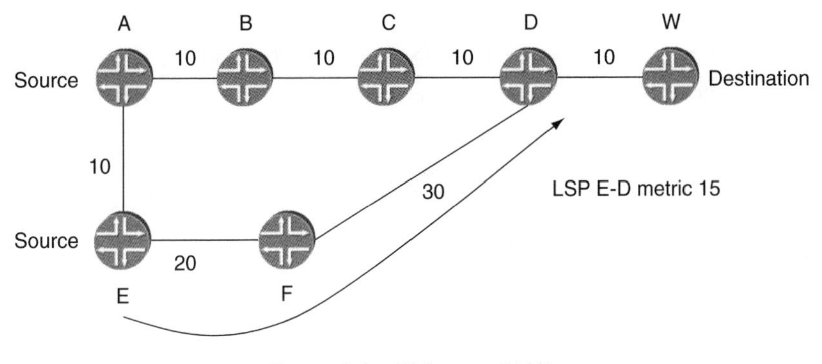

Figure 2.3 IGP use of LSPs

For source E, this means taking the LSP E–D and then the link D–W, yielding a metric of 25 (15+10). When the SPF algorithm runs at node E, in order to find this path E has to be able to take the LSP E–D into account in the SPF computation. This is the first behavior described above, called autoroute or traffic engineering shortcuts in different vendors' implementations. The concept, however, is very simple: use LSPs originating at a particular node in its SPF computation.

When source A sends traffic to destination W, the path with the smallest metric is through E and the LSP E–D, with a metric of 35 (10+15+10). However, A is oblivious of the existence of the LSP E–D, because the LSP originates at node E. For A to be able to take the LSP into account when computing its SPF, it is necessary for node E to advertise the LSP as a link in the link-state advertisements. This is the second behavior described above, called forwarding adjacency or advertise LSP in different vendors' implementations. The concept is simple: distribute the knowledge about the existence of the LSP to other nodes in the network so they can use it in their SPF computation.

Relying on LSP information distributed by other nodes can sometimes cause surprising behavior. This is because the routing decision is made based on a different router's judgment on what the shortest path should be. Let us continue the example above with a slight modification: the metric of the link E–F is 10 instead of 20, as illustrated in Figure 2.4. Because E advertises the LSP in its link-state advertisements, the node F also receives this advertisement. Consequently, F concludes that the shortest path to destination W is

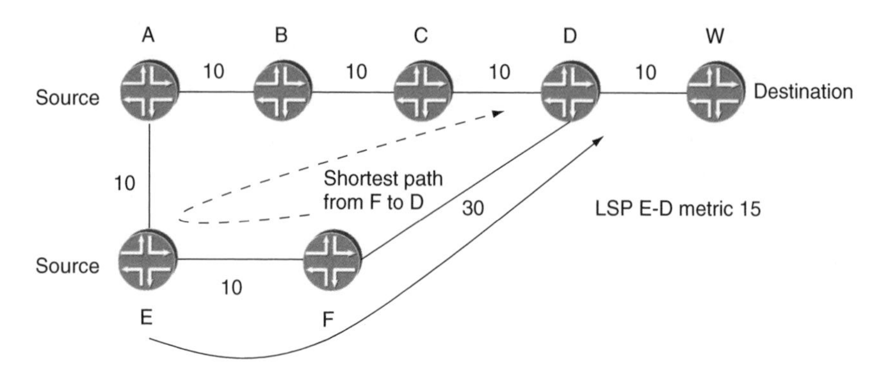

Figure 2.4 Surprising behavior when using LSPs in the shortest path computation

through E along the path F–E–LSP–D–W with a metric of 35 (10+15+10), rather than through the path F–D–W, with a metric of 40. What happens is that the traffic from F is forwarded to E and then right back to F, only to follow the same links as the pure IGP path. This happens because F has no insight into the LSP's path and relies on E's advertisement that traffic to W should be forwarded through it.

Regardless of whether the protocol used is BGP or one of the IGPs, when several LSPs are available to the same destination, most vendors allow the user the flexibility to pick one out of several LSPs for forwarding, based on various local policies. One such policy can use the class-of-service classification of the incoming IP traffic for picking the LSP. For example, best-effort traffic is mapped to one LSP, while expedited forwarding traffic is mapped to another. By manipulating the properties of these LSPs, the operator can provide more guarantees to the more important traffic. Mapping traffic to different LSPs in this way is particularly useful in the context of MPLS DiffServ-TE, as will be seen in the DiffServ-TE chapter (Chapter 4).

To summarize, the ability of the routing protocols to make use of the traffic-engineered paths set up in the network enables control over the path that transit traffic takes in a domain and allows deployment of MPLS-TE in just parts of the network. After seeing how traffic-engineered paths are computed and used, the next thing to look at are some of the considerations for deploying a traffic engineering solution.

2.6 DEPLOYMENT CONSIDERATIONS

2.6.1 Scalability

The number of LSPs that are used to implement the traffic engineering requirements is one of the major scalability concerns for MPLS-TE deployments. Two major factors impact the number of LSPs in the network:

1. The extent of the deployment. For any-to-any connectivity, a solution based on RSVP-TE requires a full mesh of LSPs. Assuming N devices meshed, this yields $O(N^2)$ LSPs. For this reason, many MPLS-TE deployments are limited to the core of the network, as explained in the Foundations chapter (Chapter 1). Solutions to this problem using LSP hierarchy are also discussed there.
2. The size of the reservations. If the size of the traffic trunk between two points exceeds the link capacity, one solution is to set up several LSPs, each with a bandwidth reservation that can be accommodated by the link capacity. Traffic can then be load-balanced between these LSPs. Although logically a single LSP is necessary, the limitation on the maximum size of the reservation causes several LSPs to be set up in this case. An increasingly popular solution to this problem is to use aggregated interfaces, where several physical interfaces are 'bundled' together and treated as a single logical interface from the routing protocol's point of view. Thus, a single LSP can be set up in this case. The downside is that failure of any interface in the bundle causes the LSP not to be able to establish.

In the context of MPLS-TE deployments, vendors typically express scaling numbers by listing the maximum number of LSPs supported. Most equipment vendors distinguish between the head end, mid-point and tail end when reporting these numbers. Typical scaling numbers are in the range of several tens of thousand LSPs. It is not uncommon that different numbers are supported for head end and transit (mid point). This is mainly because of two factors: (a) the amount of state that must be maintained is different at different points in the path and (b) the fact that the head end must perform path computation for the LSPs.

When evaluating an MPLS-TE design, an important question is whether the equipment can support the number of LSPs called for by the design. It is usually pretty straightforward to evaluate the number of LSPs for which a box is ingress or egress because this information is derived directly from the tunnel end-points.

It is less obvious how to determine the number of LSPs that may transit a particular box. The temptation is to assume that all LSPs may cross a single node in the network. This is true for networks where 'choke points' exist. Such choke points could be, for example, the PoPs connecting major regions (such as the US and Europe areas of a network). However, in most designs it is safe to assume that transit LSPs are distributed among the routers in the core. Either way, the analysis must be performed not just for the steady state but also for the failure scenarios, when LSPs reroute in the network. Finally, one factor often overlooked when computing the total number the LSPs on a box is the extra LSPs that are created due to features that are turned on in the network. One example is the bypass LSPs used for fast reroute and another example is the extra LSPs created with make-before-break on reoptimization.

The number of LSPs in the network is a concern not only because of the scaling limits of the equipment used but also because of the operational overhead of provisioning, monitoring and troubleshooting a large number of LSPs. In particular, configuring a full mesh of LSPs between *N* devices can be very labor intensive to set up and maintain. When a new LSR is added to the mesh, LSPs must be established from it to all the other *N* LSRs in the mesh. However, because LSPs are unidirectional, LSPs must also be set up from all the existing LSRs to the new LSR in the mesh. This is a problem, because the configurations must be changed on *N* different devices.

The RSVP automesh capability discussed in the Foundations chapter (Chapter 1) automates the process of LSP initiation. The IGPs are extended to carry information regarding the membership of an LSR to a particular LSP mesh. When a new LSR is added to the mesh, all the existing LSRs find out about it automatically and can initiate LSPs back to it.[4]

While RSVP automesh can help alleviate the provisioning challenge of dealing with a large number of LSPs, the burden of monitoring and troubleshooting the LSPs still remains. Operators

[4] This solution assumes that the properties of the LSPs are fairly uniform for all LSPs originating at a particular LSR or that mechanisms such as autobandwidth (discussed later in this chapter) are used to handle properties that are not uniform.

use automated tools for monitoring, e.g. sending test traffic at regular intervals, gathering statistics and doing Simple Network Management Protocol (SNMP) queries. When the number of LSPs is very large, these operations may take a large amount of resources on the router and of bandwidth in the network. Thus, a tradeoff must be made between the polling frequency and the number of LSPs supported.

The number of LSPs in the network is perhaps the most important deployment consideration for an MPLS-TE network. We have seen that the reservation size impacts the number of LSPs that must be created. However, the reservation size has other effects as well, discussed in the next section.

2.6.2 Reservation granularity

The size of the individual reservations affects not just the number of LSPs that are set up in the network but also the ability to find a path, especially following a failure and the overall utilization of the links. Link capacity is the gating factor on the size of the reservation. In the previous section we saw how this impacts the number of LSPs when the reservation requests exceed the link capacity. In a network that uses links of different capacities, using the minimum link capacity as a gating factor ensures that paths can be established across any of the links. This is especially important when rerouting following a failure. The downside is that using a smaller reservation size creates more LSPs in the network and introduces the challenge of efficiently load balancing the traffic over the LSPs.

The granularity of the reservation can also affect the efficiency of the link utilization. In a network with links of equal capacity, if all the reservations are close to the maximum available bandwidth on each link, there will necessarily be unutilized bandwidth on all links that cannot be used by any of the reservations. In this case it might have been preferable to set up several reservations of sizes such that all the available bandwidth could be used. For example, if all links are of capacity 100 Mbps and all LSPs require 60 Mbps, better utilization can be achieved if instead of a single 60 Mbps reservation, several 20 Mbps reservations are made.

The approach of setting up several LSPs rather than one is not always applicable. The LSPs may not have the same delay and jitter properties, and for this reason balancing the traffic between them may not always be possible. For example, if all 60 Mbps in the previous example are used for a large data transfer by a single application, sending the packets over the different LSPs may cause them to arrive out of order at the destination. However, if the LSP is set up between two BGP peers and is used to carry traffic to destinations advertised using BGP, then packets can be load-balanced across the LSPs based on the destination address.

The preemption regime employed in the network can also impact the MPLS-TE performance. A common rule of thumb is to preempt smaller LSPs rather than large ones. This is done not only to keep the large LSPs stable but also because small LSPs have a better chance of finding an alternate path after being preempted. This method is also useful in avoiding bandwidth fragmentation (having unutilized bandwidth on the links). This is because smaller LSPs are more likely to be able to establish with the 'leftover' bandwidth remaining after the setup of larger LSPs.

2.6.3 Routing challenges

From a routing point of view, the challenges created by having LSPs in the network stem more from different implementation decisions by different vendors' software than from the technology itself. One aspect in which implementations differ is the default behavior with regards to LSP usage by the BGP protocol. BGP route advertisements include a next-hop address, which is the address of the next BGP router in the path to the destination. Thus, to forward traffic to that destination, a path to the next-hop address must be found. This process is called resolving the route. By default, all vendors use LSP next-hops to resolve VPN-IP routes. As explained in the L3 VPN introductory chapter (Chapter 7), this is required because VPN-IP routes are used for forwarding labeled traffic. However, the default resolution regime of non-VPN-IP routes differs from one vendor to another. The issue here is not one of correct versus incorrect behavior, especially because implementations typically allow the user to control the behavior through the configuration, but rather it is an issue of being aware that such differences may exist and accounting for them properly.

So far, we have seen some of things that must be taken into account when deploying a traffic engineering solution. Next, we will take a look at one of the most popular applications for traffic engineering, namely the optimization of transmission resources.

2.7 USING TRAFFIC ENGINEERING TO ACHIEVE RESOURCE OPTIMIZATION

One of the most popular applications of traffic engineering is for the optimization of transmission resources. In this context, traffic engineering is deployed in one of two ways:

1. Selectively deployed only in parts of the network. The goal in this case is to route traffic away from a congested link. This can be thought of as a tactical application, aimed at solving an immediate resource problem.
2. Deployed throughout the entire network. The goal is to improve the overall bandwidth utilization and by doing so, delay costly link upgrades. This can be thought of as a strategic application of the technology, aimed at achieving a long-term benefit.

Both applications solve valid problems and the terms 'tactical' and 'strategic' should not be assigned any negative or positive connotations.

The classic example of a tactical MPLS-TE deployment is the problem of a scheduled link upgrade that gets delayed. What is needed is a temporary solution to move some of the traffic away from the link until the upgrade actually takes place. Another example is the requirement to optimize a particularly expensive resource, such as an intercontinental link.

The classic example of a strategic MPLS-TE deployment is traffic engineering the core of the network (the WAN links). Another example is a network spanning several geographic locations, where traffic engineering is required in only some of the regions. For example, a network with a presence in both the US and Asia may run traffic engineering only in the Asia region, where traffic rates are high and links run at high utilization.

Regardless of the type of deployment, when optimizing resource utilization using RSVP-TE, the assumption is that the following information is available:

1. The bandwidth requirement for the LSP at the head end.
2. The available bandwidth at each node in the network.

In real deployments, however, this necessary information may not always be readily accessible. The following sections discuss how to deal with missing information.

2.7.1 Autobandwidth – dealing with unknown bandwidth requirements

For both tactical and strategic deployments, the first requirement for setting up a traffic-engineered LSP is to know how much bandwidth to request for it. Estimating this information can be done by looking at traffic statistics such as interface or per-destination traffic statistics or by setting up an LSP with no bandwidth reservation and tracking the traffic statistics for this LSP. Once the traffic patterns are known, an LSP can be set up for the maximum expected demand.

The problem with this approach is that typically the bandwidth demands change according to the time of day or day of the week. By always reserving bandwidth for the worst-case scenario, one ends up wasting bandwidth rather than optimizing its utilization. A more flexible solution is to allow the LSP to change its bandwidth reservation automatically, according to the current traffic demand.

This solution is called autobandwidth. The ingress router of an LSP configured for autobandwidth monitors the traffic statistics and periodically adjusts its bandwidth requirements according to the current utilization. A new path is computed to satisfy the new bandwidth requirements, in a make-before-break fashion. Once the path is set up, traffic is switched to it seamlessly, without any loss. Autobandwidth is not defined in the IETF standards, but rather it is a feature that vendors have implemented to address the problem of traffic engineering when the bandwidth constraints are not known.

2.7.2 Sharing links between RSVP and other traffic – dealing with unknown bandwidth availability

The bandwidth reservation model works under the assumption that the reservations on a link accurately reflect the traffic that is crossing the link. This assumption can break in two cases:

1. Traffic is not kept within the limits of the reservation. The implications of not keeping the traffic within the reservation limits and the use of policers for doing so are discussed in more detail in Section 4.4.7 of the DiffServ-TE chapter (Chapter 4).

2. Not all traffic traversing the link is accounted for. This can be the case when there is a mix of IP and MPLS traffic or a mix of LDP and RSVP traffic on the links, which is usually the case for a tactical MPLS-TE deployment.

The problem with having a mix of RSVP and non-RSVP traffic on a link is that bandwidth accounting breaks. A common misconception is that RSVP traffic is somehow special because it was set up with resource reservations. This is not true. The RSVP reservation exists in the control plane only, and no forwarding resources are actually set aside for it. This fact is often overlooked in network designs, especially ones for converged networks, where some of the traffic must receive better QoS than others. The result is a solution that relies on RSVP with resource reservations to carry the QoS-sensitive traffic and uses LDP for the best-effort traffic. The problem with such a solution is that both the RSVP and the LDP traffic crosses the same links. Currently, routers take into account only RSVP reservations when reporting available resources and when doing admission control. Because the bandwidth utilized by LDP is not accounted for, the bandwidth accounting is not accurate and there is no guarantee that the RSVP reservations will actually get the required bandwidth in the data plane.

One solution to this problem is to rely on DiffServ and map RSVP traffic to a dedicated scheduler queue. In this model, the bandwidth that is available for RSVP reservation is the bandwidth pool allocated to the scheduler queue. Another solution is to estimate the amount of bandwidth used by 'other' (IP/LDP) traffic and reduce (through configuration) the link bandwidth that is available for RSVP reservations. This approach works as long as the non-RSVP traffic does not exceed the bandwidth set aside for it. Statistics monitoring can be used to estimate the traffic demand in the steady state, but no mechanism is available to react to changes in the non-RSVP traffic dynamically (e.g. following a link break somewhere else in the network). Offline tools can help evaluate better the amount of bandwidth set aside for non-RSVP traffic by simulating failure events in the network and how they impact the requirement for bandwidth for 'other' traffic on the links.

The important thing to remember is that bandwidth reservations are not a magic bullet. Unless the bandwidth consumption is correctly evaluated, bandwidth reservations do not give any of the guarantees that MPLS-TE strives to achieve. This is particularly important in networks where RSVP is used locally to route traffic around points of congestion.

2.7.3 Other methods for optimization of transmission resources in MPLS networks

The only solution presented so far for doing resource optimization in an MPLS network is traffic engineering with RSVP-TE. However, most MPLS deployments use LDP for label distribution. In an LDP network, the proposition of adding a second MPLS protocol for the sole purpose of achieving resource optimization may not be an attractive one.

An alternative way of doing resource optimization in LDP networks is based on the observation that LDP label-switched paths follow the IGP. Thus, by traffic engineering the IGP paths, the LDP LSPs are implicitly traffic-engineered. A real-world example of a traffic-engineered LDP deployment was presented at the Nanog33 conference [LDP-TE]. The goal was to achieve better resource usage by allowing a higher percentage of the link to be utilized before triggering an upgrade. Traffic engineering the IGP paths was accomplished by manipulating the link metrics.

There are two main challenges when doing traffic engineering through IGP metric manipulation:

1. Changing the IGP metric on one link in one part of the network may impact routing in a different part of the network.
2. To allow higher link utilization safely, it is necessary to prove that traffic on any given link does not exceed 100% under any kind of failure.

Being able to analyze both these factors requires the ability to simulate the network behavior with different link metrics and under different failure scenarios. Thus, an offline tool is required both for planning and for validation of the design. This means that the metrics are computed offline, based on the current traffic information and after simulating different types of failures in the network. Once the metrics are set in the network, the link utilization is monitored to detect when it becomes necessary to reoptimize the computation. It is not the intention to modify the IGP metrics on a failure, because this approach would not be feasible from an operations point of view. Instead, the IGP metrics are chosen in such a way that even under failure conditions no link gets overloaded.

Given all these constraints, the question is how good is the result obtained through metric manipulation when compared to explicit routes computed with constrained routing and signaled with RSVP-TE. The answer is that it does not matter, as long as doing traffic engineering improves the existing situation by an amount that justifies the extra work involved. When choosing a metric-based approach over explicit routing, the operator is making a conscious decision to trade off some of the benefits of explicit routing, such as unequal-cost load sharing or fine-grained traffic engineering, for the sake of a simpler network design. Test results from one vendor [IGP-TE, LDP-TE] show, not surprisingly, worse results using a metric-based approach than explicit routing, but much better results than no traffic engineering at all.

2.8 OFFLINE PATH COMPUTATION

Traffic engineering with RSVP-TE relies on explicit paths. Most of the discussion so far focused on a model where the paths are computed dynamically by the routers. As seen in previous sections, the results of this computation may not be the most optimal. Offline computation tools are used to provide better results. This model is particularly familiar to operators from an ATM PVC (permanent virtual channel) background.

Offline computation tools provide the following advantages in the context of traffic engineering:

1. Exact control of where the paths are placed. The operator knows where the traffic is going to flow. There are no surprises from dynamic computation.
2. Global view of the reservations and of the bandwidth availability. As seen in Section 2.4.3, this global knowledge enables optimal placement of the LSPs.
3. Ability to cross area and AS boundaries. The computation is not based solely on the information in the TED; therefore, the restriction to a single IGP area does not apply.[5]
4. Computation can take into account both the normal and the failure cases. One of the biggest strengths of offline tools is that they can take into account the impact of one or more link failures

[5] This assumes that the TE information for the other area/AS is available through some other means.

when computing the optimal placement of LSPs. Doing so can ensure that LSPs will always be able to reroute following a failure. Figure 2.5 shows an example of such a scenario. Assume all links are 100 Mbps and three LSPs are set up: LSP1 on C–G–D with 80 Mbps, LSP2 on C–D with 30 Mbps and LSP3 on C–E–F–D with 40 Mbps. Under this setup, a failure of the link G–D will cause LSP1 not to be able to reestablish, because none of the alternate paths has the capacity to accommodate it. If instead LSP2 and LSP3 were taking the same links, LSP1 could have rerouted under failure.

5. Optimality of the solution. The computation is done offline and can take a long time to complete. More sophisticated algorithms than CSPF can be employed to look for the most optimal solution. The solution can be optimized for different factors: minimize the maximum bandwidth utilization on all links, minimize the number of changes to existing LSPs, achieve protection in case of a single failure and so on. Perhaps the biggest advantage of an offline computation tool is that it can perform optimizations taking into account all the LSPs in the network, while CSPF can take into account only the LSPs originated at the node performing the computation.

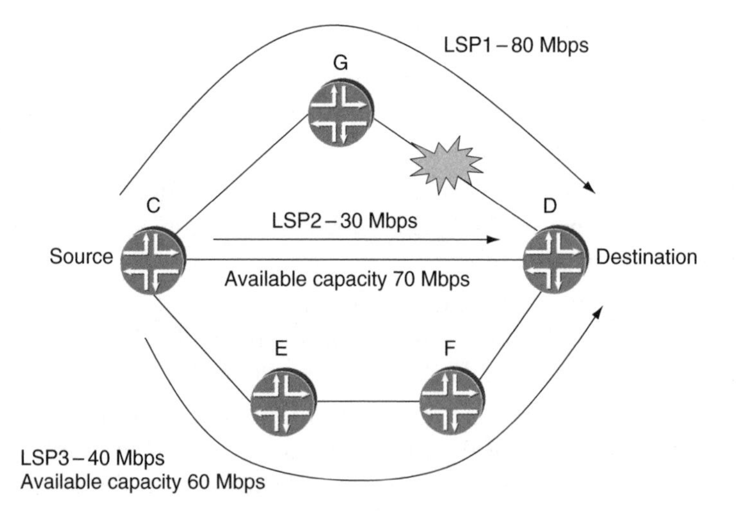

Figure 2.5 LSP placement must take into account failure scenarios

The benefits of offline computation come at a cost. Here are a few of the challenges of offline tools:

1. Input to the computation. The result of the computation is only as good as the data that the computation was based on. The traffic matrix, the demands of the LSPs and the available bandwidth must be correctly estimated and modeled.
2. Global versus incremental optimizations. As network conditions change, the computation must be repeated. The result of the new computation may require changes to a large number of LSPs and configuration of a large number of routers. To perform configuration changes, routers are typically taken offline for maintenance. For practical reasons it may not be desirable to use the result of a computation that calls for a lot of changes in the network. Instead, an incremental optimization may be more appealing: one that strives to take into account the new network conditions while leaving as many of the LSPs in place as possible. The result of an incremental optimization is necessarily worse than that of a global optimization, but the tradeoff is that fewer routers need to be reconfigured.
3. Order of the upgrade. Following a recomputation, it is not enough to know the paths of the new LSPs, but also in which order these LSPs must be set up. This is because the reconfiguration of the routers does not happen simultaneously, so an old reservation setup from router A that is due to move may still be active and take up the bandwidth on links that should be used by a new reservation from router B.
4. Limitations of the computation. The result of the computation assumes certain network conditions (such as a single failure in the network). To respond to changing conditions in a network, such as a link cut, the computation must be redone. However, the computation is fairly slow and applying its result requires router configuration, which is not always possible within a short time window. Therefore, reacting to a temporary network condition may not be practical. By the time the new computation has been performed and the changes have been applied in the network, the failure might already be fixed.

Offline computation tools can ensure optimality of the paths at the cost of the effort required to keep the simulation results and the network state synchronized. Operators have the choice of using (a) offline computation for the primary and the secondary (backup) paths, (b) offline computation of the primary and dynamic computation of the secondary paths or (c) dynamic computation for both primary and secondary paths (secondary paths are discussed in detail in the Protection and Restoration chapter, Chapter 3). Offline tools are available from several vendors, including Wandl (www.wandl.com), Cariden (www.cariden.com) and Opnet (www.opnet.com). Some operators develop their own simulation and computation tools in-house, tailored to their own network requirements.

2.9 CONCLUSION

We have seen how MPLS-TE can be used to build paths with bandwidth guarantees and how paths can be made to avoid certain links by marking such links with the appropriate administrative value and excluding them from the path computation. Using the traffic-engineered path, it is possible to achieve efficient bandwidth utilization, guarantees regarding resource allocation in times of resource crunch and control over the path that the traffic is taking in the network.

However, the traffic engineering solution presented so far has three limitations:

1. It operates at the aggregate level across all the DiffServ classes of service and cannot give bandwidth guarantees on a per-DiffServ-class basis.
2. It is limited to a single IGP area and to a single AS.
3. It provides no guarantees for the traffic during failures.

In Chapters 4 and 5 we will see how the traffic engineering solution is extended to overcome the first two limitations, using MPLS DiffServ Aware TE and interdomain traffic engineering. In Chapter 3 we will look at mechanisms available for protection and restoration, which overcome the third limitation listed above.

2.10 REFERENCES

[IGP-TE] A. Maghbouleh, *Metric-Based Traffic Engineering: Panacea or Snake Oil? A Real-World Study*, presentation at Nanog 27, http://www.nanog.org/mtg-0302/arman.html

[LDP-TE] M. Horneffer, *IGP Tuning in an MPLS Network*, presentation at Nanog 33, http://nanog.org/mtg-0501/horneffer.html

[RFC2702] D. Awduche, J. Malcolm, J. Agogbua, M. O'Dell and J. McManus, *Requirements for Traffic Engineering over MPLS*, RFC 2702, September 1999

[RFC3209] D. Awduche *et al.*, *RSVP-TE: Extensions to RSVP for LSP Tunnels*, RFC 3209, September 2001

[RFC3212] B. Jamoussi, L. Andersson, R. Callon, R. Danter, L. Wu, P. Doolan, T. Worster, N. Feldman, A. Fredette, M. Girish, E. Gray, J. Heinanen, T. Kilty, and A. Malis, *Constraint-Based LSP Setup Using LDP*, RFC3212, January 2002

[RFC3468] L. Andersoon and G. Swallow, *The Multiprotocol Label Switching (MPLS) Working Group Decision on MPLS Signaling Protocols*, RFC3468, February 2003

[RFC3630] D. Katz, K. Kompella and D. Yeung, *Traffic Engineering Extensions to OSPF*, RFC3630, September 2003

[RFC3784] H. Smit and T. Li, *IS-IS Extensions for Traffic Engineering*, RFC3784, June 2004

2.11 FURTHER READING

[Awduche Jabbari] D. Awduche and B. Jabbari, 'Internet traffic engineering using multiprotocol label switching (MPLS)', *Journal of Computer Networks* (Elsevier Science), **40**(1), September 2002

3

Protection and Restoration in MPLS Networks

3.1 INTRODUCTION

In the Traffic Engineering chapter (Chapter 2) we have seen how MPLS traffic engineering allows operators to carry traffic with stringent QoS guarantees such as voice and video. However, these applications require high-quality service, not just when the network is in a normal operating condition but also following a failure. Voice and video are referred to as 'fragile traffic' because they are real-time in nature and therefore cannot recover from traffic loss using retransmissions. Therefore, protection and restoration mechanisms are necessary to handle the failure case quickly. The ability to provide such fast protection is essential for converging voice, video and data on to a single MPLS network infrastructure.

This chapter deals with protection and restoration in MPLS networks. We will start by discussing the use of bidirectional forwarding detection (BFD) for fast-failure detection. Then we will take a look at path protection and at fast reroute using local protection and will see why MPLS-TE has become a synonym for fast reroute in MPLS networks. Finally, we will look at work currently in progress for protecting MPLS traffic that is not forwarded along a TE path, such as LDP traffic. This chapter assumes familiarity with RSVP and with basic TE concepts.

MPLS-Enabled Applications: Emerging Developments and New Technologies Ina Minei and Julian Lucek

3.2 THE BUSINESS DRIVERS

Traditionally, providers used IP/MPLS backbones to carry traffic with loose service level agreements (SLAs) and TDM networks for traffic with tight SLAs. Converging all services on to the same core is attractive because it eliminates the need to build and maintain separate physical networks for each service offering and because the flexibility of IP enables new services such as videotelephony integration. However, traffic with tight SLAs such as voice, video or ATM CBR has stringent requirements for availability and traffic loss. Thus, fast recovery following a failure is an essential functionality for multiservice networks.

One way to provide fast recovery following a link failure is to provide protection at Layer 1. This is the solution provided by SONET APS (Automatic Protection Switching). The idea is simple. Maintain a standby link that is ready to take over the traffic from the protected one in case of failure and switch traffic to it as soon as the failure is detected. Because the decision to move to the standby link is a local one, the switchover can happen within 50 ms, making any disruption virtually unnoticeable at the application layer. The quick recovery comes at the cost of maintaining the idle bandwidth and the additional hardware required for the switchover.

The goal of MPLS fast reroute (FRR) is to provide similar guarantees for MPLS tunnels. The advantage of fast reroute over SONET APS is that (a) it is not limited by the link type, (b) it offers protection for node failures and (c) it does not require extra hardware. However, the functionality is currently supported only for RSVP-TE. For a provider contemplating the deployment of a network requiring subsecond recovery (such as voice-over IP) the first question to ask is whether MPLS FRR is the only option.

Exactly how much loss can be tolerated by a particular application is an important consideration when choosing a protection method. Many applications do not really need 50 ms protection and can tolerate higher loss, e.g. up to 2 seconds for voice traffic.[1] As cell-phone users can attest, customers accept less than perfect call quality if they are compensated by either low price or increased

[1] A loss of 300 ms or more will be noticed in the phone conversation; a loss of more than 2 seconds will affect the control traffic and may cause the call to be dropped.

convenience. Given the more lax requirements, some service providers may decide to deploy pure IP networks and rely on subsecond IGP convergence (which is now available from many vendors) for the protection. The main differentiator for MPLS FRR in this context is that it can consistently provide a small recovery time, while IGP convergence may be affected by factors such as when the last SPF was run, churn in a different part of the network or CPU (central processing unit) load caused by other unrelated operations. Hence, although the average IGP convergence time might be low, the upper bound on the recovery time may be relatively high.

The amount of time during which traffic is lost depends on how fast the failure is detected and how fast the traffic is switched over to an alternate path. Most of this chapter deals with the mechanisms for quickly moving the traffic to an alternate path around the point of failure. However, no matter how efficient these mechanisms are, they are useless if the failure is not detected in a timely manner. Thus, fast failure detection, though not directly related to MPLS, is an important component of MPLS protection and is assumed throughout this chapter. In the next section we will take a look at some of the challenges with fast detection.

3.3 FAILURE DETECTION

The ability to detect that a failure has happened is the first step towards providing recovery and therefore is an essential building block for providing traffic protection. Some transmission media provide hardware indications of connectivity loss. One example is packet-over-SONET/SDH (synchronous digital hierarchy), which is widely used in the network backbones and where a break in the link is detected within milliseconds at the physical layer. Other transmission media do not have this capability, e.g. Ethernet, which is commonly used in PoPs.[2]

When failure detection is not provided in the hardware, this task can be accomplished by an entity at a higher layer in the network. Let us take a look at the disadvantages of doing so, using IGP hellos as an example. The IGPs send periodic hello packets to ensure

[2] The fast detection capability has been added for optical Ethernet.

connectivity to their neighbors. When the packets stop arriving, a failure is assumed. There are two reasons why hello-based failure detection using IGP hellos cannot provide fast detection times:

1. The architectural limits of IGP hello-based failure detection are 3 seconds for OSPF and 1 second for ISIS. In common configurations, the detection times range from 5 to 40 seconds.
2. Handling IGP hellos is relatively complex, so raising the frequency of the hellos places a considerable burden on the CPU.

The heart of the matter is the lack of a hello protocol to detect the failure at a lower layer. Based on this realization, the BFD protocol was developed jointly by Juniper and Cisco. Having rapidly gained acceptance, the BFD protocol now has its own working group (with the same name) in the IETF [BFD]. So what exactly is BFD?

BFD is a simple hello protocol designed to do rapid failure detection. Its goal is to provide a low-overhead mechanism that can quickly detect faults in the bidirectional path between two forwarding engines, whether they are due to problems with the physical interfaces, with the forwarding engines themselves or with any other component. The natural question is just how quickly BFD can detect such a fault. The answer is that it depends on the platform and on how the protocol is implemented. Available early implementations allow detection times of about 100 ms, with the possibility to improve the time in the future. While this is not perfect if recovery times of 50 ms are sought, it is a huge improvement over detection times on the order of seconds and still falls within the requirements of many applications. BFD started out as a simple mechanism intended to be used on Ethernet links, but has since found numerous applications. We will see one such application in the context of LSP failure detection in the chapter discussing management of MPLS networks (Chapter 12).

It is beyond the scope of this book to describe the details of the BFD protocol, its packet formats and processing rules, which are explained in detail in the relevant IETF drafts [BFD-BASE, BFD-MHOP]. From the point of view of MPLS protection and restoration techniques, BFD is simply a tool for solving the fast detection problem. With the knowledge that this tool exists, the problem of fast detection can be considered to be solved for all media types. Therefore, in the rest of the chapter, fast failure detection is assumed.

Let us now turn our attention to the mechanisms available for actually protecting the traffic: end-to-end (path) protection and hop-by-hop (local) protection.

3.4 END-TO-END PROTECTION

The first type of protection discussed is end-to-end protection, also known as path protection. Although not as popular as local protection using fast reroute, it is important to examine it because it highlights some of the issues solved by local protection.

A common practice in network deployments is the use of a primary backup approach for providing resiliency. Following this model, LSP protection is achieved using two LSPs: the primary, used under normal operation, and the secondary, used if there is a failure on the primary. For example, in Figure 3.1 LSP2 (S–R4–D) provides path protection for LSP1 (S–R1–D). For fastest recovery times, the secondary is presignaled and ready to take over the traffic, in effect being in hot standby mode. When a failure (such as an interface down event) is detected on the primary LSP, an RSVP error is propagated to the LSP head end. Upon receipt of this error message, the head end switches the traffic to the secondary. The problem is that

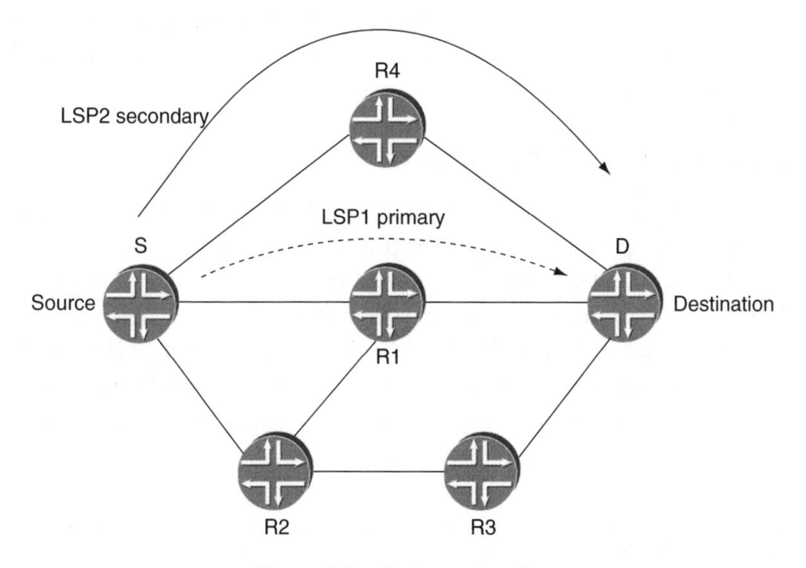

Figure 3.1 Path protection

until the error reaches the head end, traffic continues to be sent over the primary. If the secondary is not presignaled, the extra time required to set it up further increases the switchover delay.

From the example, several properties of path protection become apparent.

Control over the traffic flow following a failure

The use of a presignaled secondary path is very powerful because it provides exact knowledge of where the traffic will flow following the failure. This is important not just for capacity planning but also for ensuring properties such as low delay. Note that the same control can be achieved even if the secondary is not in standby mode, if its path is explicitly configured.

Requirement for path diversity

For the secondary to provide meaningful protection in case of a failure on the primary, it is necessary that a single failure must not affect both the primary and the secondary. Clearly, if both LSPs use a common link in their path, then they will both fail when the link breaks. To avoid this, the primary and the secondary must take different paths through the network. Path diversity is relatively easy to achieve when the LSPs are contained within a single IGP area and many implementations attempt to provide this functionality by default. However, in the chapter discussing Interdomain TE (Chapter 5), we will see that it is not trivial to ensure for LSPs that cross domain boundaries.[3]

Double-booking of resources

The secondary LSP is usually set up with the same resource reservations as the primary to ensure the same quality of service when the traffic moves from the primary to the secondary. The net result is that twice as many resources are used throughout the network if the secondary is set up before the failure. This problem could be avoided if the secondary were not presignaled, at the expense of a longer switchover time. Assuming that the secondary is presignaled and

[3] Unfortunately, path diversity alone does not guarantee that the primary and secondary will not share the same fate when a resource fails. Fate sharing is discussed in detail later in this chapter.

therefore reserves resources, an interesting situation can arise when there is a resource shortage in the network: secondary LSPs that effectively carry no traffic may reserve bandwidth while other primary LSPs may fail to establish. To prevent this situation, some providers choose to use LSP priorities and assign better values to all the primary LSPs in the network, to ensure they can always establish.

Unnecessary protection

End-to-end protection protects the entire path. Thus, even if most links in the primary paths are protected using other mechanisms (such as APS), it is not possible to apply protection selectively for just those links that need it.

Nondeterministic switchover delay

The delay in the switchover between the primary and the standby is dictated by the time it takes for the RSVP error message to propagate to the LSP head end. This is a control plane operation and therefore the time it takes is not deterministic. For example, if the CPU is busy processing BGP updates at the time of the failure, there may be a delay in the propagation of the RSVP error. Moreover, unless the secondary is set up in the standby mode, further delay is incurred by RSVP signaling of the secondary path.

The main advantage of end-to-end path protection is the control it gives the operator over the fate of the traffic after the failure. Its main disadvantages are double-booking of resources, unnecessary protection for links that do not require it and nondeterministic switchover times. They arise from the fact that the protection is provided by the head end for the entire path. Local protection attempts to fix these problems by providing the protection locally rather than at the head end and by protecting a single resource at a time.

3.5 LOCAL PROTECTION USING FAST REROUTE

The goal of protection is to minimize the time during which traffic is lost. Thus, it makes sense to apply protection as close to the point of failure as possible. The idea of local protection is simple. Instead of providing protection at the head end for the entire path, the traffic around the point of failure is rerouted. This is very similar to what happens when the highway between two cities closes somewhere between exits A and B. Rather than redirecting all the traffic away from the highway altogether, vehicles are directed on to a detour path at exit A and rejoin the highway at exit B or at some other exit down the road from B.

The use of a detour is a very intuitive concept, easily applicable to TE LSPs, as shown in Figure 3.2. An alternate path, called the detour or bypass, exists around link R1–R2. In case of a failure, traffic is shuttled around the failed link using this path and rejoins the LSP at R2. Thus, the traffic is quickly rerouted around the point of failure and for this reason this mechanism is called fast reroute. The idea is not to keep the traffic on the detour until the link recovers, but rather to keep it long enough for the LSP head end to move the LSP to a new path that does not use the failed link. There are several attractive properties to fast reroute:

1. A single resource is protected and therefore it is possible to pick and choose which resource to protect.
2. Protection can be applied quickly because it is enforced close to the point of failure.
3. Traffic is forwarded around the failure over the detour/bypass, on a path that is known before the failure happens.

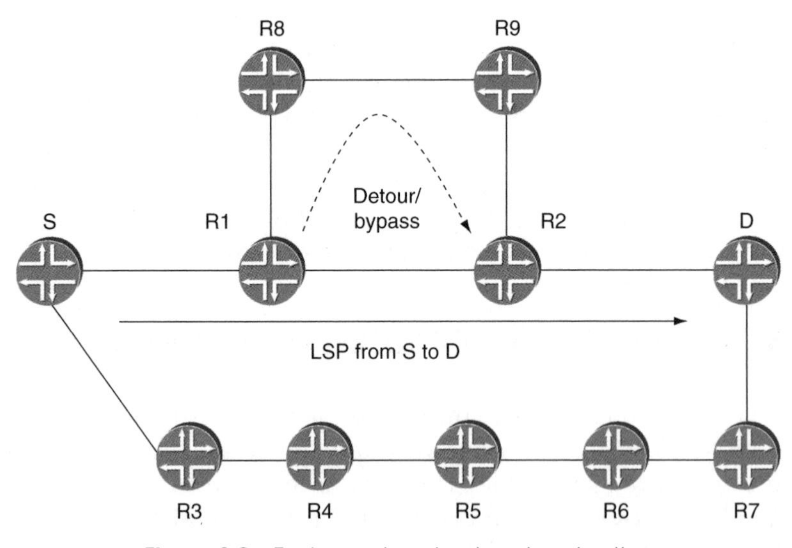

Figure 3.2 Fast reroute using local protection

If fast reroute is so intuitive and brings so many advantages, why is it not available in IP networks? The answer is because it relies heavily on source routing, where the path is determined at the source and no independent forwarding decisions are made by the individual nodes in the path. Let us see how. For local protection to work, traffic must reach the beginning of the protection path after the failure has occurred. When traffic is forwarded as IP, the forwarding decision is made independently at every hop based on the destination address. In Figure 3.2, link R1–R2 is protected by a detour along R1–R8–R9–R2. All the link metrics are equal to 1, except link R8–R9, which has a metric of 10. If after the failure node S computes its shortest path as S–R3–R4–R5–R6–R7–D and redirects the traffic towards R3, the packets will not reach the detour/bypass. Furthermore, until router R3 has also performed its path computation, its best path for destination D points back towards S (because the path R3–S–R1–R2–D is shorter than the path R3–R4–R5–R6–R7–D).

Source routing is one of the most powerful properties of TE LSPs. The LSP is set up along a path determined at the head end. Once traffic is placed into the LSP, it is guaranteed to be forwarded all the way to the tail end, regardless of the routing changes that happen in the network. Thus, traffic is always guaranteed to reach the beginning of the detour/bypass. Once it rejoins the LSP at R2, it is guaranteed to reach the tail end.

The mechanisms for providing fast reroute in MPLS networks were developed in the MPLS Working Group in the IETF and are documented in [RFC4090]. Local protection mechanisms are qualified based on two criteria:

1. The type of resource that is protected, either a link or a node. Thus, local protection is either link protection or node protection. As we will see in later sections, this influences the placement of the backup path. Regardless of the protected resource, local protection mechanisms are collectively referred to as local protection or fast reroute (FRR).
2. The number of LSPs protected by the protection tunnel, either 1:1 or *N*:1. These are called one-to-one backup and facility backup respectively. The ability to share the protection paths is not an issue of scalability alone. As we will see in later sections, it also determines how traffic is forwarded over the protection path.

Because one-to-one backup and facility backup have their respective advantages, implementations exist for both these methods. Many vendors market local protection under the generic 'fast reroute' name.

The basic mechanisms of fast reroute, one-to-one backup and facility backup are described in the section dealing with link protection. The section describing node protection focuses only on the special aspects of node protection.

3.6 LINK PROTECTION

Link protection refers to the ability to protect traffic being forwarded on an LSP when a link along the LSP fails. To protect against the failure of a link, a backup tunnel is set up around the link. This backup is called a detour in the case of one-to-one protection and bypass in the case of many-to-one protection. The head end of the backup tunnel is the router upstream of the link and the tail end of the detour is the router downstream of the link (where upstream and downstream are relative to the direction of the traffic).

Figure 3.3 shows one LSP, LSPxy from X to Y, along the path X–A–B–Z–Y. Link A–B is protected by a backup LSP taking the path A–C–D–B. When link A–B fails, traffic from LSPxy (the protected path) is forwarded on this backup around the broken link at A and delivered to B, from where it continues on its normal path to

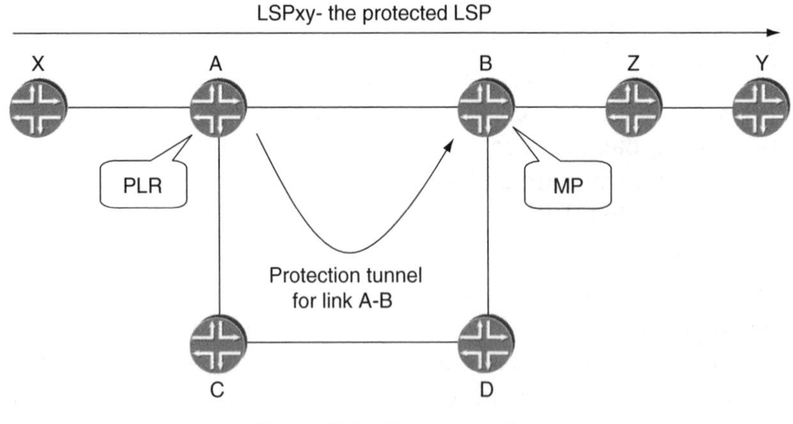

Figure 3.3 Link protection

destination Y. Node A, where traffic is spliced from the protected path on to the backup is called the Point of Local Repair (PLR) and node B, where traffic merges from the backup into the protected path again, is called the Merge Point (MP). Throughout this chapter, we will use the terms 'protected path' and 'main path' interchangeably to mean the LSP receiving protection.

Let us take a look at the different actions that need to happen before and after the failure.

3.6.1 What happens before the failure

To ensure fast protection, the backup must be ready to forward traffic when the failure happens. This means that:

1. The backup path must be computed and signaled before the failure happens and the forwarding state must be set up for it, at the PLR, MP and all the transit nodes.
2. The forwarding state must be in place at the head end of the backup tunnel (the PLR) and at its tail end (the MP) so that traffic can be forwarded into the backup at the PLR and back on to the main LSP at the MP.

Let us examine each of these separately below.

3.6.1.1 Path computation

What triggers the computation and setup of the backup path? To answer this question, let us first examine what information will cause LSR A to set up a backup path. First, LSR A needs to know that it is required to protect link A–B. Second, it must know that it is required to protect traffic flowing on LSPxy. Remember that one of the advantages of local protection is that the operator can pick and choose which resources and which LSPs to protect. For example, the operator may protect an LSP carrying voice traffic but not one carrying data. Similarly, the operator may decide not to protect a link that is already protected using APS. Thus, the operator must specify in the configuration which LSPs and which links to protect. For the link, the configuration is on the router containing the link, router A. For the LSP, the configuration is at the head end X, and therefore the information must be propagated in the RSVP Path messages for the LSP. This is done using either the 'local protection desired' flag in the Session Attribute Object or the Fast Reroute Object.[4]

Once this information is available, node A computes a protection path for link A–B by running a CSPF computation to destination B, with the obvious constraint to avoid the link A–B. The head end can signal other constraints to be applied to the backup path computation, such as the maximum number of hops that the backup is allowed to cross, the required bandwidth for the backup or its setup and hold priorities. The purpose of these constraints is to ensure that even when using the protection path, traffic continues to receive certain guarantees. These constraints are signaled from the head end to the PLR in the Fast Reroute Object. In addition, for one-to-one backup, where the backup path protects a single LSP, some of the properties of the backup path, such as bandwidth and link colors, are inherited from the protected LSP and do not require explicit signaling.

Once the backup path is computed, it is set up using RSVP. How is traffic forwarded on to it? To answer this question, let us take a look at the forwarding state that is installed.

3.6.1.2 Forwarding state installation

The goal of the backup is to carry the traffic from the protected (main) path around the failed link and merge it back into the main-path at the MP located at the other end of the failed link. Two different techniques exist for directing traffic from the backup into the mainpath, which differ in the label with which the traffic arrives at the MP. This in turn influences the number of LSPs that can be protected by a single backup tunnel, yielding either *N*:1 (facility backup) or 1:1 (one-to-one backup).

Facility backup

Traffic arrives over the backup tunnel with the same label as it would if it arrived over the failed link.[5] The only difference from the point of view of forwarding is that traffic arrives at the MP over different interfaces when arriving over the protected path and

[4] However, it is recommended that the bit (desired) in the Session Attribute should always be set if local protection is desired. Additionally, of course, the Fast Reroute Object can also be signaled.

[5] The use of a different label at the MP is not precluded in the specification. In practice, this scheme is not implemented.

over the backup. To ensure that traffic arrives at the MP with the correct label, all that needs to be done is to tunnel it into the backup by pushing the backup tunnel label on top of the protected LSP label at the PLR (label stacking) and do penultimate hop-popping for the backup tunnel label before the MP (label stacking and penultimate hop-popping are explained in the Foundations chapter, Chapter 1). Note that using this scheme, the depth of the label stack increases when traffic is forwarded over the backup tunnel.

Figure 3.4 shows the setup of the backup tunnel before the failure and the forwarding state that is installed at every hop in the path. Figure 3.5 shows traffic forwarding after a failure. In the figure, the payload happens to be an IP packet, but this is just by way of example; the packet could be anything that can be carried in an LSP. Let us take a look at some of the key properties of facility backup:

1. No new forwarding state is installed at the MP. At the PLR, the forwarding state must be set in place to push the label of the backup path (label 201 in the example) on to the labeled traffic from the protected LSP in the event of a failure.
2. Any number of LSPs crossing link A–B can be protected by the backup shown in the figure. There is no extra forwarding state for each LSP protected either at the MP or at any of the routers

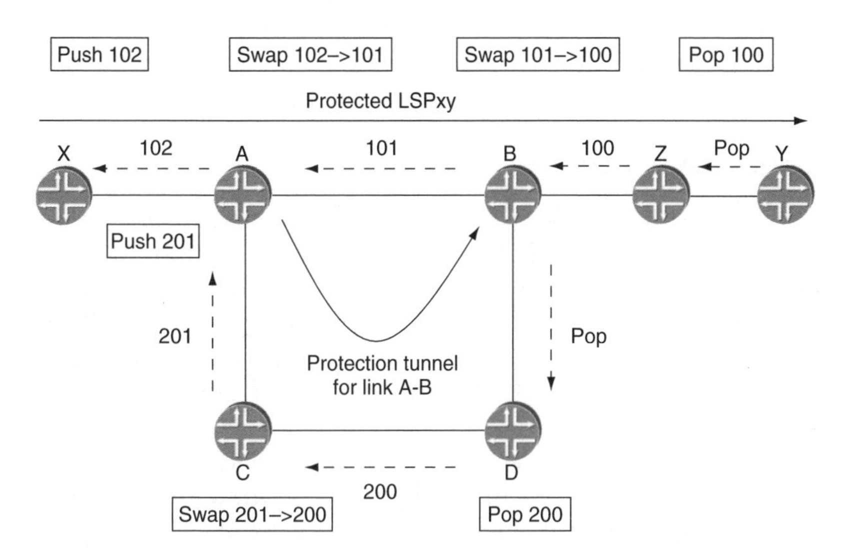

Figure 3.4 Setting up the backup tunnel for the facility backup

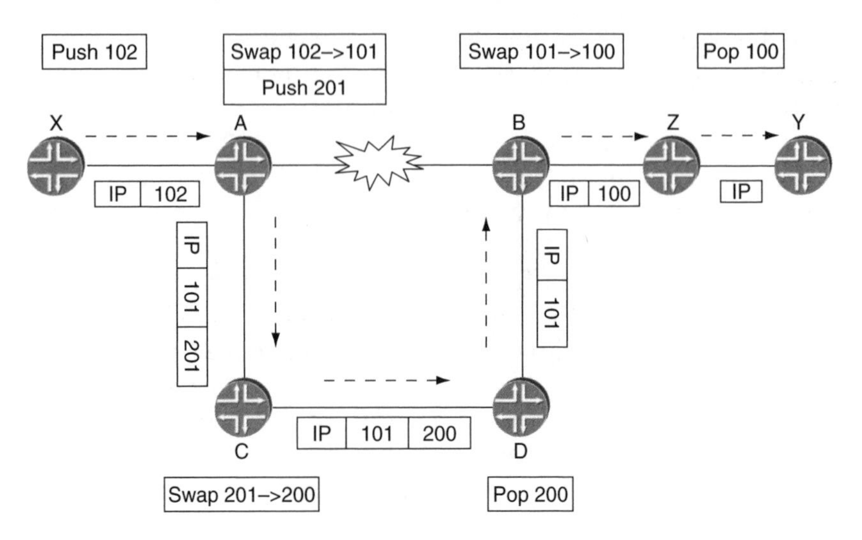

Figure 3.5 Forwarding traffic using the facility backup

in the path (e.g. routers C and D) and the action taken by the PLR is always the same: push the backup tunnel label on to the label stack of the main LSP. The ability for several LSPs to share the same protection path is an important scaling property of facility backup.

3. The label that is advertised by the MP is an implicit null label and therefore penultimate hop popping is performed for the backup tunnel. Thus, traffic arrives at the MP with the same label with which it would have arrived over the main LSP.

One-to-one backup

Traffic arrives at the MP with a different label than the one used by the main (protected) LSP. Figure 3.6 shows the setup of a one-to-one backup for the LSP from the previous example and Figure 3.7 shows forwarding over the backup following a failure. Traffic arrives at the MP with label 300, the backup tunnel label and is forwarded using label 100, the protected LSP label. Thus, the MP must maintain the forwarding state that associates the backup tunnel label with the correct label of the protected LSP. If a second LSP were to be protected in this figure, a separate backup tunnel would be required for it, and a separate forwarding state would be installed at the MP.

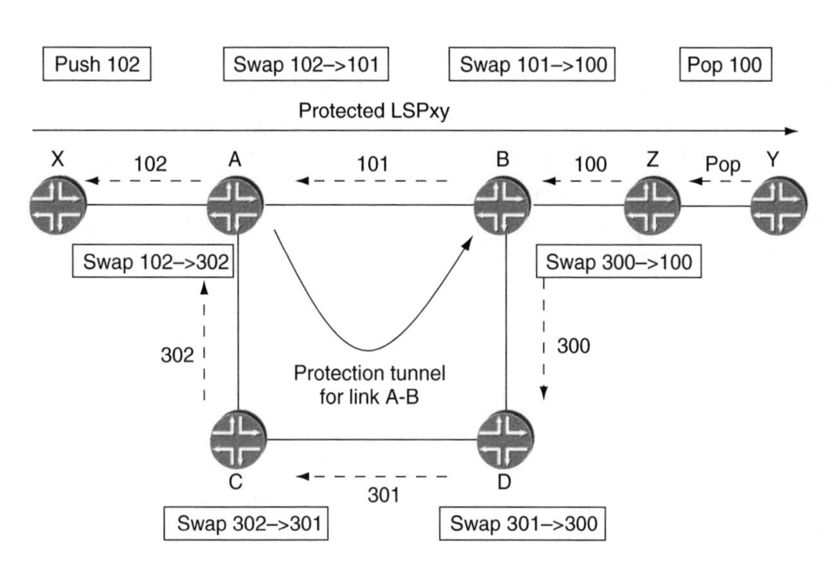

Figure 3.6 Setting up a one-to-one backup

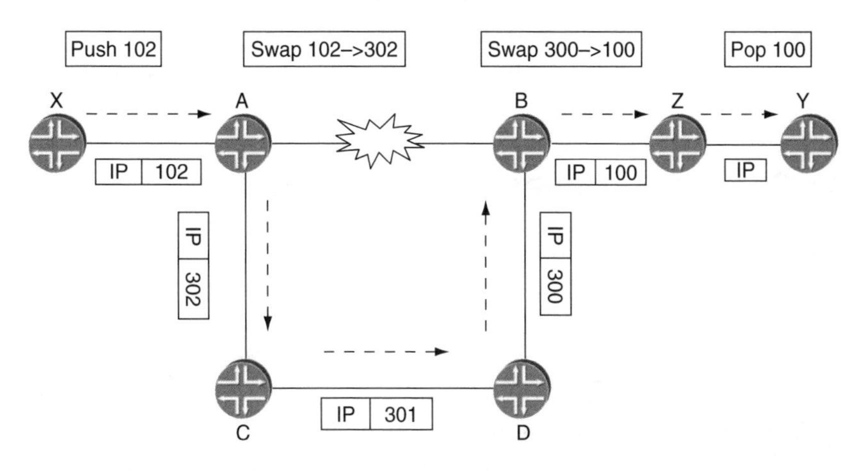

Figure 3.7 Forwarding traffic over a one-to-one backup

Similar to facility backup, the forwarding state must be set up to map traffic from the protected LSP into the backup. For example, traffic arriving with label 102, the label of the protected LSP, is forwarded over the backup using label 302, the backup tunnel label. Note that, using this approach, the depth of the label stack does not increase when packets are forwarded over the backup path, because the top label is simply swapped to the backup tunnel label.

To summarize, the use of one-to-one backup requires installing new forwarding state at both the MP and the PLR. Because the backup protects a single main LSP, the amount of state that the MP, the PLR and all the nodes in the backup path must maintain increases proportionally to the number of LSPs protected. Using separate backup tunnels means that the depth of the label stack need not increase when traffic is forwarded from the main LSP to the backup.[6] Furthermore, the fact that each backup services a single LSP allows tighter control over the backup tunnel and its properties. This is difficult with facility backup where multiple LSPs share the same backup.

Regardless of the type of protection used, the time it takes to update the state in the forwarding engine following the failure detection adds to the total protection time. This is an important, though often overlooked, component in the total protection time. To achieve fast switchover, many implementations install the forwarding state for the protection path ahead of time.

With the backup tunnel computed and signaled and the forwarding state in place to shift traffic from the main tunnel to the backup, let us now take a look at what happens after the failure of the protected link.

3.6.2 What happens after the failure

Once the failure of the protected link is detected, traffic is switched from the main LSP to the backup path. Assuming that the forwarding state was preinstalled, this operation can be done in the forwarding engine, without the intervention of the router's control plane and therefore the traffic flow is quickly restored. Would this be the end of the discussion? The answer is not quite. Following the failure, more action needs to be taken in the control plane.

Suppression of LSP teardown

Even if the protected LSP head end or tail end receives IGP notifications about the failure of the link, it must suppress any error generation that would lead to the teardown of the LSP when local protection is available. Otherwise, the purpose of local protection is defeated.

[6] Increasing the depth of the label stack was a problem in early implementations of MPLS.

Notification of the LSP head end

Remember that the purpose of the backup is to protect traffic while the LSP head end looks for an alternate path for the LSP, avoiding the failed link. For this to happen, the head end must first find out about the failure. The PLR takes care of this by notifying the head using an RSVP Path Error message with a 'Notify' error code and 'Tunnel locally repaired' subcode. In addition to this, a new flag indicating that the path is locally repaired is turned on in the Record Route Object. However, why bother with any of this when the head end will find out about the failure through the IGP anyway? Because relying on a different protocol for the failure notification will not always work. For example, when the LSP spans several IGP areas or ASs, the IGP notifications will not reach the head end.

New path computation and signaling

When the head end finds out about the switch to the backup path, it recomputes the LSP, avoiding the failed link, and sets it up in make-before-break fashion (make-before-break was discussed in the Traffic Engineering chapter, Chapter 2). This means that LSPs for which local protection is desired are always signalled as 'shared explicit', allowing the new path to share resources with the old path. It is possible for the new path to establish over the very same links used during protection. This is the case, for example, in Figure 3.4, after failure of the link A–B. Although traffic will not move to a different path, the new LSP will still be set up along this path. When the last LSP using the protection path has moved, the bypass is torn down.

RSVP message processing

The receipt of a 'Tunnel locally repaired' notification informs the head end that traffic is forwarded over a potentially suboptimal protection path. As a result, the head end attempts to re-reroute the LSP. What happens if the head end cannot find an alternate path? This can happen, for example, if the LSP is configured for an explicit path that does not allow it to move away from the failed link. The decision whether to tear down the LSP or let it stay on the protection path is one of local policy/implementation at the head end. Assuming that the policy decision allows the LSP to continue to use the protection path, the next question becomes: can the LSP stay on the protection path forever? In principle, yes, but in practice more details must be taken care of. Remember that RSVP requires periodic refreshes of its state, using Path and Resv messages. Unless these messages continue to be correctly generated and processed, the LSP will time-out. Thus, it is required to forward these messages over the backup tunnel after link failure.

To summarize, local protection is achieved through a combination of actions both before and after the failure. Before the failure, the protection tunnel must be set up and the forwarding state must be installed to switch the traffic from the main tunnel over the protection tunnel around the failed link. After the failure, actions must be taken to prevent the teardown of the main tunnel until it is rerouted. The basic mechanisms of local protection were described in the context of link protection, but they apply equally to node protection, with a few modifications, as discussed in the next section.

3.7 NODE PROTECTION

Link failures are the most common type of failure in a network. A link failure may happen because of a problem with the link itself or it may be caused by a failure of the node at the other end of the link. In the latter case, the link protection mechanisms described in the previous section will not work if they rely on the adjacent node to act as the MP. Node protection covers this case by setting up the backup tunnel around the protected node to the next next-hop in the path.

Figure 3.8 shows LSPxy from X to Y, along the path X–A–B–Z–Y. LSPxy is protected against node B's failure by a backup tunnel taking the path A–C–D–Z that merges back into LSPxy at node Z downstream from node B. When node B fails, traffic from LSPxy (the protected path) is placed on this backup at A and delivered to Z, where it continues on its normal path to destination Y.

Looking at this description and at the figure, it becomes clear that A must obtain two pieces of information to set up the backup tunnel:

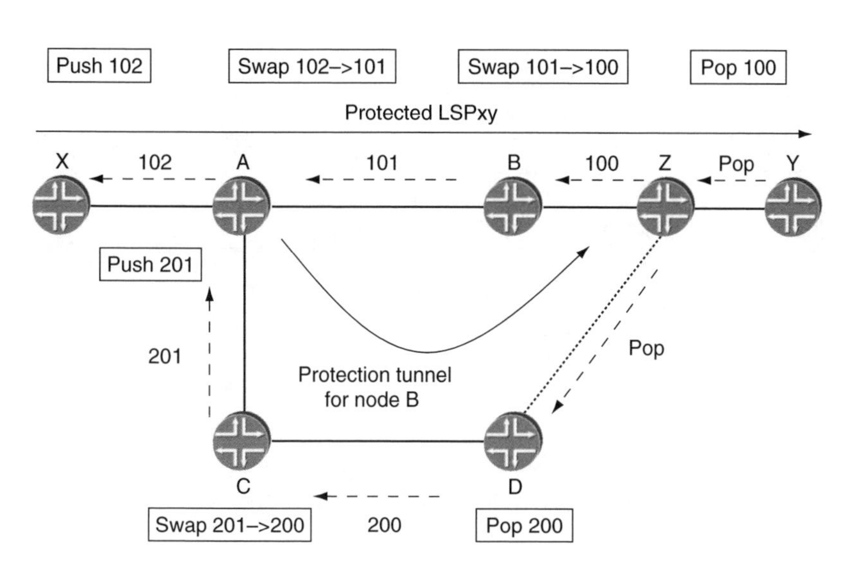

Figure 3.8 Setting up the necessary state for node protection

1. The address of node Z, the tail end of the backup tunnel. This information can be obtained from the Record Route Object (RRO). This address is used as a loose hop for reaching the MP. It can be a router ID or an interface address belonging to the MP.
2. The label used by the main LSP at node Z. Recall that when using the facility backup, traffic arrives to the MP with the same label as that used by the main LSP. Thus, A must be able to swap the incoming label 102 to the label 100, expected by node Z rather than the label 101, which is the one used in normal forwarding along the main LSP. How can A obtain this information? The answer is to use a similar approach as for the discovery of the downstream node and rely on the information in the RRO. However, the label is normally not recorded in the RRO. To solve this problem, the new flag 'label recording desired' is defined for use in the Session Attribute Object. Setting this flag indicates that the label information should be included in the RRO. As a result, labels are recorded in the RRO and becomes available to the PLR.

Given this information, the backup tunnel can be established. Figure 3.9 shows forwarding of traffic over the backup tunnel, assuming facility backup. Note that at node A traffic is already labeled with the label expected by Z before the tunnel label is pushed on to it. The rest of the mechanisms for providing node protection are very

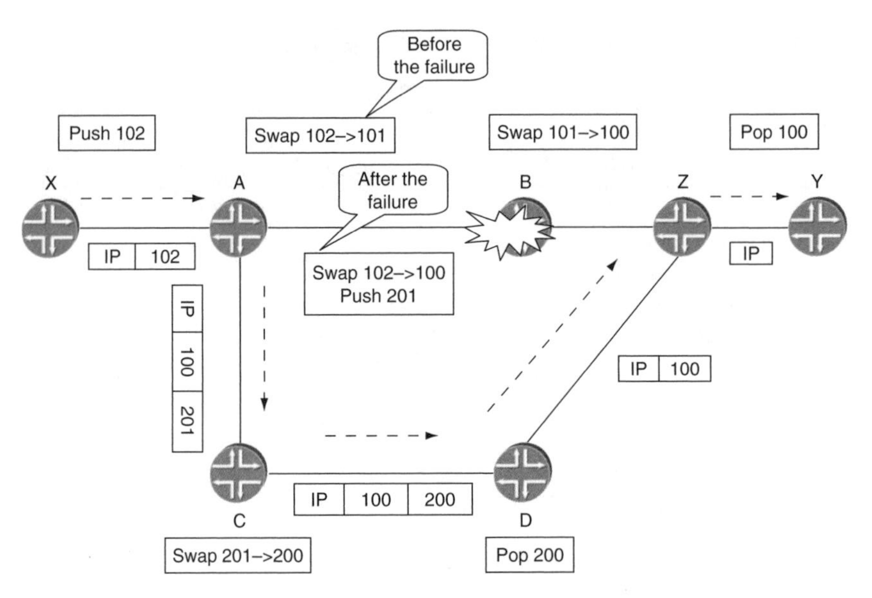

Figure 3.9 Node protection when forwarding traffic

similar to link protection and are not repeated here. Instead, let us sum up the differences between node protection and link protection:

- Node protection protects against both link and node failures.
- The MP for the backup path is a router downstream from the protected node. This backup is called the next next-hop backup.
- Label recording is required because the PLR must know the label that the MP expects the traffic to arrive with.

3.8 ADDITIONAL CONSTRAINTS FOR THE COMPUTATION OF THE PROTECTION PATH

Until now, when discussing computation of the protection path we have focused on a single constraint: avoid using the protected resource in the protection path. However, is this enough to ensure adequate protection? The answer may be 'no', on a case-by-case basis, as we will see in this section.

3.8.1 Fate sharing

Using different links or nodes in the protected and backup paths seems like a good way to ensure resiliency in the case of a failure.

However, underlying resources used in the two paths may be affected by a single event. For example, if the optical fiber for two links follows the same physical path, a single event (such as a bulldozer cutting through the fiber) affects them both. In this case, the two links are said to be in the same Shared Risk Link Group (SRLG) or fate-sharing group. Therefore, the protection path must avoid all links that are in the same SRLG with links in the main path (the protected path). This is applicable for both end-to-end path protection (discussed at the beginning of this chapter) and local protection. For example, for the path protection scenario in Figure 3.1, if links S–R1 and S–R4 belonged to the same SRLG, the secondary path would have been computed along the (less optimal) path S–R2–R3–D.

From the point of view of the CSPF computation, SRLG membership is just another constraint and it is conceptually very similar to link colors (discussed in the Traffic Engineering chapter, Chapter 2). There are two main differences between SRLG and link colors:

1. SRLG is a dynamic constraint, while link colors are static. Although SRLG membership is static, the SRLG is a dynamic constraint that changes according to the links used by the main LSP (the protected one). If the main path uses links from group x, the constraint is to avoid links from group x, but as soon as the main path reroutes and starts using links from group y, the constraint becomes to avoid links from group y. This is in contrast to link colors, where the constraint is a static one, easily expressed as a rule such as 'exclude red links', independently of the path taken by the main LSP.
2. Links from the same SRLG need not be completely excluded from the computation. CSPF does not need to exclude links that share a fate from the computation. Doing so might cause the alternate path not to establish at all if the topology is not rich enough to provide an alternate path. Instead, the cost of these links may be raised, making them less preferred in the computation. However, if no other path exists, the links can still be used, providing protection for those failures that do not affect both links. For example, in the link protection scenario in Figure 3.4 assume that links A–B and C–D are in the same risk group. Ideally, the protection path for link A–B should not cross link C–D. However, no other path exists in the given topology. In this case, it is preferable to set up the protection path through C–D anyway than not to set up any protection path at all.

So far we have seen how the SRLG information is used in the CSPF computation. How does a router find out about the SRLG membership? From an implementation point of view, there are two options on how this knowledge can be provided to the routers in the network. In both cases, the assumption is that the network operator knows which resources share what risks and can build a database of SRLG membership. The first option for distributing this information to the routers is to configure the SRLG information on the routers. The second option is to use the IGP to distribute SRLG information similarly to how other TE information is carried in the IGP. (Such extensions were defined for OSPF in the context of Generalized Multi-Protocol Label Switching, or GMPLS, extensions for OSPF.) Because this information is fairly static, carrying it in the IGP does not place a large burden on the protocol. Note that regardless of whether the SRLG information is statically configured or dynamically exchanged, the router must be configured with some measure of SRLG information: in the first case, regarding all the links in the network, in the second case, regarding membership of its own links in different SRLG groups. Both approaches are valid and the choice depends on the amount of information that must be configured and on the capabilities of the equipment used.

To summarize, taking into account fate-sharing information ensures that a single event has the least chance of impacting both the protected LSP and its protection tunnel. Fate sharing applies to both path protection and link/node protection and is implemented by computing the protection tunnel with additional constraints. The constraints remove from the computation resources that are considered to be in the same 'risk group' with those used in the protected path. Thus, the protection path is guaranteed to be able to take over traffic following a failure in the protected path. However, what if there is not enough available bandwidth on the protection path?

3.8.2 Bandwidth protection

Bandwidth protection refers to the ability to guarantee that enough bandwidth is available on the protection path. For path protection, this is achieved by setting up the secondary with the same bandwidth requirements as the primary. The consequences of doing bandwidth protection end-to-end for the entire path were discussed in Section 3.4 and are not repeated here. Instead, this section focuses on bandwidth protection in the case of local protection.

The very fact that bandwidth protection is discussed separately from the local protection mechanisms may seem counterintuitive. Without bandwidth protection, can one really ensure that no traffic is lost when switching to the backup path? And why go through all the effort of setting up a backup and switching traffic to it if packets will be lost anyway? Should bandwidth protection be an integral part of the local protection mechanisms? The answer is 'not necessarily', as we will see below.

Local protection is a short-term solution for avoiding loss by shuttling the traffic around the failed resource until the LSP reroutes to a new path. Remember that traffic is expected to stay on the protection path for a few seconds only, until the reroute happens at the head end. Even if there is not enough bandwidth on the protection path, some of the traffic will still make it through and loss will happen only for a short amount of time. This might be acceptable for some types of traffic and is an improvement when compared to total traffic loss for the same amount of time. Furthermore, links are typically not used to full capacity and some bandwidth is always available on the protection path. For example, many networks use a 50% upgrade rule, meaning that links are upgraded to higher capacity as soon as they are half full, to account for the possibility of a failure shifting traffic from a different link. For these reasons, bandwidth protection is not mandated in the local protection specification.

Bandwidth protection is an optional functionality that the head end can request for the LSP. It is signaled in the same way as link protection is signaled, by using a new flag in the Session Attribute and Fast Reroute Objects, the 'bandwidth protection' flag. The flag informs the PLR that bandwidth protection is desired. The ability of the PLR to provide the requested bandwidth protection is signaled in the same way as its ability to provide local protection, by using a flag in the Record Route Object, the 'bandwidth protection available' flag. Based on the 'local protection in-use' and the 'bandwidth protection available' flags in the Record Route Object, the head end can determine whether the LSP receives the required guarantees when it switches to the protection path. If these guarantees are not met, the head end can take action to move traffic away from this LSP at the routing level, e.g. by increasing the metric of the LSP and thus making it less desirable for use in routing.

How much bandwidth is required to protect the LSP and how should the PLR act based on this information? For one-to-one backup, the amount of bandwidth required for the backup is the same as for the protected LSP.[7] Because of the 1:1 mapping between backup paths and protected LSPs, the PLR can set up the backup path with the desired bandwidth as soon as it realizes bandwidth protection is requested.

If the same approach were used for facility backup, then separate backup paths would be built for each of the LSPs requiring protection and there would be no sharing of the backup path between several LSPs, defeating the nice scaling properties of facility backup. Many of the existing implementations deal with this problem by reversing the trigger for the setup of the backup LSP. Rather than setting up the backup based on the bandwidth of the protected LSPs, the backup is set up with a predefined bandwidth and admission control is performed for the protected LSPs into the backup based on their bandwidth requirements. For example, assume that the backup path for link A–B is set up with a bandwidth reservation of 50 Mbps and that two LSPs, LSP1 and LSP2, cross this link. If they each require bandwidth protection for 30 Mbps, bandwidth protection is provided to only one of them.

The approach of setting up the backup LSP with a bandwidth reservation and performing admission control into it is very attractive for two reasons:

1. It is easy to estimate the bandwidth that must be reserved for the backup path. For example, a safe guess for a link-protection case is to use the same value as the bandwidth of the protected link[8] because the total bandwidth of all LSPs crossing the link cannot exceed this value.
2. It is possible to achieve bandwidth protection by setting up several backup paths. When none of the paths computed satisfies the bandwidth requirements, several distinct protection paths can be set up and traffic from different protected LSPs can be spread among these paths. Traffic from a single protected LSP cannot be split or spread over different protection paths because doing so might cause packets to arrive out of order at the LSP tail end.

[7] The exact bandwidth is actually signaled and can be different from that of the protected LSP.
[8] Assuming that all of the bandwidth is available for RSVP reservations.

To summarize, bandwidth protection guarantees that enough resources are available to the traffic when it switches to the protection path, ensuring that QoS is maintained following a link or node failure. An interesting question arises with regards to the amount of bandwidth that must be reserved for the protection path, especially for facility backup, where several LSPs share the same backup. Regardless of how the bandwidth for the backup path is determined, the cost of bandwidth protection is the idle resources that are used only in case of failure.[9] Just how much bandwidth is kept idle depends on the optimality of the computation and will be discussed further in Section 3.10.3. Although bandwidth protection is expensive, not having it can impact the traffic, not just of the protected LSP but also of other LSPs as well, as we will see in the following section.

3.8.3 Bandwidth protection and DiffServ

Bandwidth protection is expensive. However, not using bandwidth protection can be very destructive in certain environments. Recall from the Traffic Engineering chapter (Chapter 2) that bandwidth reservations are done in the control plane only and no bandwidth is 'set aside' in the forwarding plane. If more traffic is forwarded than the reservation, there will be traffic loss.

In Figure 3.10, all links are 100 Mbps. Two LSPs are set up, each with a bandwidth reservation of 60 Mbps: LSP1 from A to E along the path A–B–E and LSP2 from C to B along the bath C–D–B. Link protection for link A–B is provided by the backup path A–C–D–B. LSP1 did not request bandwidth protection because it can tolerate some loss during the failure. Assume that both LSP1 and LSP2 carry 60 Mbps of traffic each. When traffic from LSP1 switches to the protection path, 120 Mbps of traffic are sent over link C–D, causing congestion and loss. The traffic that is dropped may belong to either LSP1 or LSP2, causing the very undesirable situation where LSP2 is affected by a failure in LSP1. What is needed is a way to mark the packets of LSP1 as more desirable for dropping.

One way to accomplish this is to give a different DiffServ marking to the traffic as it switches to the backup. This can be easily accomplished by manipulating the EXP bits on the label that

[9] Unless the idle resources are used by best-effort traffic at the forwarding time.

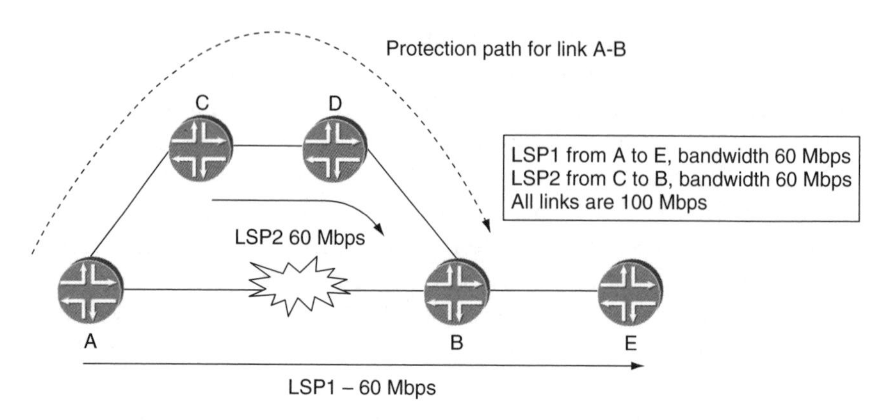

Figure 3.10 Not using bandwidth protection may be destructive

is used by A when switching traffic to the protection path. The value of the EXP bits would be such that packets have a higher drop preference and during congestion only LSP1's packets are dropped.

3.9 INTERACTION OF END-TO-END PROTECTION AND FAST REROUTE

So far this chapter has focused on local protection mechanisms using fast reroute. Fast reroute has attractive properties in terms of the speed of recovery, deterministic switchover delay and the ability to protect selected resources in the network. In contrast, using path protection to provide recovery after failure cannot offer similar guarantees, as discussed in Section 3.4.

It would seem therefore that fast reroute makes end-to-end path protection unnecessary. Why compute secondary paths when fast reroute can provide the desired level of protection? If so, are path protection and local protection mutually exclusive? The answer is 'no'; they are complementary.

Path protection allows the operator exact control over the path of the traffic after the failure. Fast reroute has the ability to limit the loss to a few milliseconds. The two can be combined by configuring LSPs with both secondary paths and local protection. Because local protection forwards the traffic around the point of failure, the main LSP can switch over to the secondary path slowly.

Furthermore, there is no need to presignal the secondary path and reserve resources for it ahead of time, because there is time to set it up after the failure has happened. On the other hand, the use of a secondary path allows tight control over the traffic patterns following a failure. The secondary can be computed offline taking into account different failure scenarios.

Having seen the different mechanisms for providing protection, let us take a look at some of the deployment considerations.

3.10 DEPLOYMENT CONSIDERATIONS FOR LOCAL PROTECTION MECHANISMS

Service providers are attracted to MPLS FRR by the promise of fast and predictable recovery times. However, the challenges of deploying it, its cost and the guarantees it can deliver in a real deployment are not always well understood. This makes it difficult to make the correct tradeoff between costs and benefits. In the following sections we discuss some of the deployment issues for MPLS FRR.

3.10.1 Scalability considerations

Local protection comes at the cost of setting up extra LSPs in the network. In the Traffic Engineering chapter (Chapter 2) we saw that the number of LSPs is a scalability concern, both because the routers themselves have limits on the number of LSPs they support and because a large number of LSPs in the network is cumbersome to manage. In the context of protection, another scaling dimension is the amount of extra forwarding state created along the protection path. Let us take a look at several of the scalability aspects of local protection.

Extra configuration work

One of the prevailing misconceptions is that local protection is operationally very difficult to deploy and manage. This belief has its origins in some of the early implementations of local protection which required listing the protection paths in the configuration. Apart from being labor intensive, this approach also required that the computation of the path be done offline, either manually or by using a specialized tool. Furthermore, the use of preconfigured protection paths can work well only if the protected LSPs were also preconfigured to ensure that they actually do use the protected resources. (This is especially true in the case of node protection, where the merge point is determined based on the current path of the LSP.) For networks relying on dynamic computation for the primary paths, the requirement to specify the protection in the configuration was therefore unacceptable.

For this reason, many of the implementations available today can compute and set up protection paths dynamically. From a network operations point of view, this can be an important factor when choosing one implementation over another. Ultimately, the complexity of deploying such a solution impacts the number of resources that can be protected in the network.

Number of LSPs created

Regardless of whether the protection paths are preconfigured or dynamically computed, they do contribute to the overall number of LSPs that are created and maintained in the network. Just how many extra LSPs are created depends on the type of protection implemented and can be an important consideration when choosing a protection method.

The easiest to analyze is the 1:1 protection. Clearly, for each LSP traversing a protected resource, a new LSP is created and a state is maintained for it. The total number of new LSPs is a function of the number of existing LSPs and the average number of protected resources for each LSP. In principle, the protected resources should be a function of the number of hops crossed by the LSP. However, remember that one of the advantages of local protection is that it can be applied selectively, so fewer resources may be protected. Deployments of 1:1 protection show an increase of a factor of 1.5 or 2 in the number of LSPs after deploying local protection. The relatively low numbers (in the context of 1:1 protection) may be attributed both to selective application of the protection and to the fact that LSPs do not cross many hops.

What about *N*:1 protection? The whole idea of *N*:1 protection is to allow sharing of the protection path. It is very tempting to think that when sharing is allowed the amount of new state created becomes solely a function of the number of resources protected. This is true for link protection, where the MP of the protection LSP is unambiguously identified as the end-point of the link. However, for node protection, the MP depends on the path of the LSP crossing the node. Several LSPs crossing the same node may 'fan out' past the protected node and require separate protection paths, as shown later in Figure 3.13, where LSP1 and LSP2 fan out past node B. Therefore, the backup paths for LSP1 and LSP2 for node protection of node B have different merge points. Thus, for node protection, both the network topology and the LSP placement influence the number of protection paths.

The discussion so far made the assumption that protection paths are implemented as a single LSP. This is not always the case. Recall from Section 3.8.2 that when bandwidth protection is required, several LSPs may need to be set up to satisfy the bandwidth requirements. This may yield a higher number of LSPs than the analysis above implies.

Increase in the forwarding state

An important although often overlooked scalability consideration is the amount of extra forwarding state that must be maintained following the deployment of local protection. No equipment has unlimited resources; therefore understanding the impact of local protection on the forwarding resources used is essential.[10] If analysis shows that the limits are reached, then alternative solutions such as adding extra routers, protecting less resources or lowering the number of LSPs can be evaluated.

There are several factors impacting the amount of forwarding state:

1. Network topology. The protection requires forwarding resources on all the routers in its path. Some network topologies (such as string-of-pearls or dual plane) yield extremely long protection paths, as shown in Figure 3.11. Therefore, the extra forwarding state consumes resources on a large number of routers.
2. Protection type. Intuitively it is easy to understand that the type of protection (one-to-one or facility) increases the amount of forwarding state in the routers in the protection path because of the different number of paths created.
3. Make-before-break. Remember that the new LSP is created using make-before-break. Thus, while the new LSP is established, twice as many forwarding resources are being used (one for the old

[10] This is especially true when forwarding is done in hardware.

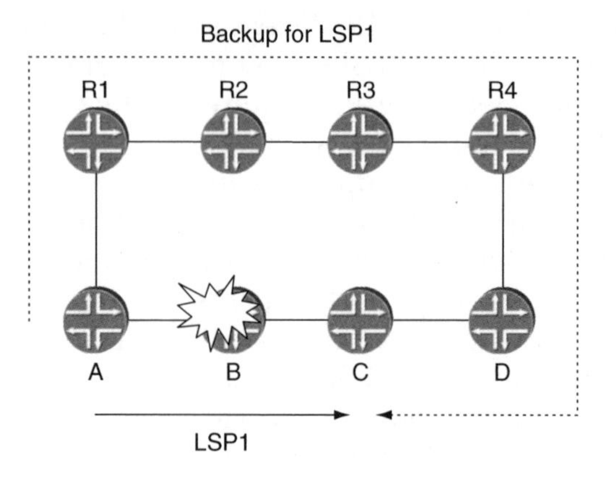

Figure 3.11 Some network topologies yield very long protection paths

path and one for the new path). If the new path triggers the creation of new protection paths, the forwarding state may triple (one for the old path, one for the new path and one for the protection path). It is true that the old path is removed very shortly afterwards, but make-before-break does temporarily increase the utilization of forwarding resources. What this means is that there must be enough free forwarding resources to accommodate the new paths. Doing so may mean that fewer LSPs are allowed to cross a particular router under steady state, which in turn may impact the network design (different LSP placements or the addition of an extra router). This is particularly important because many times at the design stage the only question that is asked is how many LSPs will cross a particular router, and no thought is given to either the protection paths or the make-before-break scenario.

To summarize, the benefits of local protection come at a cost. It is important to understand how the deployment of local protection affects the resource consumption on the routers, to determine whether the equipment can support the proposed deployment, not just in the steady state but also following a failure.

3.10.2 Evaluating a local protection implementation

The expectation is that local protection will provide very fast recovery following a failure of the protected resource. Just how fast 'very fast' is depends on the time it takes to detect the failure and the time it takes to update the forwarding state. Let us take a look at some of the ways to evaluate an implementation with regards to local protection.

Detection time

When fast detection is not available in hardware, support of BFD and the speed at which it operates directly impacts on the total recovery time. When deploying BFD, it is important to bear in mind that because BFD is a hello protocol, it must run on the routers at both ends of the link for which fast detection is required.

Switchover time for a single LSP

Vendors often express their local protection performance in terms of the time it takes to switch over one LSP. This time translates to the amount of traffic that is lost following a failure. It is typically measured by sending traffic at a known rate over an LSP set up with local protection and measuring the traffic loss following the failure of a protected resource. However, it is seldom the case that a single LSP needs to be protected.

Number of LSPs switched over in a certain amount of time

The second scaling number that vendors quote is the number of LSPs that can be successfully protected within a certain amount of time. Assuming that N LSPs cross the failed resource, the question is how much time it takes to switch over all of them, which boils down to how many forwarding entries must be changed for moving them all to the protection path. For 1:1 protection, the answer is unambiguously N, because each LSP has a different label on the protection path. What if all N LSPs were to share the same protection path? In this case, it might be possible to push the same label on to all LSPs by making a single change to the forwarding table. In any case, the maximum number of updates is bounded by the maximum number of LSPs traversing the protected resource. From a practical point of view, the number of LSPs traversing any link or node is not more than a few thousand.

Forwarding state update for IP routes resolving over a protected LSP

Even when testing the recovery of a single LSP, the location of the failure (head end, mid-point or tail end) may play a role in the protection time. This is because the use of the LSP and the forwarding state that is installed (and thus needs to be updated) is different at the head end and mid-point. Imagine a network where BGP traffic traverses the network encapsulated in MPLS and all BGP routes resolve over a single LSP. The LSP head end has a forwarding state for each and every one of the BGP destinations. The routers in the middle of the network maintain only a single forwarding entry (the one created for the LSP). When the failure happens at the LSP mid-point, protecting the traffic to any of the BGP destinations requires only fixing the underlying LSP, basically switching over a single LSP to the protection path.

However, when the failure happens at the head end, the forwarding state for each of the BGP destinations must be updated. The heart of the matter is whether these updates need to be implemented as separate forwarding state updates or not. It is easy to understand that if route 10.0.0.1 is forwarded over (L1, if1) and the protection path is through (L2, if2), then the forwarding entry for route 10.0.0.1 has to be updated to point to (L2, if2). Assume that 100 000 other routes share the same forwarding state (because they are forwarded along the same LSP).

Clearly, there need to be 100 000 distinct forwarding entries for the 100 000 distinct destinations in the forwarding table. How are they represented? One option is to create 100 000 distinct forwarding states, one for each IP prefix. When the LSP switches to the protection path, all 100 000 entries must be updated. A more efficient option is to share the actual forwarding state between all IP prefixes by introducing a level of indirection. For example, rather than maintaining the exact label and interface, maintain an abstract entity representing the LSP. Even if the LSP switches to the protection path, from the BGP routes point of view forwarding has not changed and the BGP forwarding entries need not be modified. By using indirection, fewer forwarding state updates must be made.

The issue of updating forwarding state for IP routes falls at the boundary between local-protection and routing protocol implementation. Understanding the design options provides an insight into what kind of problems to look for when testing the performance of a vendor's software.

3.10.3 The cost of bandwidth protection

Bandwidth protection requires reserving bandwidth on the protection path. The cost for doing so is quite high: idle bandwidth that is used only under failure conditions. Therefore, the goal is to minimize the overall amount of bandwidth that is reserved.

Intuitively, it is easy to understand that shorter is better in this context: the longer the path, the more resources are kept idle. Reserving bandwidth along the long paths shown in Figure 3.11 is clearly not appealing. However, the picture is not that simple. The placement of the LSPs themselves can influence the protection path, especially in setups where some links are protected using other mechanisms and do not require protection using fast reroute. Figure 3.12 shows a network where all links except the link A–B are protected using other mechanisms, such as APS. An LSP is set up between nodes A and E, with a bandwidth requirement of 100 Mbps and bandwidth protection. Let us take a look at the total bandwidth reservation in this network, counted as the sum of all reservations on all links. When the LSP is set up along the shortest path A–B–E, 300 Mbps are reserved for the protected path and another 300 Mbps for the protection path for the link A–B, a total of

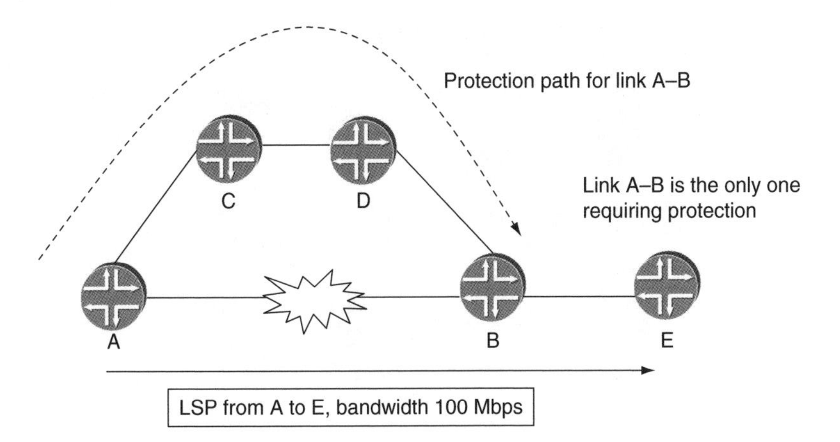

Figure 3.12 Setting up the LSP along the shortest path may not yield the best bandwidth utilization

600 Mbps. If, instead, the LSP was set up over the longer path A–C–D–B–E, no protection would be necessary and only 400 Mbps would be reserved in the network.

Another interesting scenario arises with regards to the computation of the protection paths for different links used by the same LSP. Figure 3.13 shows two LSPs, LSP1 from node A to G along the path A–B–G and LSP2 from node A to E along the path A–B–E. The capacity of all links is 100 Mbps. The requirement is to provide bandwidth protection in case of a failure of node B. Ideally a single backup would be built around B and shared by both LSPs. However, because the two LSPs diverge after node B, they require different merge points and therefore different backup paths, as shown by the dashed lines in the figure.

The next question is what should the bandwidth be for each backup. The first backup protects all LSPs crossing the path A–B–G, up to 100 Mbps. The second protects all LSPs crossing the path A–B–E, which can also add up to 100 Mbps. If the two backups are set up with 100 Mbps each, a total of 200 Mbps are needed for links A–C, C–D and D–E, and the setup of one of the backup paths with fail. However, the total bandwidth of all LSPs crossing the paths A–B–E and A–B–G cannot exceed 100 Mbps, because of the shared link A–B. This knowledge is lost when computing the backup paths and more bandwidth is required for them than necessary.

What the two examples discussed above illustrate is that to achieve optimal bandwidth utilization, more information is needed than

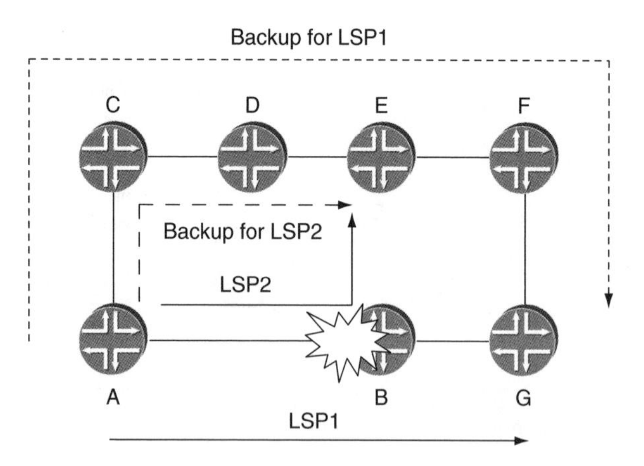

Figure 3.13 Using CSPF to compute the backup paths may not yield optimal bandwidth utilization

what is available to CSPF. Therefore, the use of an offline tool may make sense, especially in situations where bandwidth is at a premium and strict guarantees are required during failure. This could be the case for delivery of large amounts of video traffic (e.g. satellite TV channels), where loss is not acceptable and traffic volumes are very high.

To summarize, bandwidth protection effectively means paying for bandwidth that is used only when a failure happens. To minimize the cost of bandwidth protection, it is necessary to have a global view of all LSPs and protection paths and to employ more specialized algorithms than CSPF. Offline tools can provide this functionality, at the cost of increased operational complexity, as explained in the Traffic Engineering chapter (Chapter 2).

3.11 IP AND LDP FRR

Fast failure recovery is a fundamental requirement for carrying sensitive traffic such as voice or video and is an important building block for providing QoS in MPLS networks. The problem is that the local protection schemes described so far only work in conjunction with RSVP, but many MPLS deployments use LDP as the label distribution protocol.

If an LDP network is to carry voice or video traffic, it must ensure fast failure recovery. Let us see the options available:

- Move away from LDP and switch to RSVP. This is an unacceptable proposition for most providers because it requires a massive reengineering of the network.
- Use one-hop RSVP LSPs with link protection to protect selected links, and continue to run targeted LDP sessions over these RSVP tunnels, as shown in Figure 3.14. Note that both under normal conditions and following a failure, LDP traffic is tunnelled inside RSVP. When the link fails, traffic on the one-hop RSVP tunnels is protected, so the LDP traffic is also protected. This approach is attractive because it allows the operator to continue to use LDP for the setup of LSPs and does not require changes to the edge routers. However, protection for only link failures can be achieved.
- Find a mechanism that will provide fast-reroute behaviour for LDP. This is the most attractive proposition for the customers, and for this reason vendors are actively trying to engineer such

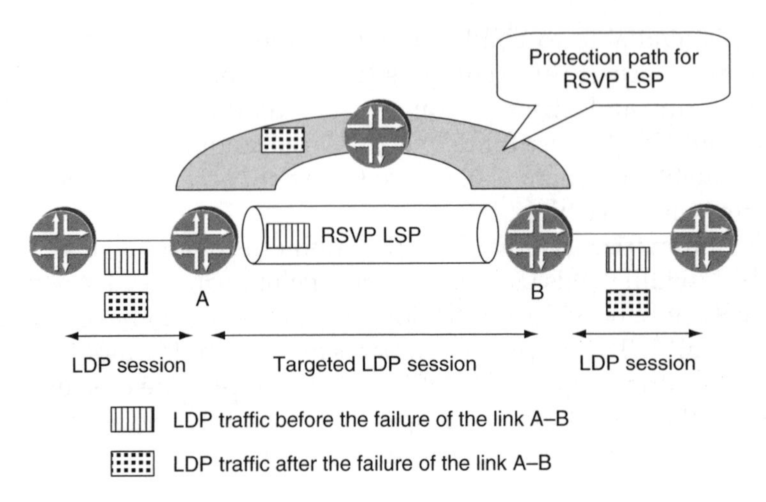

Figure 3.14 Protecting LDP traffic by using one-hop RSVP tunnels with link protection

a solution. Because this work is still in progress at the time of this writing, we will look at some of the issues that arise in the context of LDP FRR rather than present any one solution in detail.

The fundamental difference between LDP and RSVP in the context of fast reroute is that LDP LSPs follow the IGP while RSVP LSPs are set up along a predefined path and are immune to changes in the IGP. Thus, LDP is influenced by changes in IP routing while RSVP is not. Based on this observation, it can be argued that if the problem of IP fast reroute is solved, the same protection can be automatically extended to LDP. Under this assumption, this chapter discusses fast reroute for IP and LDP interchangeably.

There are several proposals for providing IP fast reroute currently under discussion by the Routing Area Working Group [RTGWG] in the IETF. The proposals fall broadly into two categories: tunnel based or alternate-path based. Let us discuss them separately below.

The tunnel-based approach

The tunnel-based approach is conceptually very similar to RSVP FRR. A bypass tunnel is set up ahead of time around the protected resource and is used to shuttle the traffic around the point of failure. In this case, the tunnel is set up with RSVP and the same mechanisms as described above for RSVP can be used. Note that this approach is different from protection using one-hop RSVP

LSPs discussed at the beginning of the section, because in this case LDP traffic is forwarded over the RSVP tunnel only following a failure, whereas in the one-hop RSVP LSP case, LDP traffic is always tunnelled in RSVP.

The tunnel-based approach can provide protection for both link and node failures. However, to provide node protection, the label used at the MP must be known to the PLR, as discussed in Section 3.7. For RSVP FRR, this label was learned from the Record Route Object. For LDP, new procedures must be set in place to advertise this label from the MP and the PLR. Different methods to do so were discussed by the MPLS Working Group in the IETF, but no solution has yet been adopted.

The tunnel-based approach is very attractive because it relies on the same elements that have been deployed and are proved to be working for RSVP FRR. However, is it really all this simple? For it to work, traffic must reach the PLR. For RSVP LSPs, this is guaranteed, because the traffic is source routed and follows a pre-established path that is not influenced by routing changes. Unfortunately with LDP this is not the case, as LDP paths track IP routing.

Changes in IP routing as the network reconvergences following the link failure may cause the traffic never to reach the beginning of the bypass tunnel, which is similar to the situation described in Section 3.5 for Figure 3.2. To recap, before the failure the route from node S to node D is through node R1 and the route from node R3 to D is through node S. In the event of a failure of the link R1–R2, the route from S to D is through node R3. If S starts forwarding traffic to R3, packets will not arrive at R1 and at the protection path. This can be a problem if the IGP on R3 has not yet performed its SPF and still points to S as its best path (the route from before the failure). In this case, traffic will loop between S and R3. This situation is referred to as microloops and there are several proposals for avoiding it. The most intuitive is to slow down the convergence of the routers so as to avoid the situation where node S reconverged faster than node R3. The solution to the microloop problem is currently being worked on by the Routing Area Working Group in the IETF.

Apart from the issue of microloop prevention, a tunnel-based approach is attractive for the following reasons:

- Most elements of the solution have been deployed in networks today in the context of MPLS FRR.
- The computations are well bounded (one computation per link for link protection, or one per next next-hop for node protection).

- The forwarding paradigm during failure is simply MPLS.
- The path of the packets during failure is known and there is control over it.

The alternate-path approach

The alternate-path approach relies on maintaining an alternate path towards the destination and taking that path when the primary fails. This is best shown in an example. In Figure 3.15 there are two paths from source A to destination D, A–B–D and A–C–D. From the IGP's point of view, path A–B–D is shorter, with a metric of 20, and normally the forwarding state for D at A would simply point to if1. When using the alternate-path approach both paths are maintained. This means two things: both paths must be computed and the forwarding state must be kept for both paths. This must be done for each of the destinations reachable from A.

When link B–D fails, traffic is sent on the alternate path over link if2 as soon as A finds out about the failure. In the simple topology in the figure, this approach works, traffic arrives to C over interface if2 and is forwarded along the link C–D to D, using the primary route at C. The solution described above is the basic approach for IP fast reroute as documented in [IPFRR], also known as loop-free alternates.

At this point, it should already be clear that this solution depends on the network topology. If the metric of link C–D was

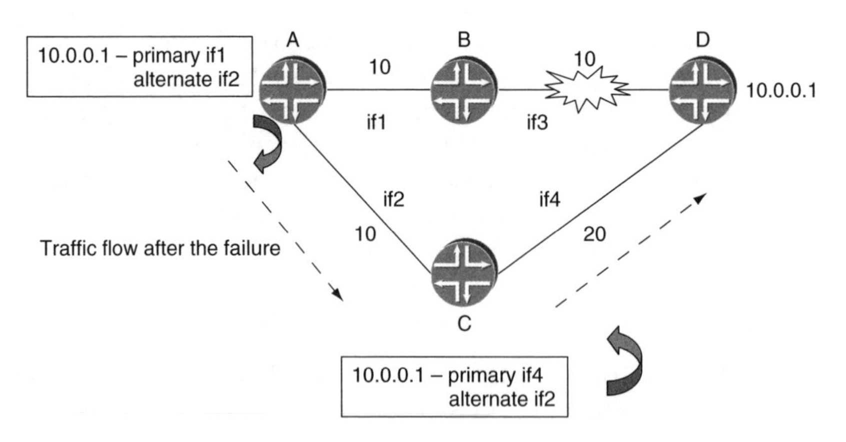

Figure 3.15 Protection using loop-free alternates

100 instead of 20, then the primary route for 10.0.0.1 at C would point back to A (through path C–A–B–D). Traffic arriving at C from A would then do a 'U-turn' and loop between A and C until the IGP on C converges. To solve this problem, the notion of U-turn alternates is introduced. If C could detect that it is causing the traffic to take a U-turn, then C could use its alternate route to forward the traffic and packets would arrive safely at D, as shown in Figure 3.16. This solution is also known as U-turn alternates and it is currently work in progress in the IETF [U-TURN].

Let us take a look at some of the properties of this solution, in contrast to a tunnel-based approach:

- Partial coverage. The solution does not work in arbitrary topologies. Neither loop-free alternates nor U-turn alternates can provide coverage for all failure cases.
- Change in the forwarding paradigm. Forwarding behaves differently in steady state and following a failure, making it difficult to debug failure scenarios.
- Computational complexity. Each router must compute two paths for every destination. The number of computations required to set up alternate paths is proportional to the number of destinations in the network, rather than being proportional to the number of neighbors or next next-hop neighbors. For U-turn alternates, more complex computations need to be performed to detect the presence of U-turns.

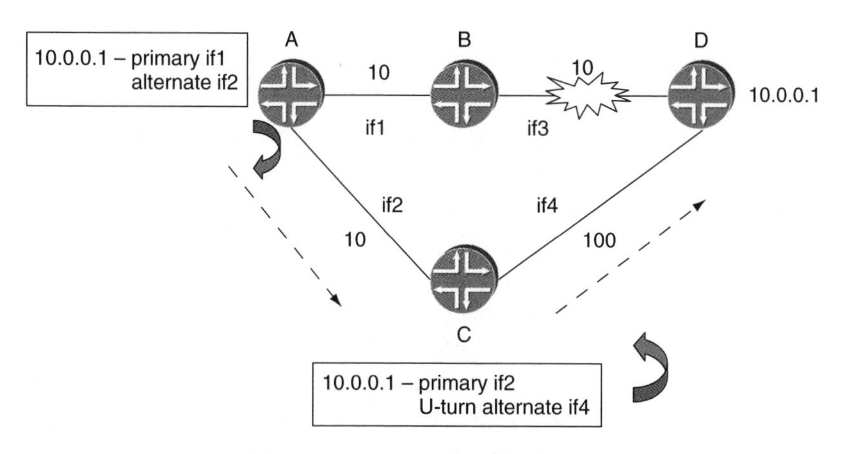

Figure 3.16 Protection using U-turn alternates

- Growth in the forwarding state. The forwarding state doubles at each and every router in the network, because a separate alternate must be maintained for each destination.
- There is no control over the path traffic will take after a failure.

To summarize, the success of MPLS fast reroute and the move towards converged networks has prompted an interest in fast-reroute solutions for both IP and LDP. The available solutions are based on either maintaining alternate paths for use during failure or on providing backup tunnels similarly to MPLS FRR. None of the proposed solutions has yet been standardized, as vendors and operators struggle to find the most optimal tradeoff between functionality and complexity.

3.12 CONCLUSION

MPLS fast reroute provides protection for the traffic following link or node failures, within times that are virtually undetectable at the application layer. This is a fundamental requirement for carrying sensitive traffic such as voice or video and is an important building block for converging all services on to a common MPLS core.

Used together with traffic engineering, fast reroute can ensure adherence to strict QoS guarantees, not just in the normal case but also following a failure, thus completing the TE solution as described in Chapter 2. In the next chapter, we will explore DiffServ Aware TE, which refines the TE solution by allowing bandwidth reservations to be carried out on a per-DiffServ class basis.

3.13 REFERENCES

[BFD] BFD Working Group http://ietf.org/html.charters/ bfd-charter.html

[BFD-BASE] D. Katz and D. Ward, 'Bidirectional forwarding detection', draft-ietf-bfd-base-03.txt (work in progress)

[BFD-MHOP] D. Katz and D. Ward, 'BFD for multihop paths', draft-ietf-bfd-multihop-03.txt (work in progress)

[IPFRR] A. Atlas, 'Basic specification for IP fast-reroute: loop-free alternates', draft-ietf-rtgwg-ipfrr-spec-base-04.txt (work in progress)

[RFC4090] P. Pan, G. Swallow and A. Atlas, *Fast Reroute Extensions to RSVP-TE for LSP Tunnels*, RFC4090, May 2005

[RTGWG] Routing Area Working Group http://ietf.org/html.charters/rtgwg-charter.html

[U-TURN] A. Atlas, 'U-turn alternates for IP/LDP fast-reroute', draft-atlas-ip-local protect-uturn- 02.txt (work in progress)

3.14 FURTHER READING

[BMWG] Benchmarking Methodology Working Group http://ietf.org/html.charters/bmwg-charter.html

[BMFRR] S. Poretsky *et al.*, 'Benchmarking methodology for MPLS protection Mechanisms', draft-poretsky-mpls-protection-meth-04.txt (work in progress)

4

MPLS DiffServ-TE

4.1 INTRODUCTION

In the MPLS Traffic Engineering chapter (Chapter 2), we saw how MPLS traffic engineering (TE) allows the user to create end-to-end paths across the network with bandwidth reservations. This guarantees that the resources are available to carry traffic of volume less than or equal to the bandwidth reservation. A disadvantage of the basic MPLS-TE model is that it is not aware of the different DiffServ classes, operating at an aggregate level across all of them.

This chapter introduces the concept of DiffServ Aware MPLS-TE, which refines the MPLS-TE model by allowing bandwidth reservations to be carried out on a per-class basis. The result is the ability to give strict QoS guarantees while optimizing use of network resources. The QoS delivered by MPLS DiffServ-TE allows network operators to provide services that require strict performance guarantees, such as voice, and to consolidate IP and ATM/FR (Frame Relay) networks into a common core.

This chapter explores MPLS DiffServ-TE and its extensions. It assumes familiarity with DiffServ in general and MPLS DiffServ in particular, discussed in the Foundations chapter (Chapter 1), as well as with MPLS Traffic Engineering, discussed in the Traffic Engineering chapter (Chapter 2).

MPLS-Enabled Applications: Emerging Developments and New Technologies Ina Minei and Julian Lucek

4.2 THE BUSINESS DRIVERS

Traditionally, IP/MPLS-based networks were used only for services with relatively relaxed requirements in terms of delay, jitter or bandwidth guarantees. Increasingly, providers have started carrying a wider range of services, such as PSTN-quality voice or providing ATM/FR or Ethernet over the MPLS core. The driver for offering these services is the cost savings achieved by eliminating the need to have several separate physical networks. Indeed, one of the most attractive promises of MPLS is the ability to converge all services on to a common core. The challenge lies in the fact that most of these services often require stricter service-level agreements (SLAs) than the previous norm on IP/MPLS networks.

The SLAs define the service quality experienced by traffic transiting the network and are expressed in terms of latency, jitter, bandwidth guarantees, resilience in the face of failure, and down time. The SLA requirements translate to two conditions: (a) different scheduling, queuing and drop behavior based on the application type and (b) bandwidth guarantees on a per-application basis.

To date, service providers have rolled out revenue-generating services in their networks using DiffServ alone. By assigning applications to different classes of service and marking the traffic appropriately, condition (a) was met. However, this approach assumes that there are enough resources to service the traffic according to the marking. If the traffic follows a congested path, traffic may be dropped, or it may experience different delay and jitter characteristics than required by the SLAs. In principle, service providers could solve this problem by using overprovisioning to avoid congestion altogether. Besides being wasteful with regards to resource utilization, this approach of 'throwing bandwidth at the problem' cannot provide any guarantees when congestion is caused by link and/or node failures.

In the Traffic Engineering chapter (Chapter 2) we have seen how MPLS traffic engineering sets up label-switched paths (LSPs) along links with available resources, thus ensuring that bandwidth is always available for a particular flow and avoiding congestion both in the steady state and in failure scenarios. Because LSPs are established only where resources are available, overprovisioning is not necessary. Further optimization of transmission resources is achieved by allowing LSPs not to follow the shortest path, if the available resources along the shortest path are not sufficient. An added benefit of MPLS is that built-in mechanisms such as link protection and fast reroute (discussed in the Protection and Restoration chapter, Chapter 3) provide resilience in the face of failure. The catch is that MPLS-TE is oblivious of the class-of-service (CoS) classification, operating only on the available bandwidth at an aggregate level across all classes.

MPLS DiffServ-TE makes MPLS-TE aware of CoS, allowing resource reservation with CoS granularity and providing the fault-tolerance properties of MPLS at a per-CoS level. By combining the functionalities of both DiffServ and TE, MPLS DiffServ-TE delivers the QoS guarantees to meet strict SLAs such as the ones required for voice, ATM and Frame Relay, thus meeting condition (b).

Note that even if resources are reserved on a per-CoS basis and that even if traffic is properly marked to conform to the CoS appropriate for the application, the SLAs still cannot be guaranteed unless further mechanisms, such as policing and admission control, are set in place to ensure that the traffic stays within the limits assumed when the resource reservation was made, as will be seen in Section 4.4.8.

4.3 APPLICATION SCENARIOS

The DiffServ-TE solution is the product of the TEWG Working Group in the IETF.[1] In [RFC3564], the Working Group documented a few application scenarios that cannot be solved using DiffServ or TE alone. These scenarios form the basis for the requirements that led to the development of the DiffServ-TE solution and are presented in this section. The scenarios show why per-traffic-type behavior is necessary.

4.3.1 Limiting the proportion of traffic from a particular class on a link

The first scenario involves a network with two types of traffic: voice and data. The goal is to maintain good quality for the voice

[1] The TEWG finished all its work items and has been closed.

traffic, which in practical terms means low jitter, delay and loss, while at the same time servicing the data traffic. The DiffServ solution for this scenario is to map the voice traffic to a per-hop behavior (PHB) that guarantees low delay and loss, such as the expedited-forwarding (EF) PHB.

The problem is that DiffServ alone cannot give the required guarantees for the following reason. The delay encountered by the voice traffic is the sum of the propagation delay experienced by the packet as it traverses the network and of the queuing and transmission delays incurred at each hop. The propagation and transmission delays are effectively constant; therefore, in order to enforce a small jitter on the overall delay, the queuing delay must be minimized. A short queuing delay requires a short queue, which from a practical point of view means that only a limited proportion of the queue buffers can be used for voice traffic.

Thus, the requirement becomes 'limit the proportion of voice traffic on each link'. In the past, service providers used overprovisioning to achieve this goal, making sure that more bandwidth was available than would ever be necessary. However, overprovisioning has its own costs and, while it may work well in the normal case, it can give no guarantees in the failure scenario. Figure 4.1 shows a network operating under such a regimen. Under normal conditions, the voice traffic takes the path A–C–D,

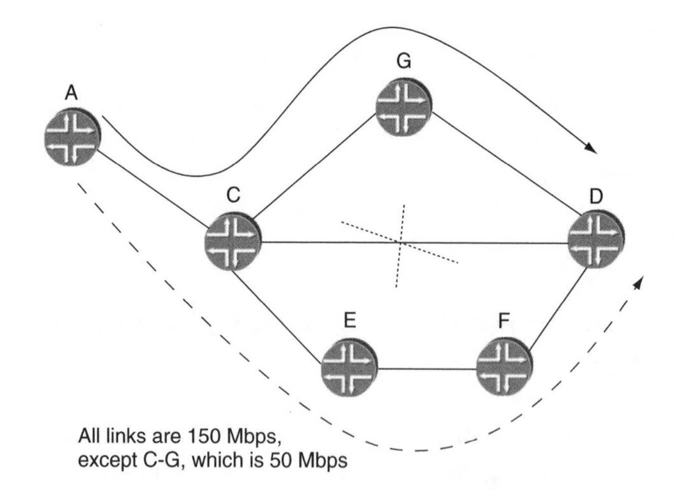

Figure 4.1 Overprovisioning cannot provide guarantees in a failure scenario

which is the shortest path. The link capacity is large, so the percentage of the voice traffic on each link is acceptable. When the link C–D fails, the traffic reroutes on the next best path, A–C–G–D. The link C–G is low-capacity and the percentage of voice traffic becomes too large. Instead, the traffic should have rerouted on the path A–C–E–F–D.

Taking the solution a step further, limiting the proportion of voice traffic on all links can be achieved by artificially limiting the available bandwidth on a link to the proportion suitable to satisfy the voice traffic requirements alone and using TE to ensure that traffic (voice and data) is mapped in such a way as to honor these artificially lower available resources. This solution provides the requested functionality but wastes resources because bandwidth that could be allocated to delay-insensitive data traffic is now idle and unavailable. The root of the problem is that TE cannot distinguish between the two types of traffic and cannot enforce allocations at a per-traffic-type granularity.

4.3.2 Maintaining relative proportions of traffic on links

The second scenario extends the previous example to a network that supports three traffic types that map to three 'classes of service'. The proportion of the different traffic types depends on the source and destination of the traffic. The challenge for the service provider is to configure the queue sizes and queue scheduling policies on each link to ensure that the correct PHB is given to each class.[2] It is impractical to configure these parameters based on the link load at a given time: changes in routing, link or node failures and preemption between LSPs make the link load a very dynamic property. Instead, from an operational and maintainability point of view, it would be ideal to fix the relative proportions of each traffic type on the links, allocate the queue sizes and scheduling policies accordingly, and use TE to make the traffic comply with the available resources. This solution requires TE to enforce different bandwidth constraints for different classes of traffic.

[2] The example uses three classes rather than the two (voice and data) from the previous scenario. This is because when just voice and data are used, it can be argued that the queue size must be set uniformly for the voice traffic.

4.3.3 Providing guaranteed bandwidth services

In this application, which is very similar to the example in Section 4.3.1, there are two types of traffic: best effort and 'guaranteed bandwidth'. The guaranteed bandwidth traffic must comply with a given SLA. The goal is to provide the required service level to the guaranteed traffic and also to be able to traffic-engineer the best-effort traffic. As in the first example, in order to enforce strict SLAs, the guaranteed bandwidth traffic must be engineered not to overflow the allotted bandwidth of the link, and TE must be employed to ensure this requirement. In addition, the best-effort traffic must also be traffic-engineered, to increase link utilization. Here again, TE must have knowledge of the type of traffic.

4.4 THE DIFFSERV-TE SOLUTION

This section examines how per-traffic-type behavior is enforced, both when setting up an LSP and when forwarding traffic.

4.4.1 Class types

The basic DiffServ-TE requirement is to be able to make separate bandwidth reservations for different classes of traffic. This implies keeping track of how much bandwidth is available for each type of traffic at any given time on all routers throughout the network. [RFC 3564] spells out the requirements for support of DiffServ Aware MPLS-TE and defines the fundamental concepts of the technology.

For the purpose of keeping track of the available bandwidth for each type of traffic, [RFC3564] introduces the concept of a class type (CT). [RFC3564] does not mandate a particular mapping of traffic to CTs, leaving this decision to the individual vendors. One possible implementation is to map traffic that shares the same scheduling behavior to the same CT. In such a model one can think of a CT in terms of a queue and its associated resources. Because the PHB is defined by both the queue and the drop priority, a CT might carry traffic from more than one DiffServ class of service, assuming that they all map to the same scheduler queue.

The IETF standards require support of up to eight CTs referred to as CT0 through CT7. LSPs that are traffic-engineered to guarantee bandwidth from a particular CT are referred to as DiffServ-TE LSPs. In the current IETF model, a DiffServ-TE LSP can only carry traffic from one CT. LSPs that transport traffic from the same CT can use the same or different preemption priorities. By convention, the best-effort traffic is mapped to CT0. Because all pre-DiffServ-TE LSPs are considered to be best effort, they are mapped to CT0.[3]

Let us revisit the application scenario from Section 4.3.1 and discuss it in terms of CTs. The voice and data network in this example supports two DiffServ PHBs, EF and BE (for voice and data traffic respectively). The goal is to provide service guarantees to the EF traffic. Two scheduler queues are configured on each link, one for BE and one for EF. CT0 is mapped to the BE queue and CT1 is mapped to the EF queue. The bandwidth available for CT1 (the voice traffic) is limited to the percentage of the link required to ensure small queuing delays for the voice traffic. Separate TE LSPs are established with bandwidth requirements from CT0 and from CT1.

In the following sections, we look at how LSPs are established with per-CT bandwidth requirements.

4.4.2 Path computation

In the Traffic Engineering chapter (Chapter 2), we discussed how CSPF computes paths that comply with user-defined constraints such as bandwidth and link attributes. DiffServ-TE adds the available bandwidth for each of the eight CTs as a constraint that can be applied to a path. Therefore, CSPF is enhanced to take into account a CT-specific bandwidth at a given priority as a constraint when computing a path. For example, the user might request an LSP of CT1 at priority 3 with a bandwidth of 30 Mbps. CSPF computes a path that meets these criteria. For the computation to succeed, the available bandwidth per CT at all priority levels must be known for each link.

This means that the link-state IGPs must advertise the available bandwidth per CT at each priority level on every link. Recall that there are eight CTs and eight priority levels, giving a total of 64 values that need to be carried by the link-state protocols. In an ideal world, all 64 values would be advertised and stored for each

[3] Pre-DiffServ-TE LSPs and DiffServ-TE LSPs from CT0 are signaled in exactly the same way.

link. However, the IETF decided to limit the advertisements to eight values out of the possible 64 [RFC4124].[4]

How are these eight values picked? TE classes are defined for this purpose as a combination of CT and priority. The IGPs advertise the available bandwidth for each of the TE classes defined. DiffServ-TE supports a maximum of eight TE classes, TE0 through TE7, which can be selected from the 64 possible CT–priority combinations through configuration. At one extreme, there is a single CT with eight priority levels, very much like the existing TE implementation. At the other extreme, there are eight distinct CTs, with a single priority level. The combinations chosen depend on the classes and priorities that the network must support. Figure 4.2 shows the 64 combinations of class type and priority, and a choice

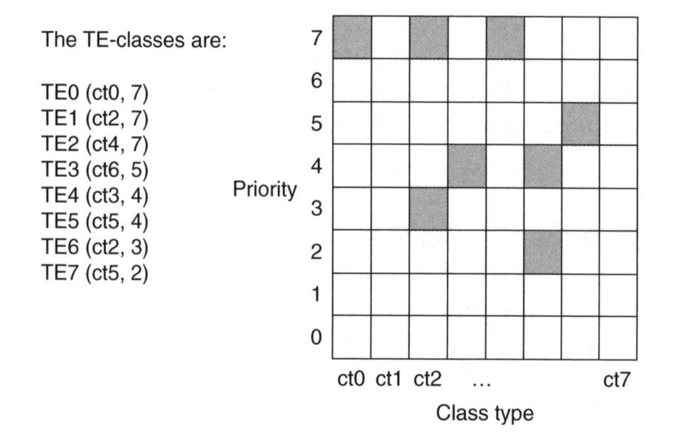

Figure 4.2 Picking eight TE classes out of the 64 possible combinations

[4] One of the most heated debates in the Working Group was around the question of whether to advertise all 64 available bandwidth values. The opponents argued that doing so would yield a very small gain, since 64 different combinations are more than anyone would deploy in a real network, and that advertising all the 64 values would place a large burden on the IGPs. This would happen because (a) the information would not fit in one link-state advertisement and would require sending several of them, greatly increasing the number of IGP advertisements in the network, and (b) new advertisements would need to be sent every time any of the values changed, creating churn. Although the concerns may seem valid, they are not entirely justified. The 64 different combinations are useful for allowing flexible configuration of CT and priorities without the need to coordinate TE class matrices across the entire network. The concern regarding the need for several link-state advertisements is also not founded given the current MTU sizes. Even without any smart packing of the data, the information would still not require more than one advertisement. If space were a concern, efficient packing could have solved the problem. Finally, the question of churn when the values change can be solved by dampening the advertisements, in the same way it is done for regular TE.

of eight TE classes, called a TE class matrix. Note that both the setup and the hold priorities used by LSPs must be used in the TE class matrix. This is because available bandwidth is reported per TE class and this information is required for both priority levels. Because TE classes are used in the IGP advertisements, all routers must have a consistent configuration of the TE class matrix. Otherwise, the advertisements will be incorrectly attributed to the wrong CT–priority combination.

The link-state IGPs advertise the available bandwidth for each TE class. [RFC4124] mandates that this advertisement be made using the existing Unreserved Bandwidth TLV, which was previously used to disseminate unreserved bandwidth for TE [RFC3784, RFC3630]. Therefore, the information that is available to CSPF through the IGPs is relevant only for the CT and priority combinations that form valid TE classes. Thus, in order for CSPF to perform a meaningful calculation, the CT and priority levels chosen for an LSP must correspond to one of the configured TE classes.

To summarize, two crucial design decisions were taken with respect to the advertisement of per-class available bandwidth:

1. Advertising information for only eight CT–priority combinations rather than for all 64 of them.
2. Overriding the semantics of an existing TLV in the IGPs to carry the available bandwidth information for the TE classes chosen.

We have already seen that decision 1 required the introduction of the not very intuitive concept of a TE class and limits the characteristics of the LSPs in the network to the classes and priorities supported by the TE class matrix. In the section discussing deployment of the DiffServ-TE solution (Section 4.4.9) we will see how decision 2 constrains the deployment of DiffServ-TE in networks that already use TE.

Despite these constraints, implementations exist today supporting this model and have been demonstrated to interoperate. As long as the classes and priorities are consistently configured, the solution is backwards-compatible with routers not supporting the functionality.

To summarize, to compute a path with per-CT bandwidth constraints, CSPF is enhanced to handle per-CT reservation requirements and the IGPs are enhanced to carry per-CT available bandwidth at different priority levels.

4.4.3 Path signaling

After the path is calculated, it is signaled, and admission control and bandwidth accounting are performed at each hop. [RFC4124] defines the necessary extensions to RSVP-TE that allow it to establish paths with per-CT bandwidth reservations.[5]

The CT information for an LSP is carried in the new Class Type Object (CT object) in the RSVP path message, and specifies the CT from which the bandwidth reservation is requested. Two rules ensure that it is possible to deploy DiffServ-TE incrementally in the network:

1. The CT object is present only for LSPs from CT1 through CT7 (if the CT object is missing, CT0 is assumed).
2. A node that does not understand the DiffServ-TE extensions and that receives a path message with the CT object rejects the path establishment.

These two rules ensure that establishment of LSPs with per-CT reservation is possible only through DiffServ-TE-aware nodes, while pre-DiffServ-TE LSPs, which are considered to belong to CT0, can cross both old and new nodes. In a mixed network, where some of the routers support DiffServ-TE and others do not, DiffServ-TE LSPs can establish through the routers that have the support.

The CT information carried in the path message specifies the CT over which admission control is performed at each node along the path. If a node along the path determines that enough resources are available and the new LSP is accepted, the node performs bandwidth accounting and calculates the new available bandwidth per-CT and priority level. This information is then passed back into the IGPs.

To summarize, for each LSP, the CT is implicitly signaled for CT0 and explicitly signaled for all other CTs. The CT is necessary to perform the calculation of the available resources. How is this calculation performed?

4.4.4 Bandwidth constraint models

One of the most important aspects of the available bandwidth calculation is the allocation of bandwidth among the different CTs.

[5] Note that although CR-LDP also supports explicit routing, no extensions are defined for it because the IETF decided in [RFC3468] to abandon new development for CR-LDP.

The percentage of the link's bandwidth that a CT (or a group of CTs) can take up is called a bandwidth constraint (BC). [RFC3564] defines the term 'bandwidth constraint model' to denote the relationship between CTs and BCs. Several bandwidth constraint models exist; the most popular are the maximum allocation model (MAM) and the Russian dolls model (RDM).

4.4.4.1 The maximum allocation model (MAM)

The most intuitive bandwidth constraint model maps one BC to one CT. This model is called the maximum allocation model (MAM) and is defined in [DSTE-MAM]. From a practical point of view, the link bandwidth is simply divided among the different CTs, as illustrated in Figure 4.3.

The benefit of MAM is that it completely isolates different CTs. Therefore, priorities do not matter between LSPs carrying traffic from different CTs. In the network shown in Figure 4.4, all links are of capacity 10 Mbps and are partitioned to 9 Mbps for data (CT0) and 1 Mbps for voice (CT1). The operator sets up two data LSPs: LSP1 with 9 Mbps and LSP2 with 1 Mbps. LSP1 is set up along the shortest path A–B–C. As a result, the available bandwidth for CT0 along this path becomes 0 and LSP2 is forced to establish along the longer path A–D–E–C. This is despite the fact that 1 Mbps is free along the path A–B–C. When the operator wants to set up a voice LSP, the resources are guaranteed to be available for class CT1 on the shortest path and no preemption of data LSPs (CT0) is necessary, or indeed possible.

The problem with MAM is that because it is not possible to share unused bandwidth between CTs, bandwidth may be wasted instead of being used for carrying other CTs. Consider the network shown in Figure 4.4. In the absence of voice LSPs, bandwidth is available on all the links on the shortest path for data traffic, but

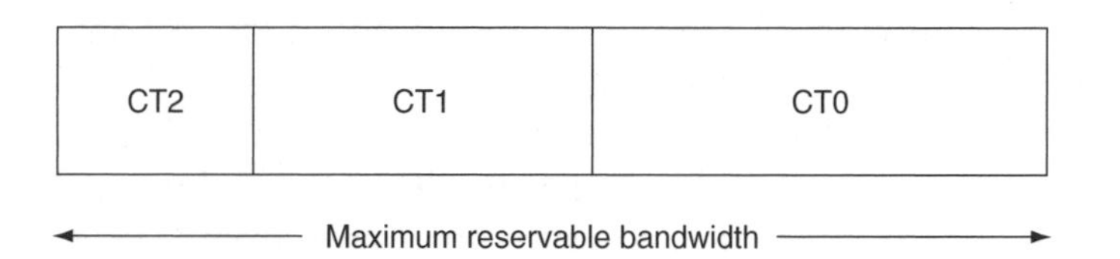

Figure 4.3 The allocation of bandwidth to CTs in the MAM model (for simplicity, only three CTs are shown)

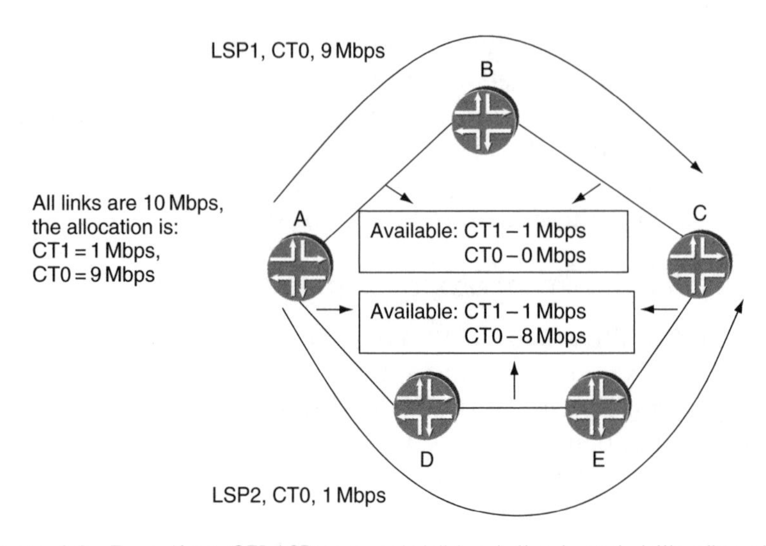

Figure 4.4 Even if no CT1 LSPs are established, the bandwidth allocated for CT1 cannot be used for CT0

this bandwidth cannot be used for setting up another data LSP. The second data LSP is forced to follow a nonoptimal path, even though bandwidth is available on the shortest path. On the other hand, after both data LSPs have been set up, if a voice LSP needs to be established, bandwidth is available for it on the shortest path.

The available bandwidth for the MAM model is accounted in a similar way as for TE, except that it is done on a per-CT basis. To calculate the bandwidth available for CTn at priority p, subtract from the bandwidth allocated to CTn the sum of the bandwidths allocated for LSPs of CTn at all priority levels that are better or equal to p.

4.4.4.2 The Russian dolls model (RDM)

The Russian dolls bandwidth allocation model (RDM), defined in [RFC4127], improves bandwidth efficiency over the MAM model by allowing CTs to share bandwidth. In this model, CT7 is the traffic with the strictest QoS requirements and CT0 is the best-effort traffic. The degree of bandwidth sharing varies between two extremes. At one end of the spectrum, BC7 is a fixed percentage of the link bandwidth that is reserved for traffic from CT7 only. At the other end of the spectrum, BC0 represents the entire link bandwidth and is shared among all CTs. Between these two extremes are various degrees of sharing: BC6 accommodates traffic from CT7 and CT6, BC5 from CT7, CT6 and CT5, and so on. This model is very much like the Russian doll toy, where one big doll (BC0) contains a smaller doll (BC1) that contains a yet smaller doll (BC2), and so on, as shown in Figure 4.5.

Figure 4.6 shows how bandwidth accounting works for the RDM model. The figure shows a network very similar to the one in Figure 4.4, which carries two classes of traffic, data (CT0) and voice (CT1). The total bandwidth available on each link is 10 Mbps;

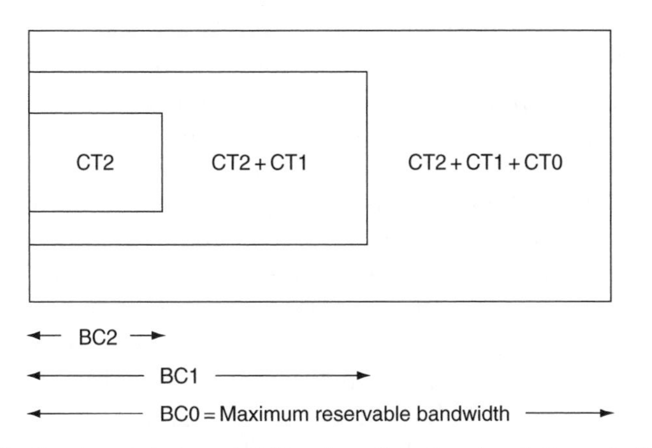

Figure 4.5 Russian dolls bandwidth allocation (for simplicity, only three CTs are shown)

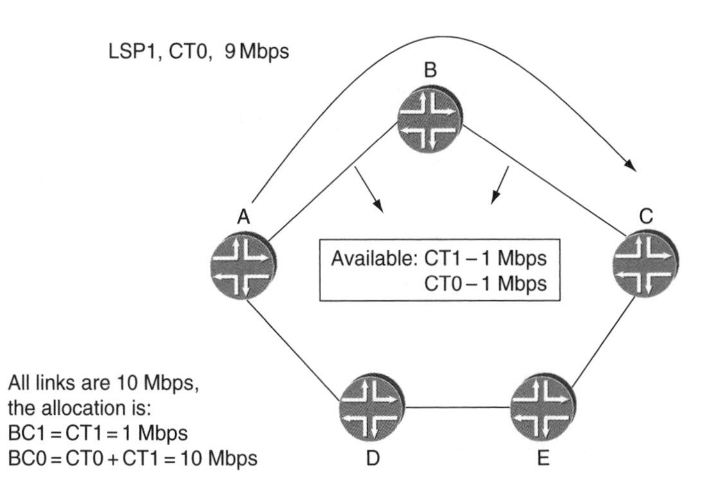

Figure 4.6 Bandwidth accounting for the RDM model

1 Mbps is allocated to BC1 and 10 Mbps are allocated to BC0. This means that each link can carry between 0 and 1 Mbps of voice traffic and use the rest for data. A data LSP, LSP1 from CT0, is already established over the shortest path A–B–C, with a reservation of 9 Mbps. Therefore, 1 Mbps remains available on this path, for use by either CT0 or CT1 traffic. Therefore, the available bandwidth for each of these classes is reported as 1 Mbps.

4.4.4.3 Comparison of the RDM and MAM models

The advantage of RDM relative to MAM is that it provides efficient bandwidth usage through sharing. In Figure 4.6, a second data LSP, LSP2, with a reservation of 1 Mbps can also be established on the shortest path to take advantage of the unused bandwidth. Another useful property that is achieved through sharing is cheap overprovisioning for real-time traffic. Because the extra bandwidth can be used by other types of traffic, allocating it to the real-time class does not affect the overall throughput of the network.

The disadvantage of RDM relative to MAM is that there is no isolation between the different CTs and preemption must be used to ensure that each CT is guaranteed its share of bandwidth no matter the level of contention by other CTs. This is shown in Figure 4.7. After establishing the second data LSP, LSP2, if the operator wants to establish a voice LSP, no resources are available

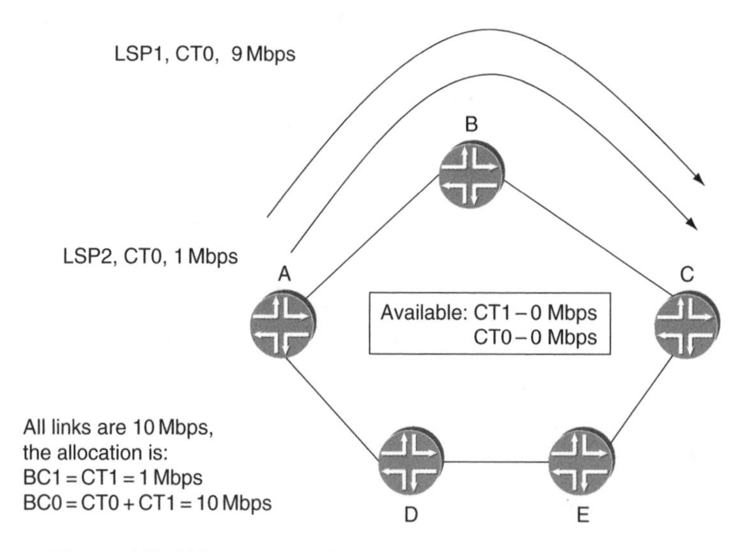

Figure 4.7 Why preemption is necessary when using RDM

for the voice traffic on the shortest path. Thus, one of the data LSPs must be preempted: otherwise, bandwidth is not guaranteed for the voice traffic. This means that voice and data LSPs must be given different priorities, because they share bandwidth resources.

Figure 4.8 shows the same network, with voice LSPs at priority 0 and data LSPs at priority 1. (Recall that the best priority is priority 0 and the worst priority is priority 7.) When the voice LSP, LSP3, is established, it preempts one of the data LSPs (LSP2) and establishes over the shortest path. LSP2 then reestablishes over the longer path A–D–E–C. Note that a voice LSP can preempt the data LSP only if the voice LSP bandwidth requirement is such that the CT1 allocation on the link is not exceeded. For example, if LSP3 had a requirement of 9 Mbps from CT1, it would not preempt LSP1. This is because the maximum bandwidth that class CT1 is allowed to reserve on any link is 1 Mbps (from the definitions of the BCs). In that case, LSP3 would simply not establish. What the example in Figure 4.8 shows is that the bandwidth-sharing capabilities of RDM come at the cost of extra planning and configuration: LSPs from different classes must be assigned different priorities, to ensure that ultimately each class gets its share of the bandwidth on a link.

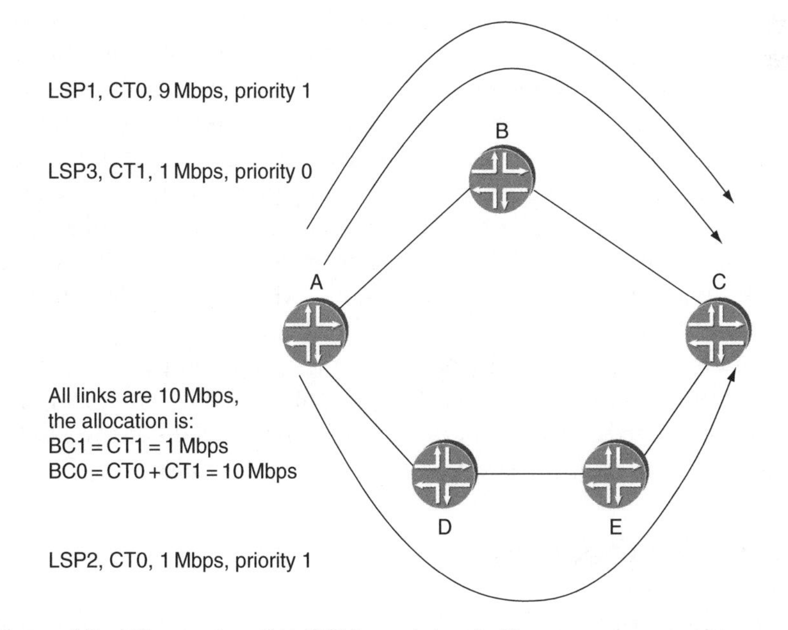

Figure 4.8 When using the RDM model, priorities are necessary to guarantee bandwidth to different CTs

Table 4.1 Comparison of MAM and RDM

MAM	RDM
Maps one BC to one CT; easy to understand and manage	Maps one BC to one or more CTs; harder to manage
Achieves isolation between CTs and guaranteed bandwidth to CTs without the need for preemption	No isolation between CTs; requires preemption to guarantee bandwidth to CTs other than the premium
Bandwidth may be wasted	Efficient use of bandwidth
Useful in networks where preemption is precluded	Not recommended in networks where preemption is precluded

The calculation of available bandwidth for the RDM model is a bit more complicated, because it must take into account LSPs at several priority levels and for all the CTs that share the particular BC. For example, the available bandwidth for an LSP from CT0 at priority p is equal to BC0 minus the allocations for all LSPs from all CTs at priorities better or equal to p. Table 4.1 summarizes the differences between MAM and RDM.

It is clear that the BC model plays a crucial role in determining the bandwidth that is available for each one of the TE classes on a link. The BC model and the bandwidth allocation for each BC are advertised by the IGPs in the BC sub-TLV. The IETF does not mandate usage of the same BC model on all links in the network. However, it is easier to configure, maintain and operate a network where the same BC model is used, and some implementations require consistent configuration of the bandwidth model on all links.

To summarize, the BC model determines the available bandwidth for each CT at each priority level. MAM and RDM are two possible BC models. They differ in the degree of sharing between the different CTs and the degree of reliance on preemption priorities necessary to achieve bandwidth guarantees for a particular CT. The IGPs advertise the BC model and the unreserved bandwidth for the CT–priority combinations corresponding to valid TE classes.

4.4.5 Overbooking

In the discussion so far, LSPs are established with bandwidth reservations for the maximum amount of traffic that is bound to traverse the LSP. However, not all LSPs are carrying the maximum amount of traffic at all times. Thus, even if a link is full from the point of view of existing reservations, there is idle bandwidth on the link. This bandwidth could be used by allowing other LSPs to establish over the link, in effect overbooking it. Several methods exist for implementing overbooking:

1. LSP size overbooking. The overbooking is achieved by reserving a lower bandwidth value than the maximum traffic that will be mapped to the LSP.
2. Link size overbooking. The overbooking is achieved by artificially raising the maximum reservable bandwidth on a link and using these artificially higher values when doing bandwidth accounting. Note that with this approach the overbooking ratio is uniform across all CTs, as shown in Figure 4.9.
3. Local overbooking multipliers (LOM). This refines the link size overbooking method by allowing different overbooking values for different CTs. Rather than 'inflating' the bandwidth for all CTs by the same factor, different factors can be used for each CT, e.g. 3:1 overbooking for CT0 but 1:1 overbooking for CT1, as shown in Figure 4.9. The per-CT LOM is factored in all local

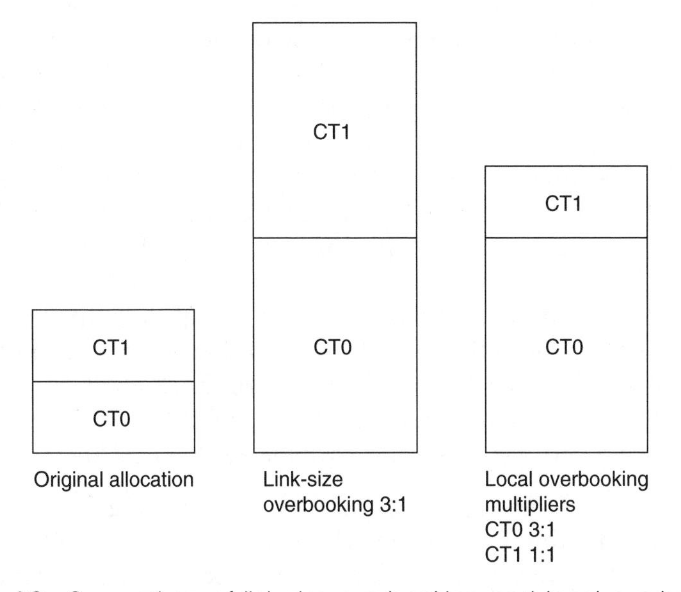

Figure 4.9 Comparison of link size overbooking and local overbooking multipliers

bandwidth accounting for the purpose of admission control and IGP advertisement of unreserved bandwidths. LOM is tightly coupled to the bandwidth model used, because the effect of overbooking across CTs must be accounted for very accurately (recall that, for example, in RDM bandwidth can be shared across classes). The details of the computation are described in [MAM-LOM] and [RDM-LOM].

4. Manual configuration of the BC. This method allows the user to specify the bandwidth constraints and by doing so to overbook a particular class. The drawback of this approach is that it is less intuitive to configure, because it does not translate easily into an overbooking percentage for a particular class.

Overbooking is useful in a multiservice network that will carry a number of different traffic classes where the statistical likelihood of congestion for each of the traffic classes varies greatly. A typical scenario is a network providing voice and data services. In this case, it is likely there will be high overbooking for the data traffic and no overbooking for the voice traffic.

4.4.6 The DiffServ in DiffServ-TE

In the previous sections, we have seen how network resources are partitioned among different types of traffic and how paths with per-traffic-type resource reservations are set up. In the solution we have presented, the traffic type equates to a desired scheduling behavior, and the available resources for a traffic type are the available resources for a particular scheduler queue. The assumption is that traffic automatically receives the correct scheduling behavior at each hop. This is achieved through DiffServ [RFC2475].

Recall from the Foundations chapter (Chapter 1) that the DiffServ CoS determines the packet's PHB and in particular the scheduling behavior at each hop. In practice, there are two ways to ensure that traffic mapped to a particular DiffServ-TE LSP maps to the correct scheduler queue, as explained in [RFC3270]:

1. Set the EXP bits appropriately at the LSP ingress (E-LSPs). Recall from the Foundations chapter that using E-LSPs at most eight PHBs are supported, so this method is good for networks in which less than eight PHBs are required. An important thing

to keep in mind is that once the packet is marked with a particular value, its QoS treatment is defined at each hop it crosses. Thus, to ensure consistent QoS behavior, it is imperative to maintain consistent EXP-to-PHB mappings.

2. Encode the scheduling behavior in the forwarding state (label) installed for the LSP and use the EXP bits to convey the drop preference for the traffic (L-LSP). The scheduling behavior associated with a forwarding entry is signaled at the LSP setup time. Any number of PHBs can be supported in this way.

A combination of both E-LSPs and L-LSPs can be used in a network, assuming that they can be identified (e.g. through configuration).

Thus, once the traffic is mapped to the correct LSP, it will receive the correct DiffServ treatment. The remaining challenge is to ensure that the mapping is done appropriately. Most vendors today provide mechanisms for picking the LSP based on flexible policies. One of the most intuitive policies is one where the IP DSCP (for IP traffic) or the EXP bits (for MPLS traffic) is used to map the packet to the correct LSP. Other policies may employ BGP communities attached to the route advertisements to pick the LSP. In that case, destinations are tagged with BGP communities, e.g. one community for a destination that requires EF treatment (such as a game server) and a different community for destinations for which traffic can be treated as best-effort (such as Internet routes). In both cases, traffic must be forwarded to the BGP next-hop, which corresponds to the address of the peer who sent this advertisement. Thus, several LSPs are set up to this destination, one per traffic class. When sending traffic to different destinations, the community is used to pick the LSP. In this way, traffic to the game server can be mapped to an LSP that gives the correct guarantees for EF.

In summary, DiffServ provides the correct scheduling behavior to each type of traffic. Vendors provide flexible policies for picking an LSP that was set up with bandwidth reservations from the correct class type. The combination of DiffServ and per-CT traffic engineering ensures compliance to strict SLAs.

4.4.7 Protection

No discussion on SLAs is complete without looking at the mechanisms available for protecting traffic following a failure.

As mentioned at the beginning of this section, the same mechanisms used to protect TE LSPs, discussed in the Protection and Restoration chapter (Chapter 3), can be used for DiffServ Aware LSPs. However, an interesting issue arises in the context of bandwidth protection.

When bandwidth protection is provided, the backup path must reserve bandwidth from the same class type as the protected path. The solution is straightforward for one-to-one backup, because separate protection paths are set up for each LSP. In case of facility backup, there are two options:

- Single backup. Use a single backup for the LSPs from all classes and treat all traffic on the backup as best-effort (this implies the backup is set up from CT0). Note that this approach is likely to cause performance degradation.
- Separate backup per-CT. Instead of a single backup, there is one backup for each class type and admission control of LSPs into the appropriate backup is performed based on both the class type and the requested bandwidth.

Because different CTs are tied to different guarantees, the operator might choose to reserve backup bandwidth for some classes but not for others or to protect LSPs from some classes but not from others. The ability to provide protection for DiffServ-TE LSPs ensures that SLAs can be guaranteed both under normal conditions and following a failure. However, is all this enough?

4.4.8 Tools for keeping traffic within its reservation limits

The carefully crafted solution presented in the previous sections would all go to waste if more traffic were forwarded through the LSP than the resources that were allocated for it. In such an event, congestion would occur, queues would be overrun and traffic dropped, with disastrous QoS consequences, not just on the misbehaving LSP but on all other LSPs from the same CT crossing the congested link.

One solution is to police the traffic entering the network at the interface between the user and the provider. Another solution is to use LSP policers to prevent such scenarios. LSP policers operate at per-CT granularity at the LSP head end and ensure that traffic

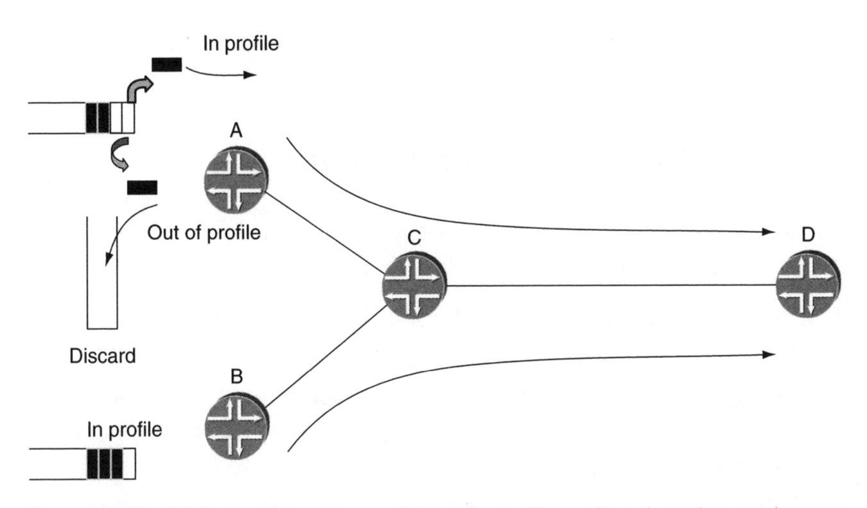

Figure 4.10 Misbehaving source A has its traffic policed and experiences QoS degradation. The well-behaved traffic from B is not affected

forwarded through an LSP stays within the LSP's bounds. Out-of-profile traffic can be either dropped or marked, affecting the QoS of the misbehaving LSP but shielding well-behaved LSPs that cross the same links from QoS degradation, as shown in Figure 4.10. LSP policers make it easy to identify the traffic that needs to be policed, regardless of where traffic is coming from (e.g. different incoming interfaces) or going to (e.g. different destinations beyond the LSP egress point). If the traffic is mapped to the LSP, it will be policed.

LSP policing provides a tool for policing traffic that is forwarded through an LSP. But how can one prevent mapping more traffic to an LSP than the LSP can carry? The answer is admission control. For example, some implementations provide admission control for Layer 2 circuits. A circuit does not establish unless the underlying RSVP LSP has enough available resources, thus avoiding oversubscription. For example, if a new Layer 2 circuit requires a 20 Mbps bandwidth, the ingress router identifies an LSP that goes to the required destination that has sufficient bandwidth, and decrements 20 Mbps from the bandwidth available for other potential Layer 2 circuits that may need to use the LSP in the future.

To summarize, LSP policing is a simple tool that ensures that traffic stays within the bounds requested for an LSP. Admission control into the LSPs prevents mapping more traffic to an LSP than the resources allocated for it. By combining admission control with policing, traffic is kept within its reservation limits and QoS can be enforced.

4.4.9 Deploying the DiffServ-TE solution

To summarize the previous sections, the following steps are required to deploy a DiffServ-TE solution:

1. Decide on a BC model and the bandwidth associated with each BC on each link.
2. Configure the buffer and bandwidth allocations on each link to be consistent with step 1 (assuming a model where CTs map to scheduler queues).
3. Decide which CTs and priorities are required.
4. Choose an IGP.
5. Configure LSPs with the desired bandwidth reservation, CT and priority.
6. Configure policers if required.
7. Decide whether the DiffServ treatment will be determined from the EXP bits or the label. If the DiffServ treatment is based on the EXP bits, configure the EXP-to-PHB mappings consistently throughout the DiffServ domain and make sure the traffic is marked correctly. If the DiffServ treatment is based on the label, make sure that all routers have a consistent view of what the PHBs are, so that when the DiffServ treatment is advertised at LSP setup time, it results in uniform behavior on all routers.

Let us briefly look at the migration of a traffic-engineered network to the DiffServ-TE model. As a first step, the network operator must decide which combinations of CTs and priorities are required in the network. Recall from Section 4.4.1 that LSPs with no per-CT requirements are mapped to CT0. Therefore, in a migration scenario, the combinations of CT0 and of the priorities used for TE LSPs that already exist in the network must be selected as valid combinations. The second step is to map the CT–priority combinations selected in the first step to TE classes. Recall from Section 4.4.2 that the Unreserved Bandwidth TLV is overwritten with the per-TE class information. Network migrations are typically done in stages, so there will be both old and new nodes advertising the Unreserved Bandwidth TLV to each other, but with different semantics. Old nodes will fill in field *i* of the Unreserved Bandwidth TLV the available bandwidth for (CT0, *i*). New nodes will fill the available bandwidth for TE*i*. To provide a consistent picture of the available resources to both old and new nodes, (CT0, *i*) must map to TE*i*. Such a definition ensures smooth interoperation between nodes that support the DiffServ-TE extensions and nodes that do not.

4.5 EXTENDING THE DIFFSERV-TE SOLUTION WITH MULTICLASS LSPs

So far we have seen that LSPs set up according to [RFC4124] carry traffic from a single DiffServ class and are set up along a path that satisfies the bandwidth constraints specified for that class. However, sometimes traffic with different DiffServ behaviors must be mapped to the same LSP and the LSP must satisfy the bandwidth constraints for each one of these classes. We will call these multiclass DiffServ Aware LSPs.

An example scenario for multiclass LSPs arises in the context of ATM trunk emulation using MPLS LSPs. To effectively emulate an ATM trunk, all the traffic classes should follow the same path in the network and should exhibit the same behavior in case of failure. If the EF class fails, so should the BE class. If traffic switches to a protection path, it should use the same path for all classes. In principle, one could argue that this behavior can be achieved by setting up a separate LSP for each class and then adding the necessary control-plane intelligence to keep them synchronized. Apart from being cumbersome to implement, such a solution also has drawbacks with regards to the number of LSPs that must be created and maintained.

This brings us to another application of multiclass DiffServ-TE LSPs: reducing the number of LSPs in a network by setting up reservations for several classes in one LSP rather than one reservation per class. When LSPs are set up with bandwidth reservations from a single class, the total number of LSPs in the network is equal to the number of classes times the number of LSPs in the mesh. With multiclass LSPs, the total number of LSPs is equal to the size of the LSP mesh. The reduction in the number of LSPs is important both from a scaling and manageability point of view, as seen in the Traffic Engineering chapter (Chapter 2).

Without a solution from the IETF, vendors developed proprietary extensions to the DiffServ-TE solution, in order to support multiclass LSPs. One such solution is documented in [MULTI-CLASS]. In this case, multiple-class types are configured per LSP and the LSP is established only if there is a path that fulfils the bandwidth requirements of each configured class-type.

4.6 CONCLUSION

Differentiated Services (DiffServ) provides QoS by dividing traffic into a small number of classes and allocating network resources on a per-class basis. MPLS-TE enables resource reservation and optimization of transmission resources. MPLS DiffServ-TE combines the advantages of both DiffServ and TE, while at the same time benefiting from the fast reroute mechanisms available for MPLS.

The result is the ability to set up traffic-engineered LSPs with per-traffic-class granularity and to guarantee resources for each particular type of traffic. Equipment vendors offer mechanisms to map traffic to the appropriate LSPs based on flexible policies, as well as tools for ensuring that traffic stays within the limits of the resources that were reserved for it. Thus, strict QoS guarantees are achieved both for the steady state and the failure cases. Based on the service guarantees that are achieved, service providers can offer services with high SLA requirements, such as voice or migration of ATM/FR on to an MPLS core.

However, as discussed so far, both TE and DiffServ-TE are limited in their scope to a single IGP area and a single AS. In the next chapter, we will see how this limitation can be overcome by Interdomain TE.

4.7 REFERENCES

[DSTE-MAM] F. Le Faucheur and K. Lai, *Maximum Allocation Bandwidth Constraints Model for Diff-Serv-aware MPLS Traffic Engineering*, RFC4125, category experimental, June 2005

[MAM-LOM] draft-ietf-tewg-diff-te-mam-00.txt, older version of the MAM draft, which includes discussion of LOM

[MULTI-CLASS] I. Minei *et al.*, 'Extensions for differentiated services-aware traffic engineered LSPs', draft-minei-diffserv-te-multi-class-01.txt (work in progress)

[RDM-LOM] draft-ietf-tewg-diff-te-russian-01.txt, older version of the RDM draft, which includes discussion of LOM

[RFC2475] S. Blake *et al.*, *An Architecture for Differentiated Services*, RFC2475, December 1998

[RFC3270] F. Le Faucheur *et al.*, *MPLS Support of Diff-Serv*, RFC3270, May 2002

[RFC3468] L. Andersson and G. Swallow, *The Multiprotocol Label Switching (MPLS) Working Group Decision on MPLS Signaling Protocols*, RFC3468

[RFC3564] F. Le Faucheur *et al.*, *Requirements for Support of Differentiated Services-Aware MPLS Traffic Engineering*, RFC3564, July 2003

[RFC3630] D. Katz, K. Kompella and D. Yeung, *Traffic Engineering Extensions to OSPF*, RFC 3630, September 2003

[RFC3784] H. Smit and T. Li, *IS-IS Extensions for Traffic Engineering*, RFC3784, June 2004

[RFC4124] F. Le Faucheur *et al.*, *Protocol Extensions for Support of Differentiated-Service-Aware MPLS Traffic Engineering*, RFC4124, June 2005

[RFC4127] F. Le Faucheur *et al.*, *Russian Dolls Bandwidth Constraints Model for Diff-Serv-Aware MPLS Traffic Engineering*, RFC4127, category experimental, June 2005

4.8 FURTHER READING

[RFC2702] D. Awduche *et al.*, *Requirements for Traffic Engineering over MPLS*, RFC2702, September 1999

[Awduche Jabbari] D. Awduche and B. Jabbari, 'Internet traffic engineering using multiprotocol label switching', *Journal of Computer Networks* (Elsevier Science); **40** (1), September 2002

5

Interdomain Traffic Engineering

5.1 INTRODUCTION

In the Traffic Engineering chapter (Chapter 2), we have seen how to compute and signal traffic-engineered paths that comply with a set of user-defined constraints. A key step in this process is acquiring the information regarding the constraints for all the links in the network. This information is distributed by a link-state IGP and is therefore confined within the same boundaries as the link-state advertisements. Because the visibility of the topology and of the constraints is limited to a single IGP area, TE LSPs dynamically computed by the head end are also limited in the same way. This becomes a problem in large networks that deploy several IGP areas for scalability or in the case of services spanning across several service providers.

In this chapter we will see how RSVP-signalled TE LSPs can extend across IGP areas and across AS boundaries. These solutions are known as interarea TE and inter-AS TE respectively and are referred to collectively as interdomain TE. They apply equally to classic TE and to DiffServ Aware TE (described in the DiffServ-TE chapter, Chapter 4). In this chapter the term 'domain' is used to denote either an IGP area or an AS.

5.2 THE BUSINESS DRIVERS

The benefits of traffic engineering were discussed in the Traffic Engineering chapter (Chapter 2). Providers use traffic-engineered paths for optimization of network resources, support of services with QoS guarantees, fast reroute and the measurement of the aggregated traffic flow between two points in the network. To achieve these functions in large networks with multiple IGP areas, the LSPs used for traffic engineering need to cross area boundaries (interarea LSPs).

Interdomain LSPs[1] are not limited to traffic engineering; they are also pivotal to the deployment of services spanning across different geographical locations. These can be services requiring assured bandwidth, such as connection of voice gateways, or they may be applications that rely on the existence of an MPLS transport tunnel, such as pseudowires or BGP/MPLS Layer 3 VPNs. When the service spans several IGP areas, the LSP is interarea; when it spans different ASs, the LSP is inter-AS.

Inter-AS LSPs exist both within the same provider and across different providers. Multiple ASs can be present within a single service provider's network, e.g. following the acquisition of another provider's network in a different geographical location. The separate ASs are maintained for reasons ranging from administrative authority to the desire to maintain routing protocol isolation between geographical domains and prevent meltdown of the entire network in the event of a local IGP meltdown.

A useful application of LSP establishment across provider boundaries is the interprovider option C of BGP/MPLS Layer 3 VPNs (discussed in detail in the Hierarchical and Recursive L3 VPNs chapter, Chapter 9). The inter-AS RSVP LSP brings two benefits: (a) the ability to traffic-engineer the path between the remote PEs and (b) the ability to simplify the configuration by not having to rely on BGP to 'glue' the LSP segments for setting up LSP between the remote PEs.

Figure 5.1 shows an interprovider VPN setup, where two customers, VPNa and VPNb, have sites attached to PE3 and PE4 (in different ASs). The loopback addresses of PE3 and PE4 are advertised as VPN routes to ensure connectivity between the PEs. Once the addresses of PE3 and PE4 are known, an External Border Gateway Protocol (EBGP) session can be set up between the two PEs and the

[1] Recall that 'domain' is used in this chapter to denote either an area or an AS.

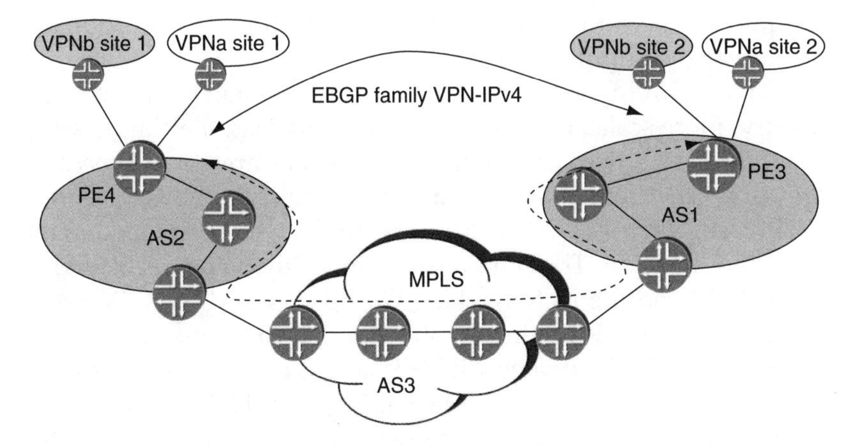

Figure 5.1 An inter-AS LSP can be used in the setup of an interprovider VPN

VPN routes of the two customers, VPNa and VPNb, are exchanged over this session. Forwarding traffic to these addresses requires an LSP between PE3 and PE4. When discussing interprovider VPNs we saw the rather drawn-out process of how this is done using BGP. If an end-to-end RSVP LSP was available, it could be used instead.

So far we have seen why interdomain LSPs are important. Next, we will look at how they can be set up.

5.3 SETTING UP INTERDOMAIN TE LSPs

As discussed earlier in this chapter, the limitation of a TE LSP to a single IGP area is caused by the limited visibility into the topology at the LSP head end. However, once a path is specified, there is no problem signaling it across IGP areas or across ASs. Therefore, the setup of interdomain LSPs is possible without any extensions, e.g. by computing the path offline and specifying the hops at the head end. The problem with this approach is that it forces a service provider to move to an operations model relying on offline computation for both the primary and secondary paths, with the implications discussed in section on offline computation in the TE chapter.[2] In addition, the issue of fast reroute is not addressed in

[2] In the case of a multiprovider environment the offline tool would also need to know the TE information of all the links of all the providers involved. That may require, among other things, that one provider discloses its internal topology to another provider, a not very attractive prospect in many cases.

this model unless the bypass tunnels protecting the interdomain links are also computed offline.

As TE and MPLS-based applications started gaining traction in the industry, more scalable and optimal solutions than the simple setup of an LSP across domain boundaries (whether area or AS) became necessary. The requirements for interdomain traffic engineering [INTER-AREA-TE-REQ, INTER-AS-TE-REQ], were developed in the TEWG[3] in the IETF. The solutions are currently being developed in the CCAMP and the PCE Working Groups [CCAMPWG, PCEWG] and at the time of this writing are still works in progress.

Setting up RSVP-signaled TE LSPs across IGPs domains is done in three steps: (a) discovering reachability and TE information, (b) computing the path and (c) signaling the LSP. However, only the latter two are modified for interdomain operation. The propagation of reachability and TE information cannot be changed to cross IGP boundaries, because this would severely impact the scalability of the IGPs.[4] For this reason, information distribution is not discussed further. Instead, the following sections focus on path computation and path signaling. For clarity of the discussion, the setup methods for interdomain LSP setup are discussed first.

5.3.1 Path setup

It may seem like a strange thing to start the discussion on interdomain TE from the setup instead of the path computation. The reason for doing so is because examining the different setup methods makes it easier to understand the choices that must be made with regards to path computation. There are three methods for setting up LSPs across domain boundaries.

Contiguous LSP

In this case, an end-to-end LSP between PE1 and PE2 is built across domain boundaries, using hop-by-hop signaling between adjacent neighbors. This method is the most intuitive, because it resembles exactly the setup of a TE LSP within one domain. Figure 5.2 shows the setup of an interarea contiguous LSP.

[3] The TEWG has in the meantime completed its work and has been closed.

[4] It may also not be feasible for the reason of privacy discussed before (as each provider may not want to disclose its internal topology to other providers).

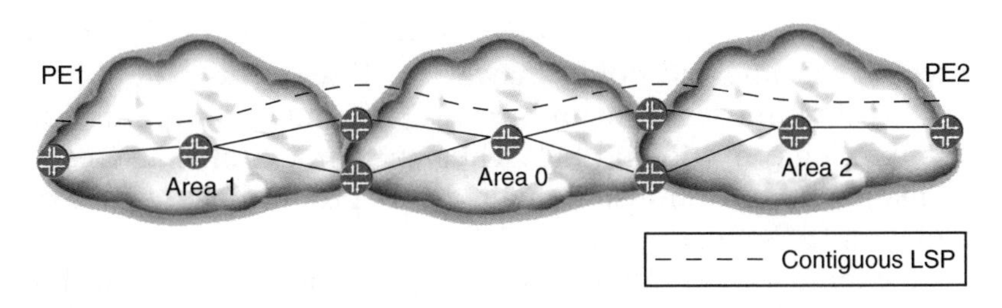

Figure 5.2 Setting up an interarea LSP using the contiguous LSP method

LSP stitching

The end-to-end LSP between PE1 and PE2 is built from several smaller LSPs (called TE LSP segments) that are set up in the different domains and 'stitched' together at locations called 'stitching points' [LSP-STITCHING]. This patching together of segments is accomplished by installing the forwarding state that takes traffic reaching the end-point of one LSP segment and maps it into the next LSP segment. A 1:1 mapping is enforced between the segments in the different domains, meaning that traffic that is mapped into any LSP segment is guaranteed to be coming from a single LSP segment. Railway cars are a useful analogy for LSP segments. They can be stitched together to allow traffic (people) to pass from one car to another and there is a 1:1 mapping between the segments because the stitching point connects exactly two cars.

Figure 5.3 shows a stitched LSP crossing three IGP areas. Separate TE LSP segments exist in each area (in this case spanning between the area border routers) and are stitched together at the ABRs to

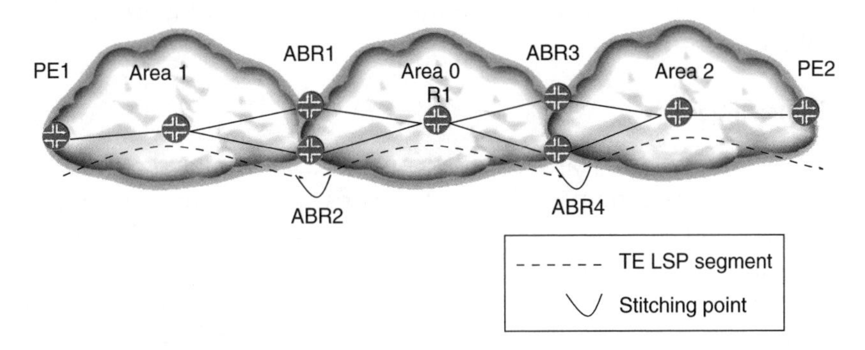

Figure 5.3 Setting up an interarea LSP using the LSP stitching method

form one LSP. If in this example a second LSP were to be set up between the same end-points, new TE LSP segments would have to be created in each domain, because a segment can participate only in a single end-to-end LSP. Thus, the amount of state created and maintained in a transit domain grows proportionally with the number of LSPs crossing it.

There are several important things to note about TE LSP segments that influence the properties of an end-to-end LSP set up using the stitching method:

- Scope. By definition, TE LSP segments span a single domain. This means that the computation of their path is limited to the domain and that functions such as reoptimization and fast reroute are also confined in the same way. The ability to perform these operations locally is a useful property of the stitching solution, as will be seen in later sections.
- End-points. TE LSP segments have a head end and a tail end, just like normal LSPs. These are usually the border routers at the entry into and exit from the domain, but can be other routers as well, depending on the topology and the provisioning used. In the section discussing LSP computation we will see why it is convenient to set up the TE LSP segments between border routers.
- Setup trigger. TE LSP segments may be preconfigured or their setup may be triggered by the arrival of an LSP setup message from a neighboring domain.

Thus, LSP stitching creates an interdomain LSP from several segments with per-domain scope. However, because any segment can be part of only a single LSP, the state created in transit domains increases with each transit LSP. LSP nesting solves this scalability limitation.

LSP nesting

An end-to-end LSP between PE1 and PE2 is tunneled inside an LSP with per-domain scope as it crosses the domain, creating a hierarchy of LSPs and hiding the details of the end-to-end LSP from the routers in the transit domain [HIER, INTER-DOMAIN-FW]. This process is called 'nesting', because one LSP is placed into another one. The LSP that acts as the 'nest' or container for other LSPs is called the Forwarding Adjacency (FA) LSP. LSP1 and LSP2 in Figure 5.4 are both end-to-end LSPs crossing three IGP areas. In the middle area, LSP1 and LSP2 are both nested into an FA LSP that spans between the area boundaries.

Nesting is accomplished by using label stacking. At the head end of the FA LSP, the label of the FA LSP is pushed on top of the label stack of the nested LSP, in a process similar to the one described for bypass tunnels in the protection and restoration chapter. Forwarding proceeds based on the top label only and routers within the FA LSP's domain are not required to maintain any state for the transit LSPs. In the example setup from Figure 5.4, router R1 is not aware of the existence of LSP1 or LSP2.

More than one transit LSP can be nested into the same FA LSP. Figure 5.4 shows two LSPs, originating and terminating on different PE routers, that share the same FA LSP in the transit area. Thus, nesting uses a 1:*N* mapping between the FA LSP and the transit LSPs.

LSP nesting brings the scaling advantages of LSP hierarchy to interdomain TE: no state needs to be maintained for the interdomain LSPs inside the transit domain and the setup and teardown requests for the nested LSPs do not place any additional burden on the routers inside the transit domain. FA LSPs also allow for easier enforcement of policies for LSPs traversing the domain. One example is the ability to control the links used by transit LSPs by limiting the links used by the FA LSP. Another example is the ability to perform admission control for LSPs traversing the domain by simply looking at the available resources on the FA LSP rather than checking each and every link that the traffic would traverse.

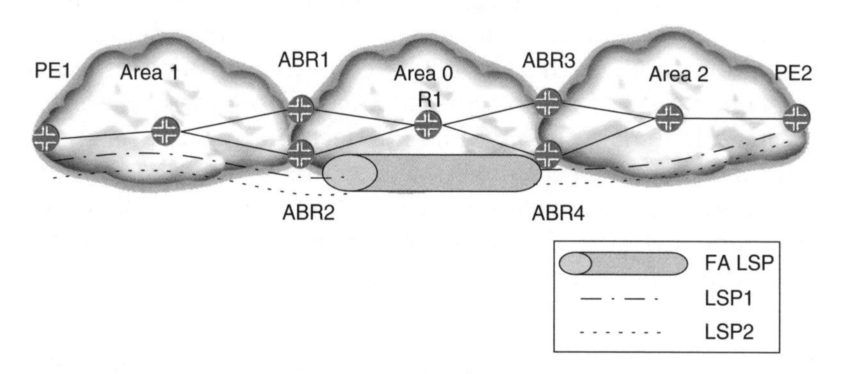

Figure 5.4 Setting up an interdomain LSP using the LSP nesting method. Several LSPs can be nested on to the same FA LSP

FA LSPs share the same properties as TE LSP segments from the point of view of scope, end-points and triggers for their setup. Thus, the useful properties of domain-local path computation, reoptimization and repair also apply to FA LSPs and end-to-end LSPs set up with stitching and nesting share similar properties.

The main difference between nesting and stitching is in the amount of state that is created in the transit domain. Stitching requires individual segments for each transit LSP while nesting can share a single FA LSP, yielding a more scalable solution. The natural question is why bother with stitched LSPs at all? To answer this, recall that interdomain LSPs are often used for traffic engineering. Let us take a look at an end-to-end LSP with a certain bandwidth requirement. When the LSP is set up using stitching, the bandwidth requirement can be easily satisfied by ensuring that all the TE LSP segments are set up with the correct bandwidth allocation. In contrast, when the LSP is set up using nesting, the same approach does not automatically work, because any number of LSPs may be mapped into the same FA LSP. To ensure adequate resources for all transit LSPs there is a need to perform admission control into the FA LSP. In addition to the admission control, one may also need to perform traffic policing at the entrance to the FA LSP, especially if such an entrance is on the administrative domain boundary between two providers.

Having seen the different LSP setup methods, the natural question is how the setup method is picked at the domain boundary. The answer is that it is chosen based on administrative policies locally configured at the border router.[5] This implies that a single end-to-end LSP may be set up using different methods in different domains: for example, it may use nesting in one domain and stitching in another. This should not come as a surprise, especially when one thinks of the interprovider case. How the LSP is set up within each domain should be a local decision.

Regardless of the setup method used, the path of the LSP must be computed. The following section discusses the challenges of computing the path for an interdomain LSP and the different methods for performing the computation.

[5] When the head end requires the setup of a contiguous LSP, it can explicitly signal this desire using a flag in the Session Attribute Object. In all other cases, the signaling is based on local administrative policies.

5.3.2 Path computation

The main limiting factor for an interdomain[6] path computation is the visibility that the computing node has into the topology. This influences both the scope of the computation (per-domain or interdomain) and the ownership of the computation (which element is performing the computation).

1. Scope of the computation. The scope of the computation is limited by the visibility that the computing entity has into the topology. This is true irrespective of the owner of the computation. Therefore, it is either confined to a single domain (per-domain path computation) or it spans multiple domains (interdomain path computation, also referred to as end to end).
2. Ownership of the computation. The entity performing the computation can be an offline tool, the LSR at the head end, a domain-boundary node or another element (such as the path computation element, which is discussed in more detail in the following sections). The visibility that the computing entity has into the topology affects its ability to perform the computation.

At first glance it may seem that the LSP setup method dictates the scope of the computation and therefore also implicitly determines which element has enough data to perform the computation. Wouldn't the setup of a contiguous LSP require that the path computation span its entire path? And if an interdomain computation is indeed required, wouldn't it have to be performed by an entity with global visibility across all domains? The answer is 'no', as will be seen in the following example discussing the setup of an interdomain LSP using the contiguous signaling method.

It is very intuitive to think of a contiguous LSP setup where all the hops in the path are precomputed and then signaled with the Explicit Route Object (ERO). In this case, the path computation must have interdomain scope and therefore must be performed by an entity that has interdomain visibility, such as an offline tool.

A less intuitive, but perfectly valid, way of setting up the same end-to-end LSP is to perform the path computation separately within each domain. Assuming that the exit points out of the domains are

[6] The IETF documents discussing path computation use the term 'domain' to denote either an area or an AS. For this reason, the same terminology is used here.

known[7] or can be determined by some means, a path can be computed up to the border router at the domain exit. Thus, the path is not known in its entirety at the LSP head end. Instead, as the LSP is signaled, the path to the next border router is computed and added to the ERO (this process is called ERO expansion). Using this approach, the path computation is always limited in scope to a single domain and the path is computed piece by piece as it traverses the different domains. The computation may be performed by the domain boundary nodes or it may be obtained through other means, as we will see in the following sections.

The above example illustrates a fundamental point regarding interdomain TE, namely that the path computation can be performed either interdomain or per-domain, regardless of the signaling method used for the LSP setup.

The discussion so far focused on finding a path for the LSP across the different domains. However, remember from the introduction that one of the main requirements for the interdomain solution was support for TE. It is important to understand that regardless of whether the path is computed per-domain or interdomain, the assumption is that the traffic engineering characteristics of the LSP are uniformly maintained across all domains. This implies a common understanding of the LSP's constraints across all domains. The problem is that different domains may be under different administrations and therefore their local definition of DiffServ-TE class types, as discussed in the DiffServ-TE chapter (Chapter 4), or link properties may not be compatible. For example, the class type (CT) suitable for voice traffic may be CT1 in one AS and CT3 in another, or the link color for high-latency links may be X in one domain and Y in the neighboring one. For this reason, when the path computation crosses from one domain to the next, the constraints must be translated appropriately, e.g. through a mapping. Note that this implies that the administrations of the two domains must cooperate by exchanging the relevant information and agreeing on such a mapping. This is particularly true in the interprovider case, where it is very likely that different constraints are used. Thus, when talking about 'visibility' into neighboring domains, both the topology

[7] The exit points out of the domain (border routers) can be configured as loose hops in the ERO or they may be discovered based on the IP reachability information for the LSP's destination address.

information and the TE characteristics of the topology (or the appropriate mapping) must be known.

Given a common understanding of the constraints, the interdomain computation assumes visibility into all the domains in the path, but does not introduce any new requirements. However, the per-domain computation raises interesting challenges.

5.3.2.1 Per-domain path computation

Per-domain path computation is performed when there is no visibility across all domains at any one single central point, irrespective of the owner of the computation. For this reason, the computation is performed separately within each domain, from one border router to the next, each such computation bringing the path one domain closer to the final destination. The assumption is that the address of a border router on the path to the LSP destination is known. The border router is either configured as a loose hop in the path or it is discovered dynamically based on the IP reachability for the LSP destination address. The result of the computation is a path to the border router. How this path is used depends on the LSP setup method. For contiguous signaling, it can be used during ERO expansion; for stitching and nesting, it can be used to set up or select the relevant TE LSP segment or FA LSP.

Thus, when using per-domain computation, the path is traffic engineered separately within each domain rather than being traffic engineered end to end. However, the fact that each piece in the path is optimal does not necessarily mean that the entire path is optimal.

Figure 5.5 shows an example of how the complete path can be nonoptimal. The goal is to set up a shortest-path inter-AS LSP from A to B, with a bandwidth reservation of 100 Mbps. There are two inter-AS links and both exit points are equally good from a routing point of view. All links are of the same capacity. However, link ASBR3-B in AS2 has no available bandwidth because LSP2 is set up over it, with a 100 Mbps bandwidth requirement. In this case, the optimal path is A–ASBR2–ASBR4–B. However, from the viewpoint of AS1, both ASBR1 and ASBR2 appear to be valid, optimal options. If A chooses ASBR1 as its exit point, then A–ASBR1–ASBR3–ASBR4–B is the most optimal path that can be found (it is, in fact, the only feasible path, so it is the 'best' one that meets the constraints). Although the computation is optimal within each domain, the end-to-end path is not optimal.

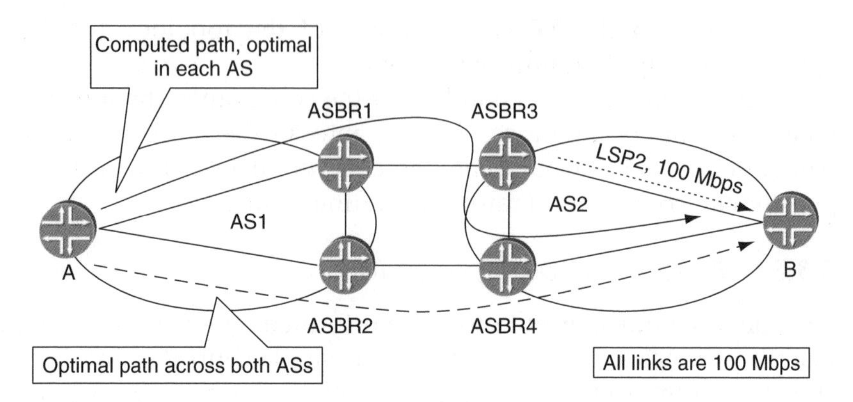

Figure 5.5 Nonoptimal resource optimization when using per-domain path computation

This example also raises an interesting question with regards to the information distribution within a single domain. Imagine that congestion occurs on the inter-AS link ASBR1–ASBR3 rather than on ASBR3–B. The problem is that the inter-AS link is not in the TED, so the congestion on the link is not detected until the path setup request is received and admission control fails. One way to alleviate this problem is to inject the TE information for links on domain boundaries (such as the ASBR–ASBR link) into the IGP TED, to improve the accuracy of the computation and minimize failures at the LSP setup time.

However, this approach cannot guarantee that admission control will succeed when the LSP is actually signaled. Of course this is no different from any other links in the TED and true for any computation method and any signaling method. The question, therefore, is how are LSP setup failures handled in the case of interdomain LSPs?

5.3.2.2 Crankback

The previous example showed an LSP setup failure caused by a computation based on inaccurate TE information. However, even if the computation is perfectly accurate, the LSP setup can still fail if, between the time the path was computed and the time that the path was signaled, one of the resources becomes unavailable (e.g. due to the setup of another LSP). The Traffic Engineering chapter (Chapter 2) describes how this situation is handled when TE is confined to a single domain. In this case, the node where admission control fails sends a path error message to the head end, indicating the location of the failure. Based on this information, the head end can compute a new path that avoids the problematic resource. In addition, the updated resource information may be advertised by the IGP to ensure that inaccurate computations are not performed by other nodes in the network.

This solution assumes that the LSP head end can use the failure information in a meaningful way when computing the new path. This may not be the case when the path crosses several domains into which the head end does not have visibility. Figure 5.6 shows a network with three IGP areas (which are labeled areas 1, 2 and 3 for the sake of clarity). An LSP must be established between node A in area 1 and node B in area 3. Imagine that the path computation yields the path A–ABR1–ABR3–R1–B and the setup fails because of unavailable resources on link R1–B (we will call link R1–B the blocked resource). In this case, propagating the failure to the LSP head end is not very useful, as there is not much that it can do with the information identifying the blocked resource as link R1–B. Instead, it makes more sense to forward the failure information to the first node that can put it to good use, in this case border router ABR3, which can look for an alternate path within area 3 that avoids the offending link (e.g. ABR3–ABR4–B).

However, what if border router ABR3 cannot find such a path? In the example, this can happen if the link ABR3–ABR4 along the path ABR3–ABR4–B does not have enough resources. In this case, ABR3 is treated as the blocked resource and an error is forwarded

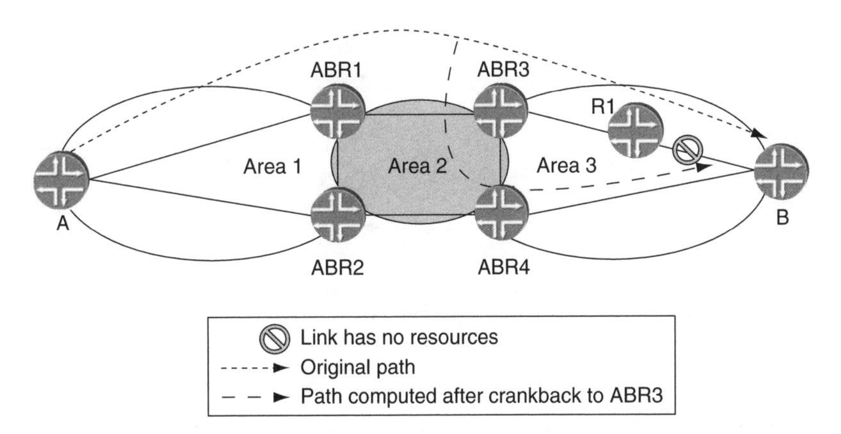

Figure 5.6 Using crankback to deal with path setup failures

to the first router that can use this information in a meaningful way, border router ABR1. What is effectively happening is that the computation is cranked back one computation step at a time, away from the failure point. This process is called crankback and is a popular technique in TDM-based networks.

Crankback is a natural fit for LSPs made up of nested or stitched segments. When there is a setup failure in one domain, rather than recomputing the entire LSP the computation can first be redone in the failed domain. If the computation fails, the error is reported to the domain upstream and a path to an alternate border router or alternate domain can be evaluated. This local repair of the path computation shields the LSP head end from recomputation requests caused by failures in domains over which it has no control.

The desire to shield upstream domains from unnecessary computations is one of the main goals of crankback. However, containing the computation within a particular domain is not enough. In the previous example, imagine that there is no feasible path within area 3 and that the computation has been cranked back to border router ABR1 in area 2, as shown in Figure 5.7. At this point, any setup request from ABR1 will fail. What is to stop border ABR1 from continuously toggling between the two blocked resources and trying to set up paths through ABR3 and ABR4 alternatively? What is needed to avoid such a situation is a way to inform ABR1 that there is no point in trying to continue the search and that it should crankback the computation. Thus, two pieces of

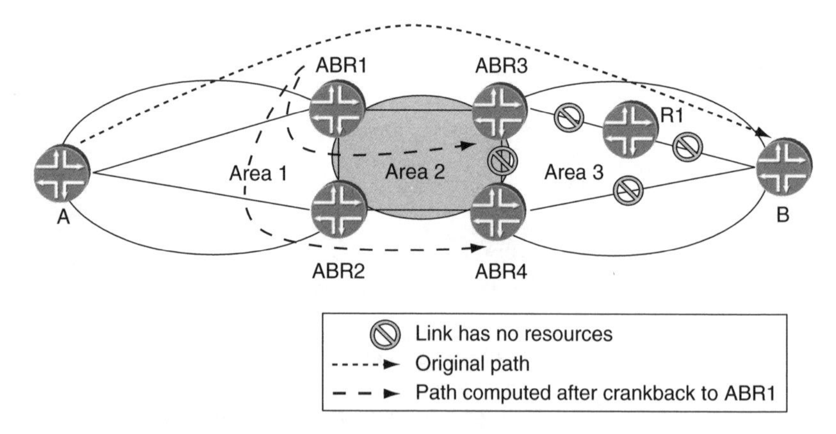

Figure 5.7 Crankback when no feasible paths exist in a downstream domain

information must be carried in the failure notification: the location of the failure and whether to continue the search or crankback. In addition to this mechanism, routers can maintain a history of failed computation attempts to improve the accuracy of computations, and a hard limit can be set for the recomputation attempts of any path.

Note that crankback does not provide any guarantees regarding the time it takes to find a path. Furthermore, because of its per-domain nature, it cannot ensure optimality of the path either. In fact, because of the limit imposed on the recomputation attempts, crankback cannot even ensure that any path will be found. Having said all this, it may seem that crankback is not a very good idea at all. However, remember that crankback does provide an efficient solution in nonpathological situations. In an engineering world, an imperfect solution may sometimes be better than no solution at all. As with anything else, the decision whether to use crankback becomes a question of whether the benefits outweigh the costs for a particular deployment. For example, think of an LSP crossing 10 ASs across the globe, when there is some congestion in the destination AS. Without an attempt to local repair, the errors would need to be propagated back all the way to the head end LSR, 10 ASs away.

As a protocol, RSVP lends itself readily to crankback signaling because it already includes the failure notification. The IETF defined further extensions to RSVP for crankback support [CRANKBACK]. In particular, the ability to do crankback and the node that should perform the recomputation (head end, border router or other router in the path) can be signaled using flags in the Session Attribute Object in Path messages. New objects are added to carry more detailed failure information in the Path Error message issued when the LSP setup fails.

Although by no means perfect, crankback is an important tool when the path is computed separately within each domain. From the discussion so far, it may seem that the entity performing the computation is always one of the routers in the network. However, in an interdomain setup, the entity performing the computation may need more information than is available to a router in the network. For this reason, a path computation element may be used.

5.3.2.3 Path computation element (PCE)

Recall from earlier discussions that path setup across domain boundaries is not a problem when the path is specified in its entirety at the head end. For this reason, it is tempting to want to offload the path computation to an all-knowing, all-seeing entity that can deliver the necessary information on demand. Path computation elements (PCEs) first started out as such entities, and thus in many people's minds became inextricably tied to the interdomain solution. From the discussion so far it should already be clear that this is not the case. In the previous sections we have talked about setting up LSPs using per-domain computation without the aid of PCEs, but the same computations performed by the border routers could have been performed by PCEs in each domain.

Let us take a look at a few of the challenges of doing path computation in an interdomain setup, as this will shed some light on the benefits of PCEs:

1. *Constraint communication.* Recall from the Traffic Engineering chapter (Chapter 2) that not all the constraints are signaled when the LSP is set up. The assumption is that some of the constraints (e.g. link colors) need to be known only at the head end, which is the node doing the path computation. This assumption breaks in an interdomain setup where the computation is performed per-domain and the constraints must be communicated to all nodes participating in the path computation.
2. *Need for extra visibility.* Intuitively, it is easy to think of an interdomain LSP as one whose head end and tail end are in different domains. However, this may not always be the case. For example, for optimization or policy reasons it may be necessary to set up an LSP whose head end and tail end reside in the same domain but crosses into a different domain at some point along its path. To compute the path of such an LSP, more information is required than is available in a single domain.
3. *Constraint translation.* Earlier in this chapter we discussed the issue of translating constraints such as class types or priorities at domain boundaries. Although mappings may solve the problem, a different approach is to have a different entity do the translation.
4. *Optimality of the solution.* For an optimal solution, it may be necessary to run more specialized algorithms than CSPF or to have more information than just the resource availability, as discussed in the Traffic Engineering chapter.

From looking at these requirements, it looks like an offline path computation tool could almost fit the bill. However, thinking of the way the path is used during LSP setup, it is clear that the computing entity should be able to service requests dynamically in real time. The notion of a clever entity that can compute paths based on more information than what is available to a router in the network, using more sophisticated algorithms than simple CSPF, was first introduced in the early interdomain drafts. The entity was called the path computation element (PCE).

Subsequently, the task of defining PCE operation and protocols became the task of a new working group in the IETF, the PCE Working Group [PCEWG]. This group started its work around the time of this writing. For this reason, we will look at the open questions and challenges facing the definition of the PCE rather than at solutions. First, however, let us see how the PCE is envisaged to operate in the interdomain setup.

The PCE can either be a module on the router or a separate entity that the router can communicate with. Its role is to service path computation requests. The resulting path may be confined to a single domain or may cross several domains. For example, when setting up an interdomain path using the ERO expansion technique, the border router can query a PCE for the path to the next loose hop instead of simply running CSPF on the router itself. To compute such a path, the PCE must have at least the same information as is available in the TED, but to provide added value it may store additional information not available to the routers (such as global knowledge of all LSPs set up in the domain). Other ways in which the PCE can improve the path computation are by running more sophisticated algorithms than CSPF and by collaborating with other PCEs to find the most optimal path across several domains.

Based on the description of the PCE operation and requirements, let us list the different functions of the PCE that need to be defined and standardized. Here is a (partial) list of the work that will be done in the PCE Working Group:

1. Router–PCE communication protocol. The PCE operates in a request/reply mode, where the router requests a computation and the PCE replies with a result. Thus, there is a need to define the protocol for this communication. The router–PCE relationship is a client–server one and therefore the protocol used must be appropriate for client–server interactions. At the time of this writing, the choice of communication protocol is still under

discussion in the PCE Working Group. One of the big debates is whether to extend RSVP for this purpose. The argument is that because the result of the computation will be handed off as an ERO, it makes sense to reuse RSVP. This argument is not convincing, for two reasons. First, the encoding of the result should not dictate a choice of protocol. Second, RSVP is not a client–server protocol and does not lend itself naturally to client–server interactions.

2. PCE–PCE communication protocol. Inter-PCE collaboration is envisaged as one of the options either for computing an end-to-end path or for minimizing the risk of having to run crankback when computing per-domain paths. Ideally, the same protocol used for router–PCE communication should also be used for PCE–PCE communication. However, the PCE–PCE protocol may be more of a peer-to-peer protocol than client–server protocol, and thus the protocol for router–PCE communication may not be suitable for PCE–PCE interactions.
3. PCE discovery. The question of how a router finds out which PCE to query seems a simple one: why not simply configure the address of the PCE? A statically configured PCE becomes cumbersome to maintain in a large network and does not solve the problem of switching to a backup PCE when the primary fails or is too busy servicing other requests. Autodiscovery may provide a solution to at least some of these problems. One of the proposals is to use the IGP as a vehicle for distributing PCE information, similar to the way it carries RSVP automesh information (as seen in the Foundations chapter, Chapter 1).
4. Acquiring the TE database. The TED is the minimum information that the PCE must have in order to provide meaningful computation. For the computation to be as accurate as possible, the TED on the PCE must be at least as accurate as the one on the router. This is not a problem when the PCE is part of the router. For PCEs that are separate entities, the TED can either be built dynamically by 'sniffing' the IGP advertisements or it can be requested from the routers. Sniffing IGP advertisements implies that the PCE is part of the network (which in turn means that the operator must qualify the equipment appropriately). Synching-in large TED databases in an efficient and reliable way requires an appropriate database exchange protocol built into both the

router and the PCE. At the time of this writing, the PCE Working Group had not yet decided on the preferred method of acquiring the TED.

5. Statefull versus stateless PCE. Clearly if the PCE has a global view of all the current reservations, its computation can be much more accurate. Maintaining extra information also allows the PCE to perform more interesting functions such as computing disjoint paths for the primary and secondary or avoiding links that failed in a previous computation. For these reasons, a statefull PCE is attractive. The price for doing so is not just maintaining large amounts of information on the PCE but also synchronizing this information between several PCEs in the network (either between the primary and the secondary or between collaborating PCEs) and possibly maintaining this state across failures.
6. Computation algorithms. One of the earliest and least controversial decisions of the PCE Working Group was that the path computation algorithms used in the PCE are not an area of standardization. Instead, they are left to the individual implementations as a differentiator between vendors. Although the algorithms themselves are not standardized, both the speed of the computation and the quality of the result are important when evaluating PCE performance and the Working Group is defining methods for measuring this performance.

The notion of PCE is not foreign either to vendors or to large network operators. Some large operators have home-grown tools that provide a lot of the functionality required by the PCE (such as gathering TE information from the network or computing paths) and that are used as offline computation tools. Therefore, the standardization work is driven equally by service providers and vendors.

The PCE is a tool that can be used to improve and ease path computation, both within a single domain and across domains. Although PCEs are often equated with interdomain solutions, they are not a requirement, regardless of whether the computation is done per-domain or interdomain.

So far we have described the different path computation methods. It is important to understand that the path computation methods can be used with the different path setup methods and are not tied to the LSP setup method.

5.3.3 Reoptimization

Reoptimization refers to the process of finding a more optimal path for an LSP and moving to it in a seamless fashion. The trigger for doing so may be an operator request, the expiration of a timer or the availability of updated information for the path computation. The point of reoptimization is to move the LSP to a better path if such a path can be found. In the Traffic Engineering chapter (Chapter 2) we saw how this can be done in a make-before-break fashion by setting up the LSP along the new path before tearing down the old path.

The important thing to understand with regards to reoptimization is that it is done in two steps: path computation and path signaling. Within a single domain, reoptimization is driven by the LSP head end and requires recomputation and resignaling of the entire path. For interdomain LSPs the situation is different: both the path computation method (per-domain or interdomain) and the signaling method (contiguous, stitching or nesting) influence how reoptimization happens.

When per-domain computation is used, it is possible to compute a new path in just one domain without disturbing segments in other domains. If, in addition, the LSP is set up using stitching or nesting, it is also possible to signal the new path within the domain without disturbing the head end or other domains. Thus, the entire reoptimization process is contained within a single domain. This is important for two reasons:

1. Locality. Remember that the reasons for reoptimization are usually local ones: new TE information or a decision on the part of the operator. A local decision in one domain should not impact the neighboring domain. This is especially true for interprovider situations, where the administrative decision to perform optimization of a segment in one domain should not create control plane operations (and thus load on the routers' CPUs) in the neighboring domain.
2. Scalability. Containing the optimization work to the domain where it was triggered is important for scalability. The head end does not need to be aware of the path changes happening in domains downstream from it and does not have to be involved in carrying out the reoptimization process (make-before-break) for every event in a domain several AS hops away. This approach also shields intermediate domains from the extra activity that

would be triggered were the head end to initiate the reoptimization. Therefore, the ability to contain reoptimizations to a single domain is important for scalability.

However, is per-domain reoptimization always used? The answer is 'no'. In some cases, per-domain reoptimization is not desirable. For example, if an LSP is set up with tight constraints, allowing local reoptimization can cause violation of the constraints. This is similar to the situation in Figure 5.5, where locally optimal paths yield a nonoptimal end-to-end result. In other cases, per-domain reoptimization is not possible; e.g. if the LSP is set up as a contiguous LSP. In such cases, the head end must be involved in the reoptimization process. There are two questions to be answered: is it possible for the head end to initiate reoptimization requests for nodes downstream and is it desirable to allow it to do so?

The answer to the first question is straightforward, because it involves only the mechanics of signaling. RSVP can be extended to provide the necessary signaling capabilities [LOOSE-PATH-REOPT]. The head end can signal a reoptimization request to the nodes that perform the per-domain computation, using a bit in the path message and, conversely, these nodes can inform the head end that a better path is available using a path error message.

The answer to the question of whether such a mode of operation is desirable is not as straightforward. In an interarea setup it might be acceptable to hand over control to the head end LSR, but in an interprovider scenario (as seen earlier) it might not be desirable to do so.

To summarize, the reoptimization of a contiguous LSP requires head end intervention, while for stitched/nested LSPs the process can be restricted to the routers in the domain where the path is optimized. Thus, the LSP setup method impacts the scaling properties of the reoptimization process, and must therefore be taken into account when choosing whether to set up a contiguous LSP or a stitched/nested one.

So far, we have discussed how to compute, set up and reoptimize interdomain TE LSPs. The last important part of TE is the protection and fast reroute aspects, discussed in the next section.

5.3.4 Protection and fast reroute

As seen in the Protection and Restoration chapter (Chapter 3), the first type of protection is end-to-end protection. This is accomplished by setting up an alternate (also called secondary) path that can be used in case of failure of the primary path. For this approach to provide protection, the failure of a link or node in the primary path should not affect the secondary path. Simply put, the primary and secondary paths must be routed differently in the network. In an interdomain setup, when the computation is done per-domain, finding diversely routed paths is not trivial. Even if the domain exit points chosen for the primary and secondary paths are different, this does not necessarily ensure diversely routed paths. For example, an LSP from A to B is set up in Figure 5.5. Imagine that the primary path enters AS2 through ASBR3 but, because of unavailable resources on link ASBR3–B, it establishes through ASBR3–ASBR4–B. Choosing a different entry point (ASBR4) for the secondary path does not ensure the path diversity that was desired.

The second type of protection is local protection. This is accomplished by setting up protection paths around the failed link or node, as explained in the Protection and Restoration chapter. Within each domain, link/node protection operates in the same way for interdomain LSPs as for single-domain LSPs: a backup tunnel is built around the protected resource between the point of local repair (PLR) and the merge point (MP), and traffic is forwarded through it when the protected resource fails. When the LSP is set up using stitching, the protection path is applied to the TE LSP segment. When nesting is used, protection is applied to the FA LSP. Doing so implicitly protects the traffic of all LSPs nested on to it. No special actions need to be taken to protect the nested LSPs, because no control-plane state is maintained for them. To summarize, local protection within a domain operates in the same way for interdomain LSPs and for intradomain LSPs. For this reason it will not be discussed further.

The interesting situation for local protection of interdomain LSPs is when the PLR and the MP fall in different domains. Regardless of whether the protected resource is a link or a node, there are two challenges in this case: how to identify the MP and how to compute the path to it. These challenges are not limited to any particular LSP setup method and they apply equally to LSPs set up as contiguous, stitched or nested. Let us take a look at a link protection scenario, where the failure of a link at the domain boundary requires a backup tunnel between the two border nodes, around the interdomain link.

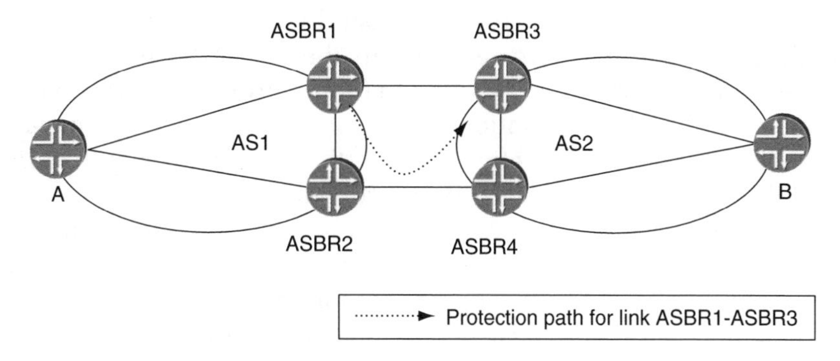

Figure 5.8 Link protection for an inter-AS link

In Figure 5.8, the link between ASBR1 and ASBR3 is protected by the tunnel ASBR1–ASBR2–ASBR4–ASBR3. How does ASBR1 identify the MP? Recall from the Protection and Restoration chapter (Chapter 3) that typically the MP address is taken from the RRO and that the FRR specifications recommend using interface addresses in the RRO. In an interdomain setup, interface addresses are not advertised in the IGP so the MP cannot be identified. To solve this problem, the FRR specification was extended to advertise node ids along with interface information [NODE-ID]. Node ids are usually loopback addresses. Once the MP is identified, the challenge is to find a backup path to it that does not cross the protected link. Because the PLR does not have visibility into the neighboring domain, it must rely on the same path computation methods described earlier for interdomain LSP computation.

Identifying the MP is not as easy as described in the previous example. An interesting challenge arises in the node protection scenario. The failure of a node requires computing a backup path to a node on the LSP path that lies beyond the failed node. When the LSP is set up as a contiguous LSP, the MP can be any node in the LSP path. However, when the LSP is set up with stitching or nesting, the MP can only be the end-point of the TE LSP segment or of the FA LSP.

In Figure 5.4, protecting node ABR2 can be accomplished by setting up the bypass tunnel to any node beyond ABR2. When the LSP is set up as a contiguous LSP, R1 is a good MP candidate. However, when the LSP is set up using stitching/nesting, the MP can only be node ABR4, resulting in a much longer protection path. The drawback of a long protection path is that when bandwidth protection is ensured by reserving bandwidth on the protection path, more resources are reserved in the network. In this case, instead of reserving extra bandwidth for the protection path on the links up to R1, the reservation is made on all the links all the way to ABR4. Thus, the LSP setup method the affects choice of the MP and thus the properties of the protection path.

To summarize, the same protection mechanisms available for intradomain TE LSPs can be extended to apply in the interdomain case. This is an important property because protection has come to be equated with MPLS-TE and is a requirement for most QoS guarantees.

5.4 INTERPROVIDER CHALLENGES

So far we have focused on the technical details of setting up interdomain LSPs. However, when LSPs span several administrative boundaries, additional concerns arise, in particular over security and compensation agreements, which translate to additional requirements from the interdomain TE solution.

Let us start by looking at the security concerns. Any kind of interprovider interaction requires a level of trust. However, operators seldom rely on trust alone to prevent accidental or malicious impact on their networks because of interprovider relations. Interprovider LSPs are no exception.

The use of RSVP for path signaling creates an interesting problem in interprovider interactions. The path of the LSP is recorded in the Record Route Object (RRO) that is propagated all the way to the head end. This means that the addresses of the links/nodes in one domain become visible in the neighboring domain. Providers are wary of exposing the internal addressing outside their networks, because by doing so their routers become vulnerable to attacks (the reasoning is that if the router address is not known, the router cannot be attacked). Therefore, the ability to conceal the hops in the path at the exit from a domain, by either filtering them out or modifying the addresses used in the RRO, becomes a requirement for interprovider LSPs. A similar requirement exists for PCEs collaborating in an interdomain path computation that exchange information regarding path segments.

Another security concern is the fact that excessive LSP setup or reoptimization requests can be sent by the upstream domain, with the same effect on the router control plane as a denial-of-service attack. Therefore, the ability to rate-limit such requests at the domain boundary becomes a requirement for interprovider LSP implementation. Furthermore, because an upstream domain can create both control and forwarding state in the network, it is necessary to ensure that LSP setup requests come from an authentic peer and cannot be faked by an attacker. This can be done by using authentication of the protocol messages and by negotiating ahead of time the type of requests accepted at a domain boundary (e.g. accept LSP setup requests only from a certain range of sender addresses).

Negotiation is necessary not just for security purposes but also for compensation agreements between the two administrative domains. As part of such agreements, the exact profile of the interprovider interaction is defined. For example, the two providers negotiate ahead of time how many LSP setup requests can be sent per unit of time, what LSP priorities are acceptable and whether FRR requests are honoured. This implies that equipment vendors can provide tools to enforce the terms negotiated in such an agreement (e.g. the ability to reject setup requests based on configured policies).

To summarize, deployments spanning several providers' networks place additional requirements on the interdomain TE solution discussed so far. The extensions are necessary for providing the additional security guarantees needed in such setups and for enforcing compliance with the negotiated interaction profiles between providers.

5.5 COMPARISON OF THE LSP SETUP METHODS

The LSP setup method is one of the important decisions in an interdomain deployment and like any other design choice it involves tradeoffs. For example, contiguous LSPs are more intuitive but they have less desirable scaling properties when compared to nested or stitched ones. The question is not which LSP setup method is better, but rather which one is better for a particular deployment. For example, the fact that a stitched LSP can be reoptimized locally is not an advantage in a setup where reoptimization will never be run. Table 5.1 presents a summary comparison of the different setup methods.

Table 5.1 Comparison of the different LSP setup methods

	Contiguous	Stitching	Nesting
Number of LSPs in the transit domain; assuming *N* LSPs in the head end domain	*N*	*N*	Smaller than *N*; depends on the number of FA LSPs
Support of per-domain path computation	Yes	Yes	Yes
Requires protocol extensions	Yes	Yes	Yes
Reoptimization in the transit domain affects other domains	Yes	No	No
Control over reoptimization	Head end	Local (head end if desired)	Local (head end if desired)
MP when protecting a boundary entry node	Any node in the path	TE LSP segment end-point	FA LSP end-point

5.6 CONCLUSION

Interdomain TE enables setting up TE LSPs across different areas and different ASs both within a single provider's network and across providers, with the same TE properties and features as intradomain TE. Along with DiffServ Aware TE, interdomain TE completes the traffic engineering solution presented in the Traffic Engineering chapter (Chapter 2). Interdomain TE tunnels are important, not just for interprovider deployments but also for enabling MPLS applications such as Layer 3 VPNs in large networks encompassing several IGP areas when the transport tunnel is RSVP-signaled.

Before we can start exploring the different MPLS applications there is one more piece of functionality that is useful for some of the advanced applications. This is point-to-multipoint LSPs, discussed in the next chapter.

5.7 REFERENCES

[CCAMPWG] CCAMP working group in the IETF http://ietf.org/html.charters/ccamp-charter.html

[CRANKBACK] A. Farrel *et al.*, 'Crankback signaling extensions for MPLS and GMPLS signaling', draft-ietf-ccamp-crankback-05.txt (work in progress)

[HIER] K. Kompella and Y. Rekhter, 'LSP hierarchy with generalized MPLS TE', draft-ietf-mpls-lsp-hierarchy-08.txt (work in progress)

[INTER-AREA-TE-REQ] J. Le Roux, J.P. Vasseur, J. Boyle *et al.*, *Requirements for Inter-Area MPLS Traffic Engineering*, RFC4105, June 2005

[INTER-AS-TE-REQ] R. Zhang and J.P. Vasseur, 'MPLS inter-AS traffic engineering requirements', draft-ietf-tewg-interas-mpls-te-req-09.txt (work in progress)

[INTER-DOMAIN-FW] A. Farrel, J.P. Vasseur and A. Ayyangar, 'A framework for inter-domain MPLS traffic engineering', draft-ietf-ccamp-inter-domain-framework-04.txt (work in progress)

[LOOSE-PATH-REOPT] J.P. Vasseur, Y. Ikejiri and R. Zhang, 'Reoptimization of multiprotocol label switching (MPLS) traffic engineering (TE) loosely routed label switch path (LSP)', draft-ietf-ccamp-loose-path-reopt-01.txt (work in progress)

[LSP-STITCHING] A. Ayyangar and J.P. Vasseur, 'LSP stitching with generalized MPLS TE', draft-ietf-ccamp-lsp-stitching-01.txt (work in progress)

[NODE-ID] J.P. Vasseur, Z. Ali and S. Sivabalan, 'Definition of an RRO node-id subobject', draft-ietf-mpls-nodeid-subobject-06.txt (work in progress)

[PCEWG] PCE working group in the IETF, http://ietf.org/html.charters/pce-charter.html

5.8 FURTHER READING

[INTER-DOMAIN-PATH-COMP] J.P. Vasseur and A. Ayyangar, 'Inter-domain traffic engineering LSP path computation methods', draft-vasseur-ccamp-inter-domain-path-comp-00.txt (work in progress)

[INTER-DOMAIN-SIG] A. Ayyangar and J.P. Vasseur, 'Inter domain MPLS traffic engineering – RSVP-TE extensions', draft-ayyangar-ccamp-inter-domain-rsvp-te-02.txt (work in progress)

[PCE-ARCH] A. Farrel, J. Vasseur and G. Ash, 'Path computation element (PCE) architecture', draft-ietf-pce-architecture-01.txt (work in progress)

[PCE-COMM] G. Ash and J. Le-Roux, 'PCE communication protocol generic requirements', draft-ietf-pcecomm-protocol-gen-reqs-00.txt (work in progress)

[PER-DOMAIN-PATH-COMP] J.P. Vasseur, A. Ayyangar and R. Zhang, 'A per-domain path computation method for computing inter-domain traffic engineering (TE) label switched path (LSP)', draft-ietf-ccamp-inter-domain-pd-path-comp-00.txt (work in progress)

[RFC2702] D. Awduche *et al.*, *Requirements for Traffic Engineering over MPLS*, RFC2702, September 1999

6

Point-to-Multipoint LSPs

6.1 INTRODUCTION

In the Foundation chapter of this book (Chapter 1), we discussed how MPLS is used to establish LSPs in the network and how the form of the LSP depends on the signaling protocol used. We saw that when RSVP is the signaling protocol, each LSP is point to point in nature, carrying traffic from one ingress point to one egress point. In contrast, when LDP is the signaling protocol, each LSP is multipoint to point in nature, carrying traffic from several ingress points to a single egress point.

In this chapter we will see how RSVP or LDP can be used to create point-to-multipoint (P2MP) LSPs which carry traffic from one ingress point to several egress points, thus enabling multicast forwarding in an MPLS domain. Using P2MP LSPs, traffic is multicast from one source to multiple destinations in a bandwidth-efficient manner, without the ingress having to send separate copies to each receiver.

The use of RSVP-based P2MP traffic engineering gives the ingress router control over the path taken by the traffic and allows bandwidth guarantees to be made. As described later in this chapter, this unification of traffic engineering and multicast enables applications that were previously difficult to support on an IP or MPLS network, such as the distribution of broadcast-quality television.

MPLS-Enabled Applications: Emerging Developments and New Technologies Ina Minei and Julian Lucek

In later chapters, we discuss the use of P2MP LSPs in the context of L3 VPN and VPLS, for the transport of customers' IP multicast traffic. The P2MP LSPs are set up in the service provider's core using either RSVP or LDP, depending on the needs of the service provider and its customers. This chapter assumes an understanding of RSVP, LDP and TE and some basic knowledge of multicast.

6.2 THE BUSINESS DRIVERS

Without P2MP LSPs, many networks use MPLS for unicast traffic and IP multicast for multicast traffic. Therefore, separate control and forwarding planes for unicast and multicast traffic operate concurrently and independently in the network, without knowledge of each other. This is sometimes referred to as a 'ships-in-the-night' situation. When using P2MP LSPs for multicast distribution, the control plane for all traffic within the core of the network is based on RSVP or LDP and the forwarding plane for all traffic is based on MPLS encapsulation. This reduction in the number of protocols used in the core of the network, and the reduction in the number of encapsulations in the data plane, results in simplified network operations.

IP multicast enables the distribution of traffic to multiple receivers without the need to send separate copies to each one of them, but it allows no control over the path the traffic takes and provides no guarantees about the bandwidth availability on the path so it cannot make any QoS guarantees. However, some applications require multicast distribution in conjunction with QoS guarantees such as reserved bandwidth and low loss. The most notable example is professional real-time video transport, which is discussed in more detail in Section 6.7.1. Other applications include core distribution infrastructure for video-on-demand services and large database downloads to multiple remote sites.

It is useful to compare [P2MPWC] some of the properties of IP multicast to those of P2MP TE. As described later in this chapter, hybrid schemes are possible in which IP multicast operates in conjunction with P2MP TE. The list below does not consider such schemes, and instead compares IP multicast in its native form to P2MP TE:

- *Failover mechanisms*. For IP multicast traffic, the failover mechanisms are relatively slow (on the order of seconds), the timescale being partly dependent on IGP convergence times.

This makes IP multicast unsuitable for real-time video distribution applications in which an interruption of this timescale would be unacceptable. In contrast, as described in Chapter 3, RSVP-TE fast-reroute mechanisms are fast (millisecond timescales) because the switchover to a back-up path is a local decision taken by the router upstream from the point of failure.

- *Control of path taken by the traffic*. With IP multicast, it is difficult to control the path taken by the traffic. The multicast tree that is built is a shortest-path tree, the path being determined by the IGP. Some implementations allow the use of static multicast routes to override this behavior, but it is a cumbersome process.[1] RSVP-TE allows control of the path taken by the traffic, according to where bandwidth resources are available or user-defined constraints. Rather than having a shortest-path tree, which minimizes latency, the user may want a minimum-cost tree (also known as a Steiner tree) which minimizes the bandwidth utilization. The difference between a shortest-path tree and a minimum-cost tree is discussed in Section 6.3.2.1.1.
- *Bandwidth guarantees*. IP multicast protocols (such as PIM, or Protocol Independent Multicast) do not have the ability to perform bandwidth reservations and hence there are no guarantees that resources will be available for the traffic to reach its destination. Even if they did have the mechanisms to perform bandwidth reservations, the path of the multicast tree is fixed, so if the required bandwidth resources were not available along that path, there is no way to change the path of the tree. RSVP-TE, in contrast, has mechanisms for reserving the bandwidth and the path computation can take bandwidth availability into account.
- *Control over receivers permitted to join the tree*. With IP multicast, there is no overall control over the extent of the tree or the receivers allowed to join it, and receivers can splice themselves on to any existing tree, unless prevented from doing so through the use of tools such as PIM Join filters. In contrast, with P2MP TE, the set of receivers to which the tree extends is determined at the ingress node (e.g. through configuration).

[1] Alternatively, one can use a different IGP topology for multicast traffic to that for unicast traffic, but this does not give control over the path followed by multicast traffic with per-flow granularity.

6.3 P2MP LSP MECHANISMS

This section examines the forwarding and control plane mechanisms associated with P2MP LSPs. First we discuss how data are forwarded along a P2MP LSP. This is independent of the signaling protocol used to create the P2MP LSP. Then we discuss the two control plane mechanisms by which P2MP LSPs can be created: the RSVP-based scheme and the LDP-based scheme. We then discuss LAN procedures for P2MP LSPs, the coupling of traffic into P2MP LSPs and fast-reroute for P2MP LSPs.

6.3.1 Forwarding plane mechanisms

A P2MP LSP [P2MP REQ, P2MP TE] has a single ingress router and multiple egress routers. This is illustrated in Figure 6.1. PE1 is the ingress router for the P2MP LSP. The egress routers are PE2, PE3, PE4 and PE5. As can be seen in the figure, PE1 creates two copies of each packet arriving from the data source. One copy having the MPLS label value L1 is sent to P1 and another copy having the label value L5 is sent to P2. Routers PE1, P1 and P2 are called branch nodes. As can be seen, replication of MPLS packets occurs at these nodes.

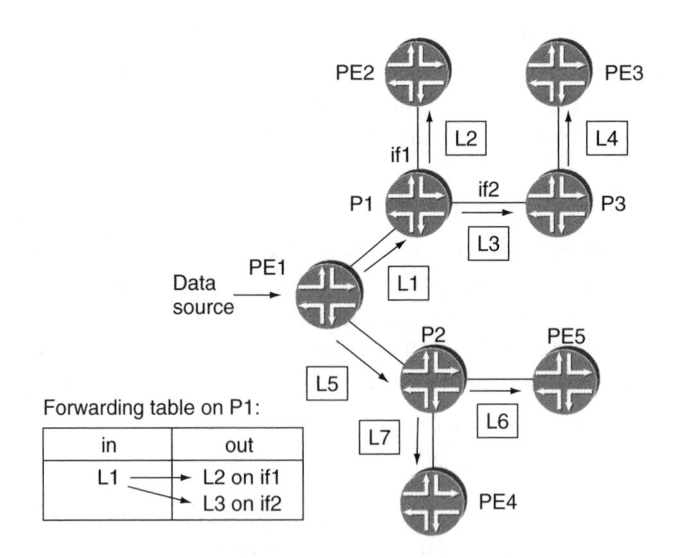

Figure 6.1 P2MP LSP forwarding operation

Let us consider the packet forwarding process on P1. For each incoming packet belonging to the P2MP LSP, P1 makes two copies, one of which is sent to PE2 and the other to P3. Let us look at this process in more detail. Packets arrive at P1 having label L1. Looking at the inset of the figure, which shows the forwarding entry corresponding to the P2MP LSP installed on P1, it can be seen that P1 has an entry in its forwarding table for label L1 saying that one copy of the packet should be sent out on interface if1 with label L2 and another copy should be sent out on interface if2 with label L3. Hence P1 is performing a packet replication process in the MPLS domain. The copy of the packet arriving at P3 is forwarded to PE3. No branching occurs at P3, so P3 is just a transit node for this particular P2MP LSP, rather than a branch node.

A key advantage of the P2MP scheme is its bandwidth efficiency. For example, let us suppose that a flow of 100 Mbps is forwarded using the P2MP LSP. On the link between PE1 and P1, only 100 Mbps of bandwidth is used, rather than 200 Mbps if PE1 had to send separate copies of the traffic to PE2 and PE3. As with point-to-point LSPs, the flow of traffic in a P2MP LSP is unidirectional, so no traffic can flow from the egress routers to the ingress routers along the P2MP LSP.

To summarize, the key property of P2MP forwarding is the ability to construct a distribution tree that replicates packets at the branch points. This is done based on the forwarding information maintained by those branch points. How is this information built? To answer this question, we need to turn our attention to the control plane mechanisms.

6.3.2 Control plane mechanisms

This section describes the control plane mechanisms underpinning P2MP LSPs. First we describe how RSVP creates a P2MP traffic-engineered LSP and discuss how the path computation can be performed. Then we discuss how LDP can create (non-traffic-engineered) P2MP LSPs.

6.3.2.1 Use of RSVP for P2MP traffic engineering

One of the design principles behind the P2MP scheme was to minimize the changes to RSVP-TE needed to accommodate P2MP operation. This section describes how a point-to-multipoint LSP is signaled using RSVP-TE and the changes that were made to RSVP-TE to achieve this. It is useful to refer to the Foundation chapter (Chapter 1) and the Traffic Engineering chapter (Chapter 2) of this book as a reminder of how (point-to-point) traffic engineering works. Figure 6.2 shows the same network as in Figure 6.1 and illustrates how the point-to-multipoint LSP that was shown in Figure 6.1 is signaled by RSVP. It should be noted that the ingress of the P2MP LSP is assumed to know the identity of the egress nodes. The way in which the ingress acquires this information is outside the scope of RSVP-TE, but could be via manual configuration, or could be discovered via PIM, as described in Section 6.5.3 of this chapter. The figure shows the flow of RSVP Path messages (solid arrows) and Resv messages (dotted arrows). The label values associated with the Resv messages in the diagram (L1, L2, etc.) are those contained in the Label Object in the Resv messages. Bear in mind that as with point-to-point LSPs, downstream label allocation is used. Therefore, the control messages (the Resv messages) containing the label for each link shown in Figure 6.2 travel in the opposite direction from the actual MPLS data packets.

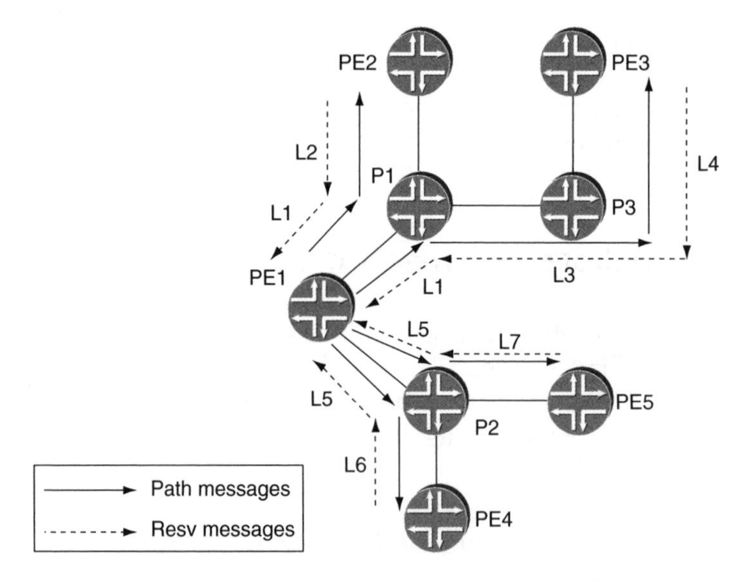

Figure 6.2 RSVP-signaling operation for P2MP LSP

A key point to note is that from the control plane point of view, a P2MP LSP is regarded as a set of point-to-point LSPs, one from the ingress to each of the egress nodes of the LSP. Each of the LSPs within the set is known as a sub-LSP. Recall that for normal point-to-point traffic engineering, an LSP is signaled by sending Path messages that flow from the ingress to the egress and Resv messages that flow from the egress to the ingress. The Path messages contain an Explicit Route Object (ERO) that determines the path followed by the LSP and the Resv messages at each hop contain the label to be used for forwarding along that hop. In the point-to-multipoint case, each sub-LSP is signaled using its own Path and Resv messages, the Path messages containing the ERO of the sub-LSP in question. The Path and Resv messages contain a new object, the P2MP Session Object, so that the routers involved know which P2MP LSP a particular sub-LSP belongs to. This knowledge is essential for creating the replication state in the forwarding plane. A branch node must realize that two or more sub-LSPs belong to the same P2MP LSP in order to treat them correctly.

Let us see how this works in the example network shown in Figure 6.2. The P2MP LSP has four egress nodes, so it is composed of four sub-LSPs, one from PE1 to PE2, another from PE1 to PE3, and so on. Because each sub-LSP has its own associated Path and Resv messages, on some links multiple Path and Resv messages are exchanged. For example, the link from PE1 to P1 has Path messages corresponding to the sub-LSPs to PE2 and to PE3. Let us examine in more detail how the P2MP LSP in Figure 6.2 is signaled, looking at sub-LSP PE1 to PE3 whose egress is PE3:

1. A Path message is sent by PE1, the ingress router, containing the ERO {PE1, P1, P3, PE3}. This can contain a bandwidth reservation for the P2MP LSP if required.
2. PE3 responds with a Resv message that contains the label value, L4, that P3 should use when forwarding packets to PE3. Similarly, the Resv message sent on by P3 to P1 contains the label value, L3, that P1 should use when forwarding packets to P3.
3. In a similar way, for the sub-LSP whose egress is PE2, P1 receives a Resv message from PE2 containing the label value, L2, that P1 should use when forwarding packets to PE2. P1 knows that the Resv messages from PE2 and P3 refer to the

same P2MP LSP, as a consequence of the P2MP Session Object contained in each.

4. P1 sends a separate Resv message to PE1 corresponding to each of the two sub-LSPs, but deliberately uses the same label value for each, L1, because the two sub-LSPs belong to the same P2MP LSP.
5. P1 installs an entry in its forwarding table such that when a packet arrives with label L1, one copy is sent on the link to PE2 with label L2 and another copy on the link to P3 with label L3. If a bandwidth reservation is being created for the P2MP LSP, the shared explicit (SE) reservation style is used. This ensures that when the Resv messages are sent from P1 to PE1 corresponding to the two sub-LSPs, no double-counting occurs of the bandwidth reservation.
6. PE1, knowing that the two Resv messages received from P1 refer to the same P2MP LSP, a consequence of the P2MP session object contained in each, forwards only one copy of each packet in the flow to P1, with the label value L1 that had been dictated by PE1 in those two Resv messages.

The section of the P2MP LSP from PE1 to PE4 and PE5 is set up in an analogous way to the section from PE1 to PE2 and PE3.

In addition to the scheme described above, in which each sub-LSP is signaled using its own Path message, at the time of writing, the IETF draft [P2MP TE] also discusses other mechanisms for signaling P2MP LSPs. For example, in one of them, each Path message contains details of all the sub-LSPs, including explicit routes for each. However, the existing implementations use the scheme described above.

An interesting question is: what should happen if it is not possible to bring up all of the sub-LSPs belonging to a P2MP LSP? This could be because one of the egress routers is down or there is a loss of connectivity to one or more egress routers due to link failures in the network. Should the entire P2MP tree be torn down? The IETF draft that covers the requirements for P2MP-TE [P2MP REQ] leaves this decision to the local policy in the network, because for some applications a partial tree is unacceptable while for others it is not. For example, for an application such as broadcast TV distribution, the typical requirement is that the P2MP LSP should still stay active so that the reachable egress nodes still receive traffic.

In some networks, it may be necessary for a P2MP LSP to cross nodes that do not support P2MP operation. This could happen at the time of the initial deployment of P2MP capability in the network, when some of the nodes support it and other legacy nodes do not. This is a problem because the RSVP messages travel hop by hop, so a sub-LSP will not be established if a node sees an unsupported object (e.g. the P2MP session object) in the RSVP message.

If a node does not support P2MP operation in the control and forwarding planes, a workaround is to use LSP hierarchy (see the Foundation chapter of this book, Chapter 1, for an explanation of LSP hierarchy). In this scheme, sub-LSPs pertaining to a P2MP LSP are nested within an outer LSP, so that the transit nodes of the outer LSP are not aware that they might be carrying a P2MP LSP. Naturally, such nodes cannot act as ingress, branching or egress nodes of a P2MP LSP, which may mean that the overall path taken by the P2MP LSP is further from optimum than if those nodes could support branching. As an example of the use of LSP hierarchy, let us refer to Figure 1.9 of the Foundation chapter. Suppose that P2 does not support P2MP operation. It is required to set up a P2MP LSP for which PE1 is the ingress node and PE4, PE5 and PE6 are the egress nodes. The three corresponding sub-LSPs are nested within the core LSP that passes through P2. Hence P2 is unaware of the existence of those sub-LSPs. P3 acts as a branching node so that the traffic is received by the three receiving PEs. Note that the same core LSP can also be used to carry normal point-to-point LSPs at the same time. Another scenario is where P2 is semi-compliant with P2MP TE in that it supports the P2MP control plane, but does not support the branching operation in the forwarding plane. In this situation, it is not necessary to use LSP hierarchy as P2 can process the RSVP messages associated with the three sub-LSPs, but the network administrator needs to bear in mind that the node cannot be expected to act as a branch node.

So far we have looked at the signaling aspect of the P2MP setup and assumed that the ERO is known at the head end. Next we will look at some of the challenges of computing the path of a P2MP-TE LSP.

6.3.2.1.1 Path computation in P2MP traffic engineering

It is interesting to explore the path computation of a P2MP-TE LSP. The task is to perform a computation of a P2MP tree taking into account the criteria that define an optimum path from the point of view of the user. For example, if the main requirement is to minimize the latency experienced by the traffic, a shortest-path tree would be appropriate. If, on the other hand, the requirement is to minimize the bandwidth utilisation, a minimum-cost tree (Steiner tree), as measured in terms of bandwidth utilisation, would be appropriate.

Figure 6.3 compares the path of a P2MP LSP in the shortest-path tree case to the minimum-cost tree case, with the assumption that each link in the network has equal cost and latency and that any link in the network can meet the bandwidth requirement of the LSP. In the case of the shortest-path tree, each egress node is two hops from the ingress node, and the total bandwidth utilization is six units, because the P2MP tree structure uses six links in total. In contrast, for the minimum-cost tree case, the bandwidth utilization is only four units but with the downside that two of the egress nodes are three hops from the ingress node rather than two. A variation not shown in the figure is a delay-bounded minimum-cost tree in which the minimum cost tree is computed for which the propagation delay to any egress is less than a specified maximum.

This freedom to define the path of the P2MP tree according to the user requirements contrasts to traditional IP multicast, in which there is no such flexibility: the tree is a shortest-path tree (either rooted at the source or the rendezvous point) and there is no way of changing that. As with point-to-point LSPs, potential methods of computing the path of a P2MP LSP are as follows:

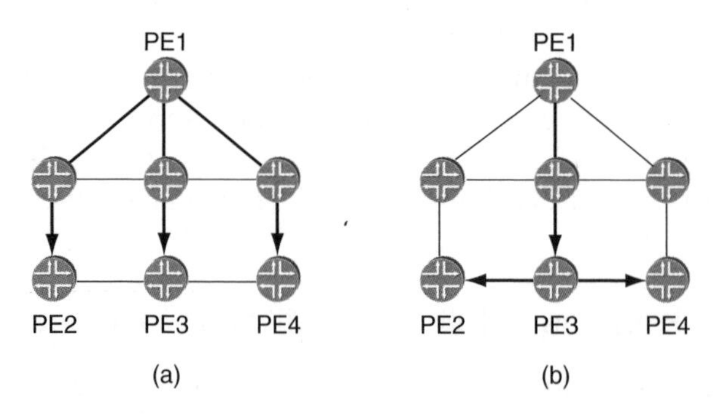

Figure 6.3 Path of P2MP LSP: (a) shortest-path tree and (b) minimum-cost tree

- Manual determination of path by inspection
- Online computation by the ingress node
- Computation by an offline tool

The considerations about which to use are similar to those discussed for point-to-point LSPs in the Traffic Engineering chapter of this book (Chapter 2). An additional factor to consider is that in some applications of P2MP TE, application level redundancy is sometimes used. This is done by having two P2MP LSPs carry the same datastream. The two LSPs originate at separate ingress routers and follow diverse paths through the network to the receivers, to prevent loss in case of a failure along the path of one of those LSPs. In such cases, it is often easier to use an offline tool to compute the paths of the LSPs, as it can be difficult to ensure that paths of the two LSPs do not overlap if the paths are computed by two different ingress routers.

The amount of computation required to calculate an optimum P2MP tree depends on which type of tree is required. In the case of a shortest-path tree, the path to any egress node is independent of the location of other egress nodes, so the computation of the shortest path tree can be decomposed into the computation of each individual sub-LSP. However, in the case of a minimum-cost tree and the delay-bounded minimum cost variant, the optimization problem is more complex, as the path of a sub-LSP to an egress node depends on the location of other egress nodes. In fact, the optimization problem can be shown to be NP-hard (nondeterministic polynomial-time hard). As a consequence, depending on the size of the tree, there may need to be a tradeoff between identifying the optimum tree, which might take an unacceptably long time, and identifying an acceptable, but not necessarily optimum, tree in a shorter period of time. In order to achieve the latter, there exist approximate algorithms that reduce the optimization task from one of NP-hard complexity to one of polynomial complexity.

An interesting question is what to do if one wishes to add or remove a branch from an existing P2MP LSP. In the case of the minimum cost tree (and its delay-bounded variant), should the branch simply be spliced on to or removed from the existing tree, without changing the path taken to any of the egress points already present? This may mean that the tree as a whole is no longer the optimum. Or should the entire tree be reoptimized?

The answer may depend on the application and how often egress nodes come and go. Although make-before-break procedures analogous to those for point-to-point LSPs exist for P2MP LSPs, as with the point-to-point case, there is the possibility of transient reordering of traffic. For example, returning to Figure 6.3(b), let us suppose that PE2 and PE3 are no longer to be required to be egress nodes of the P2MP LSP. If the path to the remaining egress node PE4 is reoptimized from PE1–P2–PE3–PE4 to PE1–P3–PE4, the first packets to travel along the new path may reach P4 before the last packets to travel along the old path. Whether this is an issue or not depends on whether the application is sensitive to mis-sequencing of packets. Hence the best course is for implementations to give some degree of choice to the user, e.g. by allowing the user to request a recomputation of the tree on an on-demand basis or on a periodic basis.

6.3.2.2 LDP signaling for P2MP LSPs

So far we have seen how P2MP LSPs can be created using RSVP. However, many MPLS deployments currently use LDP as the label distribution protocol. For such networks, if P2MP LSPs are required but the service provider does not need the traffic engineering advantages of RSVP-signaled P2MP LSPs, the possibility of using LDP as the signaling mechanism for P2MP LSPs is attractive. At the time of writing, work is under way in the IETF MPLS Working Group to define the necessary mechanisms for the P2MP LSP setup using LDP.

Recall from the Foundations chapter (Chapter 1) that LDP-signaled LSPs are initiated by the egress router. The label propagation is initiated by the receiver and is propagated hop by hop throughout the entire network. All LSRs in the network maintain the forwarding state towards the receiver following the IGP path, and any LSR can act as an ingress to this LSP. In effect, a multipoint-to-point LSP is built with several senders and one receiver. The goal when setting up P2MP LSPs, in contrast, is to have a single sender and several receivers, so the question is how to modify LDP to accommodate such a scheme. Since at the time of writing the solutions are still under discussion in the IETF, we will look into the design process of one such solution [P2MP LDP], as this will shed some light on how such solutions are developed.

One of the fundamental questions is who initiates the signaling of the LSP. In previous sections, we saw that in the RSVP case, the signaling of a P2MP LSP is initiated by the ingress router. However, in the LDP case, requiring the ingress router to initiate the LSP setup requires fundamental changes in the way labels are distributed and therefore is not an attractive option. Instead, the problem of discovering the source and destinations can be decoupled from the actual signaling of the P2MP LSP via LDP. (The discovery problem is also decoupled in the RSVP case, in that the source learns the identity of the receivers by some means outside of RSVP.) This allows the LDP solution to be developed to be receiver initiated rather than sender initiated if required.

Assuming that the receivers know that they must establish a P2MP path towards the sender, the second fundamental question is how to identify the P2MP LSP. Similar to the RSVP case, this is necessary to be able to install the correct forwarding state at the branch nodes. Clearly the ingress router of the LSP must be identified. The ingress router alone is not enough, because several P2MP LSPs may originate at the same ingress. Thus, it is necessary to identify not just the source but also the tree. LDP does not need to be aware of the semantics of the tree identifier; from its point of view the identifier is opaque. To set up the LSP, a label must be assigned by the receivers and associated with the entity of {source, tree identifier}. We will call this the P2MP forwarding equivalence class (FEC).

Recall that LDP LSPs follow the IGP. As we saw in the Foundations chapter, for an FEC corresponding to an IP address, this is accomplished by using for forwarding only those labels received over sessions that lie in the IGP path for that IP address. In the case of with P2MP FECs, the procedure is different. The rule for distribution is to advertise a label only towards the neighbor that lies on the IGP best path towards the source. Thus in the regular LDP case, the receiver of the label determines the best path towards the egress, but in the P2MP case, the sender of the label determines the best path towards the ingress.

Figure 6.4 shows an example of how a P2MP LSP is signaled by LDP. PE4 is the ingress router of the P2MP LSP and PE1 and PE2 are the egress routers. PE2 advertises a label, L2, for the P2MP FEC only towards P1, and not towards P3, because P1 lies in the best path towards the ingress. The P2MP FEC contains the address of PE4, the ingress of P2MP LSP to be built and the P2MP tree identifier.

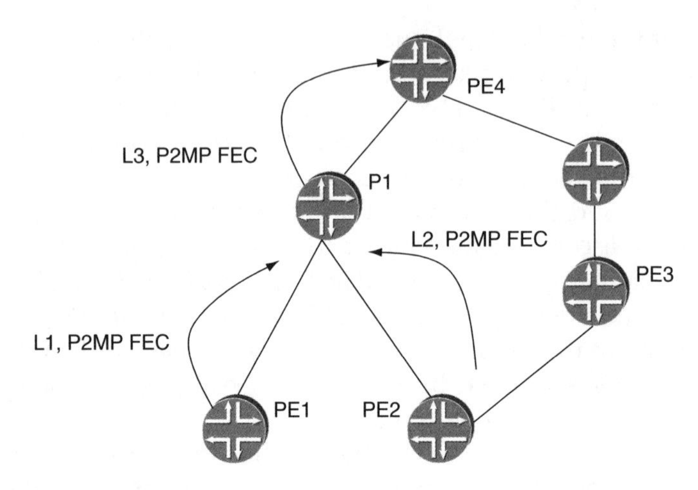

Figure 6.4 Signaling of a P2MP LSP using LDP

PE1 advertises a label L1 for the same P2MP FEC towards P1. At node P1, the labels L1 and L2 are identified as belonging to the same P2MP FEC. As a result, a single label L3 is advertised towards the source, PE4, and the forwarding state is installed to replicate packets arriving with label L3 on each of the interfaces towards PE1 and PE2 with labels L1 and L2 respectively.

In this way, the signaling for the P2MP LSP can be done from the receivers towards the source. Similar procedures have been set in place to define the behavior for label withdrawals.

6.4 LAN PROCEDURES FOR P2MP LSPs

One of the main goals of P2MP LSPs is to minimize the bandwidth used to distribute the content from the source to all the receivers. Thus, one of the fundamental requirements is to send every packet at most once over any given link. Let us take a look at an interesting problem that arises when dealing with multiaccess links, e.g. Ethernet. Figure 6.5 shows a simple network topology where source S is required to send traffic to three destinations, R1, R2 and R3. The destinations are connected to three transit routers, P1, P2 and P3, which are all on the same LAN.

To achieve the optimum bandwidth utilization, S sets up a P2MP LSP to the three receivers, according to the procedures described so far. During the setup of the branch LSPs, each of the

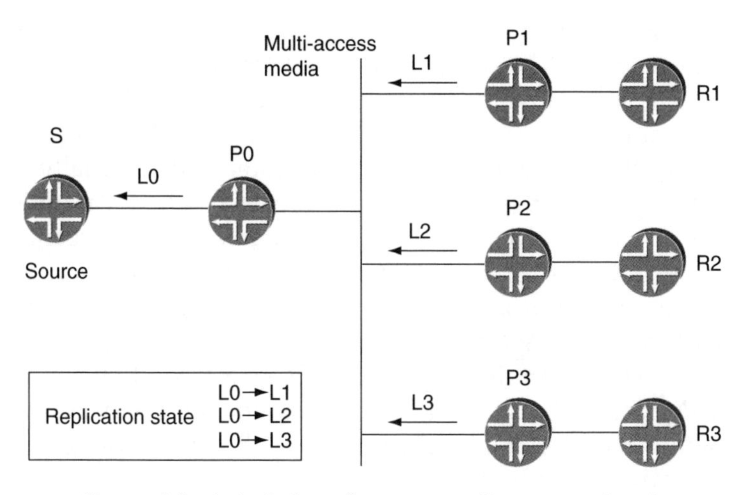

Figure 6.5 Label allocation on a multiaccess network

routers P1, P2 and P3 assigns a label and advertises it to P0. As a result, a single packet is sourced at S towards P0, but three separate copies are sent by P0 towards P1, P2 and P3, although these routers are connected to a shared media and a single packet could have reached all three of them. Indeed, if the three routers had assigned the same label, replication at P0 would not be necessary, and a single packet could be sent over the LAN.

Unfortunately, there can be no guarantee that P1, P2 and P3 assign the same label because they each assign the labels independently from their global label space. One possibility in principle could be to implement a scheme to coordinate between P1, P2 and P3. Alternatively, one could devise a scheme in which router P0 is given control over the label allocation. The latter approach is one that has been followed in the current proposals. Thus, the same label, L4, is advertised by P0 to P1, P2 and P3, which in turn install the correct forwarding state to carry the packets towards R1, R2 and R3. A single copy of the packet is sent over the shared media, labeled with the label assigned by P0 and reaches all the P*x* routers, which in turn forward a copy of the packet to the correct P*x*–R*x* interface with the correct label.

However, what are the implications of reversing the control over the label allocation? Before examining these, let us stop for a moment and revise some terminology. Recall from the Foundations chapter (Chapter 1) that the routers are referred to according to their location relative to the traffic flow. For example, in Figure 6.6, traffic flows from left to right, from Ru1 towards R2. Router R1 is performing downstream label allocation because it is assigning a label that it expects router Rd to use. Thus, the allocation is done by a router that is 'downstream' of the router that is actually going to put the label on the packet. Downstream label allocation is the scheme that is used by both RSVP and LDP today. Upstream label allocation is the scheme that was proposed as a solution to the multiaccess media problem in the previous paragraph. The label is assigned by the same router that is putting the label on the packet. In Figure 6.6, router Ru1 advertises label L1 for FEC 1.1.1.1 to router Rd, meaning that Ru1 intends to send labeled traffic destined to 1.1.1.1 using label L1. (Although we use the LDP notation, the same is applicable to RSVP.)

If you look carefully at Figure 6.6, the first problem with upstream label allocation becomes immediately evident. Router Ru2 advertises the FEC 2.2.2.2 and by coincidence chooses the same label, L1, that Ru1 had chosen for FEC 1.1.1.1. This can happen because Ru1 and Ru2 assign the labels independently from their global label space. When the labeled traffic is received, how can Rd determine if it is destined towards R2 or towards R3? Clearly, Rd must be able to identify the neighbor from which it receives the traffic, because the label has meaning in the context of that particular neighbor. In the example from Figure 6.6, the obvious answer is to use the incoming interface to distinguish between the two neighbors, but the general answer may be different under different situations. For example, if Ru1, Ru2 and

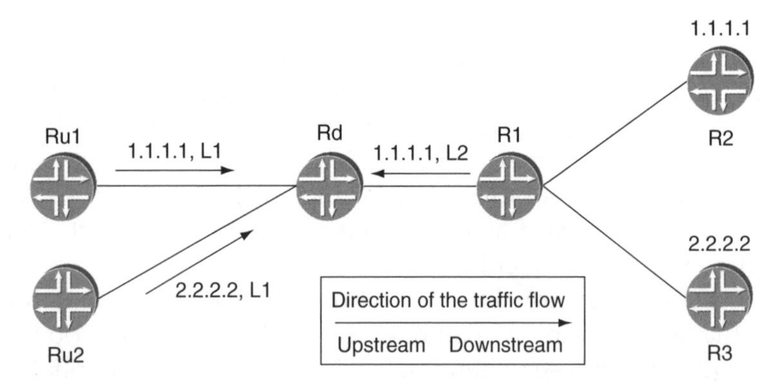

Figure 6.6 Illustration of an issue with upstream label allocation

Rd were on a shared interface, the MAC address of the sender could be used instead.

Solving the problem of packet replication using upstream label allocation is currently under development in the MPLS Working Group [UPSTR, MCST] and the mechanisms for the label distribution or context evaluation have not yet been laid out.

Having seen how P2MP LSPs are set up, the next section describes how they can be used.

6.5 COUPLING TRAFFIC INTO A P2MP LSP

The previous sections described how a P2MP LSP is created. Let us now examine how traffic can be coupled into a P2MP LSP at the ingress node. We consider three categories of traffic: Layer 2 traffic, IP traffic having a unicast destination address and IP traffic having a multicast destination address. All three categories apply to video applications, because for each there exist examples of commercially available video equipment that encapsulate video flows into packets of that format.

6.5.1 Coupling Layer 2 traffic into a P2MP LSP

One application for P2MP LSPs is to carry Layer 2 traffic such as ATM. For example, some encoders encapsulate digital TV signals into ATM AAL1 frames. With a native ATM network, point-to-multipoint VCs are often used to distribute the traffic to multiple destinations. When using an MPLS network, P2MP LSPs provide the analogous function, allowing the Layer 2 traffic to be distributed to multiple receivers in a bandwidth-efficient manner.

An existing implementation achieves this by using a point-to-multipoint version of the Circuit Cross Connect (CCC) [CCC] scheme described in the Layer 2 Transport chapter (Chapter 10). In this scheme, a binding is created, through configuration, between an incoming Layer 2 logical interface (e.g. an ATM VC or an Ethernet VLAN) and a P2MP LSP at the ingress router. Similarly, at the egress routers, a binding is created between the P2MP LSP and the outgoing Layer 2 logical interface. Note that because CCC depends on RSVP signaling, this scheme applies only to P2MP LSPs that are signaled by RSVP. The detail of how the Layer 2 frames are encapsulated for transportation across the MPLS network is exactly the same as for the point-to-point CCC case described in the Layer 2 Transport chapter. For example, in the ATM case the user can choose how many ATM cells should be carried by each MPLS packet.

6.5.2 Coupling IP unicast traffic into a P2MP LSP

Another type of traffic is a flow of packets having an IP unicast destination address. This would be case in a scenario where the source generates a stream of IP packets that have a unicast destination address but nevertheless need to be distributed to multiple video receivers. Each receiver would typically be directly connected to one of the egress nodes of the P2MP LSP. The coupling of the IP traffic into the point-to-multipoint LSP at the ingress router could be carried out using a static route, with the P2MP LSP as the next-hop. At the egress routers, if the destination address of the packet is on a subnet to which the egress router is attached, the packet is automatically routed correctly. Alternatively, a static route could be used to direct the packet to the appropriate output interface. Although this scheme may sound odd in that multiple receiving hosts are configured with the same IP address, and potentially multiple subnets around the network are configured with the same address and mask, the scheme is useful for expediency because some commercially available video-to-IP encoders currently generate packets having a unicast IP destination address.

6.5.3 Coupling IP multicast traffic into a P2MP LSP

In this section, we discuss two methods by which IP multicast traffic could be coupled into a P2MP LSP, using static routing and using PIM. This section assumes some degree of knowledge of IP multicast mechanisms.

Let us first take the case where no multicast protocols are in use, such as IGMP or PIM, but the application, such as a video encoder, generates packets with a multicast destination address and the receivers are configured to receive packets with that destination address. This scheme would be applicable in a scenario where the multicast source is directly connected to the ingress router of the P2MP LSP and the receivers are directly connected to egress routers of the P2MP LSP. In this case, a static route at the ingress router can be used to direct the packet into the appropriate P2MP LSP. At the egress nodes, again a static route is used to direct the packet to the appropriate receiver.

Another variation in the IP multicast case is a hybrid one in which P2MP LSPs provide a core distribution capability but multicast trees formed through PIM procedures are used for local distribution beyond the egress routers of the P2MP LSPs. At the time of writing, no implementations of these schemes exist but they are under discussion in the IETF. We will discuss two cases, one in which the P2MP LSP is fixed and another in which PIM triggers the addition of new branches to a P2MP LSP.

Fixed P2MP LSPs and dynamic PIM trees

The P2MP LSP tree is fixed but the local PIM trees are formed dynamically using normal PIM procedures in response to IGMP reports generated by receivers wishing to join particular multicast groups. This scheme might be appropriate for video-on-demand scenarios where all the channels are distributed by the P2MP LSP to local head ends, but multicast group membership determines the onward distribution of those channels from a local head end (i.e. from a P2MP LSP egress point) to a multicast receiver. In this situation, the PIM Joins triggered by IGMP reports received by the router attached to the receiver only extend as far as the egress router of the P2MP LSP.

Figure 6.7 illustrates such a scheme. A P2MP LSP extends from the ingress router, PE1, to the egress routers, PE2, PE3 and PE4. Each egress PE is attached to a local distribution infrastructure. The egress PEs and the local distribution routers (e.g. R8, R9 and R10 in the case of PE2) have PIM enabled. PE1 is attached to the sources of the multicast groups G1, G2, G3 and G4. The P2MP LSP distributes traffic belonging to these multicast groups to the egress PEs. Let us suppose receiver Rx1 (which could be a set-top box) needs to receive multicast group G1. It sends a corresponding IGMP message to R5. This triggers R5 to generate a PIM Join which propagates hop by hop in accordance with normal PIM procedures to PE3 (but not beyond). This results in traffic for group G1 to be forwarded by PE3 towards Rx1.

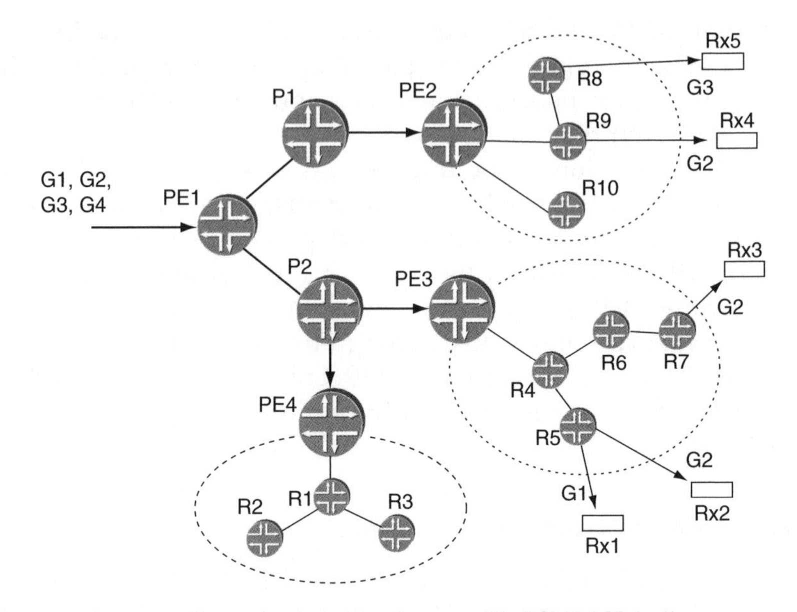

Figure 6.7 Illustration of a hybrid scheme with P2MP LSP in the core and PIM islands at the edge

P2MP LSPs in conjunction with PIM discovery

In this case, PIM is being used as a discovery mechanism. This enables the ingress router of a P2MP LSP to discover, through a PIM Join, the leaf nodes requiring to receive that traffic, rather than the identity of those nodes having to be configured manually on the ingress router. Sub-LSPs are added or removed from a P2MP LSP in accordance with PIM Join and Prune messages arriving from the edge of the network [TE-PIM].

The scheme requires changes to the way PIM operates. PIM as it stands today assumes that neighbors are directly connected, so a PIM Join is propagated hop by hop through the core of the network. However, this leads to an undesirable situation where the P routers in the network need to run PIM and hold an associated multicast state. Hence a more useful scheme would be one in which a router X can send a PIM Join directly to the ingress router of a P2MP LSP in order to trigger the ingress router to add a sub-LSP with egress X to an existing P2MP LSP that is carrying the multicast group in question (or to set up a P2MP LSP if one does not already exist).

Current IETF work proposes a scheme [DIR-PIM] in which a PIM adjacency can be formed between two routers that are not directly connected neighbors. The scheme uses 'directed' PIM Hello messages, i.e. Hello messages whose destination IP address is an IP address (usually a loopback address) belonging to the remote neighbor. Once an adjacency is formed, PIM Join and Prune messages can be sent directly to the remote neighbor rather than being propagated in a hop-by-hop fashion. In this way, the P routers in the network are not required to run PIM.

Let us look again at Figure 6.7 and see how the scheme described above operates. PE3 has receivers in its domain that require multicast groups G1 and G2. PE3 receives PIM Joins corresponding to those groups that have propagated hop by hop through its local domain. This triggers PE3 to send a directed PIM Hello to PE1. Once a PIM adjacency is formed with PE1, PE3 can send the joins corresponding to groups G1 and G2 directly to PE1. This triggers PE1 to add PE3 to the P2MP LSP being used as the distribution tree in the core of the network (or to create a P2MP LSP if one does not already exist). Similarly, PE2 builds a PIM adjacency with PE1 to be added to the P2MP tree. Note that if a single P2MP tree is being used for the core distribution, then PE2 and PE3 both receive groups G1, G2 and G3, even though PE2 does not require G1 and PE3 does not require G3. On the other hand, PE4, not having any active multicast receivers in its domain, is not an egress point of the P2MP LSP and so does not receive any unnecessary traffic.

A variation on this scheme is to use a separate P2MP LSP for each multicast group, to avoid the situation where PEs receive unnecessary traffic at the expense of requiring more P2MP LSPs in the core. In this case, PE3 would be an egress point on two P2MP LSPs, one corresponding to G1 and the other to G2. Likewise, PE2 would an egress point on two P2MP LSPs, one corresponding to G2 and the other to G3.

6.6 MPLS FAST REROUTE

A key attraction of P2MP LSPs signaled using RSVP is that MPLS fast reroute can be used for traffic protection, giving low failover times. In contrast, in normal IP multicast, the failover mechanisms are relatively slow (on the order of seconds), which is unacceptable for applications such as real-time video. In the Protection and

Restoration chapter (Chapter 3), the following variants of fast reroute were described in the context of point-to-point LSPs:

- Link protection
- Node protection

In principle, either of these variants could be used in the point-to-multipoint case. The most straightforward case, for implementation and operations, is the link protection case, because the overall topology of the P2MP LSP in terms of the location of branch nodes remains unchanged. In this case, the task of protecting sub-LSPs related to P2MP LSPs is the same as that involved in protecting normal point-to-point LSPs. In the node protection case, the protection paths end downstream of the next-hop node that is being protected, which would result in the location of branch points changing if the node being protected is a branch node. In the facility protection schemes, where a single bypass tunnel protects multiple LSPs, the same bypass tunnel can be used to protect point-to-point LSPs and sub-LSPs of P2MP LSPs. This is illustrated in Figure 6.8. In the figure, there is a P2MP whose ingress is PE1 and whose egress points are PE3, PE4 and PE5. There is also a point-to-point LSP from PE2 to PE5. The link between P1 and P3 is protected by a bypass tunnel that follows the

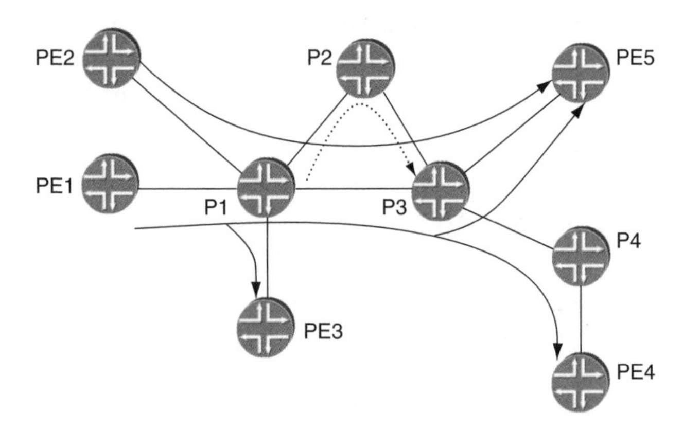

Figure 6.8 Illustration of link protection for a P2MP LSP

path P1–P2–P3. This is shown as a dashed line in the figure. If the link between P1 and P2 fails, the bypass tunnel protects the PE2–PE5 point-to-point LSP and the P2MP sub-LSPs from PE1 to PE4 and PE5.

6.7 APPLICATIONS OF POINT-TO-MULTIPOINT LSPs

This section describes some of the main applications of P2MP LSPs. We first discuss how P2MP TE is being used for the purposes of broadcast TV distribution. We then describe proposals for how P2MP LSPs can be used as infrastructure tools to enable service providers to carry their customers' multicast L3VPN traffic and VPLS multicast traffic more efficiently. These proposals are discussed in more detail in the L3VPN and VPLS chapters (Chapters 7, 8 and 11).

6.7.1 Application of P2MP TE to broadcast TV distribution

An interesting application of P2MP TE is for professional real-time broadcast TV distribution [IBC, MPLS-VID]. This application should not be confused with Internet video streaming applications, which typically involve the sending of low-bandwidth video streams to end users' PCs without any quality guarantees. In contrast, professional real-time broadcast TV distribution requires exacting performance guarantees from the network. Customers of such a service are TV broadcast companies who transport real-time video between studios, from a studio to head ends of distribution infrastructure (terrestrial, cable or satellite) or from an outside broadcast location to studios. The ability to offer broadcast TV distribution services is attractive to service providers because of the high-value revenue streams that can be generated. The demand for such services is likely to grow as the number of TV channels increases as a consequence of the extra capacity available through the growth of satellite, digital terrestrial and cable infrastructure.

Traditionally, such networks have been based on TDM transport (in the form of PDH or SDH, or SONET) or on ATM. However, there is increasing interest in moving to IP/MPLS networks for the following reasons:

1. As well as transport of real-time video, broadcast TV companies and production houses increasingly demand the ability to

transport non-real-time video with file transfer using IP-based protocols (e.g. FTP), as opposed to the traditional method of physically transporting a tape between one location and another. Cost savings can be achieved by using the same network for the real-time and non-real-time transfers. A packet-switched network is more suitable for this than a TDM network because of the statistical multiplexing advantages offered when dealing with the bursty data flows associated with the transfer of the non-real-time video.

2. Higher interface speeds are available for IP/MPLS networks than for ATM networks.
3. It is easier to build a shared network on which multiple TV broadcast companies can book bandwidth. The lead times for making bandwidth available to new customers or for existing customers requiring extra capacity are much less than for TDM-based networks.
4. The service provider can go one step further than in item 2 above. Rather than building a dedicated network for the purpose of broadcast TV distribution, this can be just one service among many carried over an IP/MPLS network.

The transport of broadcast quality real-time video places stringent requirements on the network, even more so than voice transport. The nature of the application is that there is no opportunity to resend data that failed to reach the destination and even very short-lived interruptions to the data flow can have a noticeable impact. The key requirements are as follows:

1. *Bandwidth guarantees*. Once a booking for a particular video flow has been accepted, the traffic must be transported without loss of data. There cannot be any contention for the bandwidth from other data flows.
2. *Low delay variation*. The tolerance of the flow to delay variation depends on the nature of the decoding equipment, but on the order of milliseconds is a typical target.
3. *High network availability*. The disturbance to the datastream must be minimal in the event of link failure or failure of components within the network equipment. Hence a high degree of component redundancy and schemes for rapid recovery from link failures are very desirable.

4. *Distribution from a single source to multiple destinations.* It is a common requirement for particular real-time video flows to be transported to multiple destinations. It is important to be able to add or remove a destination corresponding to a particular flow without interruption to the flow of data to the other destinations.

Let us see how the use of P2MP TE on an MPLS network allows the above requirements to be met. The requirements for low delay variation and bandwidth guarantees can be met as follows. If the network is to be shared with other traffic, on each link the real-time video packets are placed into a dedicated queue that has high scheduling priority. This means that the latency experienced by the packets in that queue is minimized, as long as the queue is not oversubscribed. Oversubscription of that queue is avoided by using traffic engineering mechanisms: bandwidth is reserved on each P2MP LSP and admission control is performed so that the sum of the bandwidth reservations does not exceed the bandwidth available to that queue. If the video traffic is the only form of traffic in the network that requires bandwidth guarantees and admission control, then RSVP-based traffic engineering can be used as described in the Traffic Engineering chapter (Chapter 2), with the maximum available bandwidth being set to the size of the queue assigned to the real-time video. If other forms of traffic also require bandwidth guarantees and admission control, RSVP-based DiffServ Aware Traffic Engineering can be used, as described in the DiffServ Aware Traffic Engineering chapter (Chapter 4).

The service provider can make the most efficient use of bandwidth when meeting the customer's requirement of distributing the traffic to multiple destinations by building P2MP LSPs in the form of minimum-cost trees. Bandwidth efficiency is especially important, bearing in mind that the bandwidth of a single uncompressed standard definition video exceeds 300 Mbps and that of an uncompressed high definition video exceeds 1.5 Gbps. In some cases, compression, e.g. based on MPEG-2, is used to reduce the bandwidth requirement. The requirement of being able to add or remove egress points of a P2MP LSP without affecting traffic travelling to the other egress points of that LSP can be met through a careful router forwarding plane design and implementation.

The use of fast-reroute mechanisms for P2MP LSPs means that the disturbance to traffic is minimized should a link in the network fail, although even when using fast reroute a visible disturbance can be noticed on the TV screen. Note that this is also the case when using SONET or SDH protection when carrying a video over a TDM network rather than an MPLS network. This sensitivity to short interruptions is in contrast to voice, where an interruption of a few tens of milliseconds would be unnoticed.

For the most critical broadcast TV traffic, application-level redundancy is sometimes used. In this scheme, two copies of the real-time video stream are sent into the network, the replication required to create the two copies being carried out within the video domain. The two streams follow different paths through the network such that they do not share the same fate (e.g. not following the same fibre or duct). At the receiving end, the two streams are sent to a receiver that is capable of seamlessly switching from one stream to the other should the first stream be interrupted. This scheme increases the end-to-end availability of the video flow because traffic remains uninterrupted in the event of a single failure of a link or component within the network. When using a pair of P2MP LSPs to transport the traffic, the fact that the path of each branch of each of the two LSPs is under the full control of the user makes it straightforward to ensure that the two LSPs do not share the same fate.

Codec equipment is commercially available that can convert a video feed into a stream of IP packets or ATM cells and vice versa. Either of these traffic types can be coupled into a P2MP LSP, as described in Section 6.5 of this chapter.

When used for the application of broadcast TV distribution, in many cases each P2MP LSP is dedicated to a single video stream. Some P2MP LSPs may be relatively short-lived, e.g. existing for perhaps a couple of hours to transmit footage from a sports event to multiple destinations. This is in contrast to traditional point-to-point traffic engineering in service provider networks, in which a typical LSP is very long-lived (on the order of months or years) and would carry a large number of end-to-end traffic flows.

6.7.2 Application of P2MP LSPs to L3 VPN multicast

An application of P2MP LSPs is an ingredient of a next-generation scheme [VPN-MCAST] for carrying IP multicast traffic belonging to Layer 3 VPN customers. At the time of writing this scheme is under discussion in the IETF. As discussed in the Layer 3 VPN chapter of this book (Chapter 7), currently Layer 3 VPNs are the largest deployed applications of MPLS today. Some Layer 3 VPN customers, in addition to using IP unicast applications in their enterprise, also use IP multicast applications. Hence there is a need for service providers to transport this traffic between sites on behalf of their customers. In the Advanced L3 VPN chapter, we discuss the proposed next-generation scheme for carrying L3 VPN multicast traffic, of which P2MP LSPs are an important component. The P2MP LSPs are used to distribute IP multicast traffic arriving at a PE from an attached CE to other PEs in the network for onward distribution to their attached CEs.

This is illustrated by the two scenarios in Figure 6.9. Figure 6.9(a) shows how a P2MP LSP (or more than one P2MP LSP) rooted at each PE is used to distribute the traffic. Figure 6.9(b) shows how a P2MP LSP (or more than one P2MP LSP) rooted at a central point is used to distribute the traffic, each PE using a unicast tunnel in order to send multicast traffic to that central distribution point. These schemes are discussed in more detail in the Advanced Layer 3 VPN chapter. Depending on the requirements of the service provider and the customers, the P2MP LSPs can either be LDP-signaled or RSVP-signaled.

6.7.3 Application of P2MP LSPs to VPLS

In current Virtual Private LAN Service (VPLS) implementations, multicast traffic arriving at a PE from a customer site is sent to all PEs

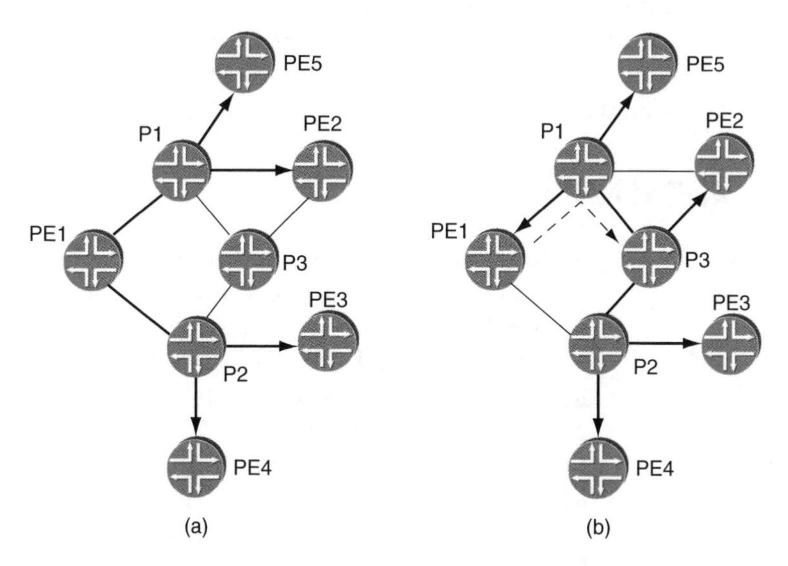

Figure 6.9 Use of P2MP LSPs as distribution trees for L3 VPN multicast traffic: (a) distribution tree routed at the PE router and (b) central distribution tree

having members of that VPLS instance attached. Ingress replication is used by the PE. This is wasteful of bandwidth, because in many cases links within the core of the network carry multiple copies of the same packet, each destined to a different egress PE. The advantage, however, is that no multicast state is required in the core of the network. The current scheme may be fine if the volume of multicast traffic is relatively low, but if not then it could be advantageous to the service provider to use a more bandwidth efficient scheme. Such a scheme is currently under discussion in the IETF [VPLS-MCAS]. As with the next-generation scheme for L3 VPN multicast discussed above, P2MP LSPs are a key component, being used to perform the distribution of multicast VPLS traffic. More details of this scheme are discussed in the VPLS chapter of this book (Chapter 11).

6.8 CONCLUSION

This chapter has discussed a solution to a missing piece of the converged network jigsaw puzzle, namely point-to-multipoint LSPs. Previously, MPLS has not interacted comfortably with multicast, typically coexisting via a 'ships in the night' approach. The advent of P2MP-TE means that multicast traffic can enjoy the traffic engineering advantages already offered by MPLS in the unicast case, such as bandwidth guarantees and fast-failover mechanisms. As a consequence of the 'grand unification' of the two worlds of MPLS and multicast, MPLS networks are now being used for professional broadcast TV distribution, a very exacting application that was previously difficult to support on an MPLS network.

6.9 REFERENCES

[CCC] K. Kompella, J. Ospina, S. Kamdar, J. Richmond and G. Miller, 'Circuit cross-connect', draft-kompella-ccc-02.txt (work in progress)

[DIR-PIM] R. Aggarwal and T. Pusateri, 'PIM-SM extensions for supporting remote neighbors', draft-raggarwa-pim-sm-remote-nbr-01.txt

[IBC] M. Firth, 'Challenges in building a large scale MPLS broadcast network', in International Broadcasting Convention (IBC 2004), Amsterdam, September 2004

[MCST] T. Eckert, E. Rosen, R. Aggarwal and Y. Rekhter, 'MPLS multicast encapsulations', draft-rosen-mpls-multicast-encaps-00.txt (work in progress)

[MPLS-VID] A. Rayner, 'Real time broadcast video transport over MPLS', Paper D1–13, in MPLS World Congress, Paris, February 2005

[P2MP LDP] I. Minei, K. Kompella *et al.*, 'Label distribution protocol extensions for point-to-multipoint label switched paths', draft-minei-mpls-ldp-p2mp-01.txt (work in progress)

[P2MP REQ] S. Yasukawa (ed.) *et al.*, 'Signaling requirements for point to multipoint traffic engineered MPLS LSPs', draft-ietf-mpls-p2mp-sig-requirement-03.txt (work in progress)

[P2MP TE] R. Aggarwal, D. Papadimitriou, S. Yasukawa *et al.* (eds), 'Extensions to RSVP-TE for point to multipoint TE LSPs', draft-ietf-mpls-rsvp-te-p2mp-02.txt (work in progress)

[P2MPWC] R. Aggarwal, 'Point to multipoint MPLS TE solution and applications', Paper D1–12, in MPLS World Congress, Paris, February 2005

[TE-PIM] R. Aggarwal, T. Pusateri, D. Farinacci and L. Wei, 'IP multicast with PIM-SM over a MPLS traffic engineered core', draft-raggarwa-pim-sm-mpls-te-00.txt

[UPSTR] R. Aggarwal, Y. Rekhter and E. Rosen, 'MPLS upstream label assignment and context specific label space', draft-raggarwa-mpls-upstream-label-00.txt (work in progress)

[VPLS-MCAS] R. Aggarwal, Y. Kamite and L. Fang, 'Multicast in VPLS', draft-raggarwa-l2vpn-vpls-mcast-01.txt (work in progress)

[VPN-MCAST] E. Rosen and R. Aggarwal, 'Multicast in MPLS/BGP IP VPNs', draft-ietf-l3vpn-2547bis-mcast-00.txt (work in progress)

Part Two

7

Foundations of Layer 3 BGP/MPLS Virtual Private Networks

7.1 INTRODUCTION

BGP/MPLS IP VPNs, referred to in short as MPLS L3 VPNs or simply L3 VPNs throughout this book, are one of the most widely deployed applications enabled by MPLS. When talking about MPLS, it is not fast reroute or traffic engineering that springs to mind, but rather VPN support. In fact, traffic engineering and fast reroute are most often thought about in terms of the benefits that they can provide in the context of a particular service. Perhaps the most popular service is provider-provisioned IP VPNs and the L3 VPN solution described in this chapter is the way this service is realized in MPLS networks. For many providers, L3VPNs is the major and sometimes the only driver for deploying MPLS in the network.

VPNs existed long before MPLS. The success of L3 BGP/MPLS VPNs is owed to the scaling and simplicity advantages that the combination of BGP and MPLS brings to VPN scenarios. The L3 BGP/MPLS VPN solution was extended to the Layer 2 space as well, as we will see in the chapters discussing Layer 2 Transport and VPLS (Chapters 10 and 11).

MPLS-Enabled Applications: Emerging Developments and New Technologies Ina Minei and Julian Lucek

In this chapter we will see how the MPLS VPN solution emerged, introduce its basic principles and concepts and shed light on some of the design decisions taken. We assume the reader has a basic understanding of both MPLS and BGP. In the next chapters we will look at more advanced topics that arise in the context of L3 VPNs. Readers familiar with the BGP/MPLS VPN concepts and basic operation can skip over this chapter and go directly to the advanced topics (Chapters 8 and 9).

7.2 THE BUSINESS DRIVERS

In the simplest scenario, a customer has geographically dispersed sites and requires connectivity between them, in order to run his day-to-day business. The customer does not want to invest in the infrastructure for connecting the sites, nor in the effort of administering this infrastructure. In a competitive world, he or she would rather concentrate on the core business and outsource the task of providing connectivity between sites to the networking expert, the service provider.

From the customer's point of view, the goal is to achieve connectivity with minimum hassle. First of all, connecting the dispersed sites should have the same QoS and privacy guarantees as a private network, and should not require changes to the way the customer's network is configured or run. For example, the customer should be able to use a private address space if he or she chooses. Secondly, the operations that affect connectivity should be easy. For example, adding connectivity to a new site, changing the connectivity between sites or increasing the bandwidth between sites should not require many configuration changes and should be achievable at short notice. Finally, the solution should not require complex routing configuration at the customer's sites.

From the provider's point of view, the goal is to fulfil the customer's expectations while maximizing profits. To fulfil the customer's expectations, the provider must be able not just to provide connectivity but also to extend the service easily and allow customers to use private (and thus possibly overlapping) address spaces. To maximize profits, the provider must support a large number of customers, as well as to be able to support a wide range of customers with respect to the numbers of sites, from customers with a handful of sites to customers with thousands or even tens of thousands of sites. Furthermore, the provider must be able to provide customers with value-added services that can be charged at a premium. Finally, the resources used in providing the service must be shared among the customers.

Based on these goals, let us see why the solution is called a Virtual Private Network (VPN). First, it is a network because it provides connectivity between separate sites. Second, it is private because the customer requires it to have the same properties and guarantees as a private network, both in terms of network operations (addressing space, routing) and in terms of traffic forwarding. Third, it is virtual because the provider may use the same resources and facilities to provide the service to more than one customer.

In the real world, it is seldom that the goals are crisp and clear from the beginning. What happens instead is that a solution is developed for a given problem. As experience is gained from existing deployments the drawbacks of the solution become apparent and more requirements are added to the 'goals' section, yielding a new and improved solution, in an iterative process. Thus it makes sense to look back at the VPN solutions that existed before the BGP/MPLS solution, as they will help us to understand how the current MPLS VPN model emerged and will highlight some of its advantages. In our discussion, we will concentrate on VPNs for which the service provider (SP) participates in the management and provisioning of the VPNs. This type of VPN is known as a provider-provisioned VPN (PP VPN).

7.3 THE OVERLAY VPN MODEL

The overlay model is the most intuitive VPN model. If it is connectivity that the customer wants, what can be simpler than connecting the customer sites via point-to-point links between routers at the various sites? The point-to-point links could be Frame Relay or ATM circuits, leased lines or IP-over-IP tunnels such as Generic Route Encapsulation (GRE) or IP Security (IPSec). What is provided is a virtual backbone for the customer's network, overlaid on top of the provider's infrastructure. Designated routers at the different customer sites (the customer edge routers, or CE routers) peer with each other and exchange routing information, thus allowing traffic to flow over the links between the different sites.

In this model, the provider is oblivious of the internal structure and addressing in the customer's network and provides only a transport service. Provisioning the circuits between the customer sites assumes knowledge of the traffic matrix between sites. However, in most cases it is not the traffic matrix that is known but the average traffic sourced and received, thus making it difficult to estimate the bandwidth required. After the circuits are set up, the bandwidth that is not used is wasted, making the solution expensive. One more interesting note on provisioning involves the case where Frame Relay or ATM is used. In this case, increasing the bandwidth between sites may require provisioning of new circuits, which can take a long time to set up.

In the overlay model, the VPN service is provided by the customer routers. A VPN where the intelligence and control are provided by CE routers is called a CE-based VPN. When customers are responsible for configuring and running the CE routers, they are in fact designing and running their own VPN, a task they may not always have the expertise or desire to be involved in. As a result, the provider may take over the management of the customers' virtual backbone (thus providing a managed VPN service). However, managing the networks of many VPN customers requires managing a large number of CE devices and places a burden on the provider, thus limiting the number of customers that he can service.

Regardless of who manages the customer routers, a model where routers at the customer sites exchange routing information with each other has limitations. Let us take a look at a scenario where there are many sites and a fully meshed configuration. In such a scenario, the number of routing peerings can be very large. This can be a scaling problem for the IGPs due to the large amount of information that may be exchanged when routing changes. Another limitation concerns the amount of configuration that must be done when a new site is added to the VPN. Obviously, the customer router at the new site must be configured to peer with the routers at the other existing sites. Unfortunately the routers at the existing sites must also be reconfigured to establish peering to the new site.

The overlay model achieves the fundamental goals of a VPN. It provides connectivity between customer sites, allows the use of a private address space and ensures the privacy of the traffic between the sites. The functionality is provided by the CE routers and comes at a cost: difficulty in evaluating the bandwidth requirements between sites in cases where the bandwidth must be pre-provisioned, the need to manage a large number of customer routers, complex configuration when adding a new site and the need for a large mesh of routing peering.

7.4 THE PEER VPN MODEL

The problems of the overlay model stem from the fact that customer routers peer directly with each other. The peer model attempts to overcome the drawbacks of the overlay model by lifting the requirement for direct routing exchanges between the customer routers. Instead of peering with each other and forming an overlay on top of the service provider's network, CE routers now peer only with directly attached PE routers. As a result, the large mesh of routing peerings between CE routers disappears. From the customer's point of view, routing becomes very easy. The burden of managing the route distribution between the customer sites is now passed on to the provider and the intelligence moves out of the CE routers into the PE routers.

Moving from a CE-based solution to a PE-based one has other benefits as well:

- Adding a new customer site to a VPN requires configuration of the CE and PE for the new site only, rather than configuration of all the customer's CEs.
- The number of points of control in the network (i.e. the number of intelligent devices that make meaningful routing decisions) does not necessarily increase for each new customer site added (assuming that more than one CE can attach to the same PE and that the CE can simply run static routing).
- A single infrastructure is used to service all VPN customers.
- The exact traffic matrix between customer sites is not required in order to provision bandwidth between customer sites. Instead, it is enough to know the amount of traffic flowing in/out of a site, since the provider's infrastructure is used to carry the traffic.
- Increasing the amount of bandwidth between sites requires increasing the bandwidth between the CE and PE, rather than upgrading several circuits or leased lines.
- Simple routing from the CE point of view. Each CE advertises to the PE reachability information for the destinations in the site

to which the CE belongs. Optimal routing between the CEs is ensured by the fact that the routing protocols in the provider network ensure optimal routing between the PEs to which these CEs attach.

- Different routing protocols can run within each one of the different customer sites.

Clearly the PE-based solution is very attractive, assuming it can meet the connectivity and privacy requirements of a VPN; traffic must flow between sites of the same VPN, but is not allowed between sites of different VPNs. Thus the requirement is to constrain the flow of traffic. This can be done either by constraining the traffic at forwarding time or by constraining the distribution of routing information (which implicitly constrains the destinations to which traffic can flow). Let us take a look at two of the early PE-based solutions, as they will highlight some of the problems that MPLS solves in the VPN context.

One of the earliest PE-based VPN solutions ensured traffic isolation between VPNs by constraining traffic at the forwarding time using access lists on the CE–PE links. Access lists operate at the forwarding time on IP packets and allow/disallow forwarding based on fields in the IP header such as source and destination addresses. While conceptually intuitive, a solution based on access lists quickly becomes unmanageable in practice. Implementing complex intersite access policies becomes a challenging task because it is driven by source/destination addresses. As the number of sites and the number of customers grow, the number of access lists increases. In some vendor's implementations, processing large numbers of access lists impacts the forwarding performance of the routers, thus making the solution even less attractive. One last but crucial point is that since access lists operate based on source/destination address information, the model assumes distinct addressing spaces in each VPN. Therefore, the access-list-based solution cannot service a customer network that uses a private address space, limiting the usefulness of the solution.

The logical next step was to get rid of the access lists by ensuring that traffic arriving at a PE is only destined for one particular VPN. This can be accomplished by connecting every VPN customer to its own dedicated physical or virtual PE router. A dedicated PE router is not enough to ensure that traffic cannot be forwarded from one VPN to another. One must also make sure that there is no routing state on this PE that would allow traffic to be forwarded towards a PE that services a different VPN. The second early PE-based solution used constrained route distribution (based on BGP communities) coupled with dedicated virtual PE routers to ensure traffic isolation between VPNs. In this model, the PE accepts and installs only routes belonging to the VPNs that it services. This model is the basis for the current BGP/MPLS-based VPN solution, and the mechanisms for constrained distribution of routes will be discussed in detail in later sections. However, in its early incarnation, this model also suffered from the limitation regarding private address spaces because of the way routes were advertised in BGP.

An important thing to note about both the early PE-based solutions discussed so far is that forwarding in the provider's network is based on the IP header. Therefore the routers in the provider's network must know how to forward traffic to all destinations in all the VPN customer sites. Forwarding based on IP is a fundamental difference between the early PE-based VPNs and the BGP/MPLS VPN solution. Let us take a look at the impact IP forwarding has on the solution:

- The use of private address spaces is precluded. Forwarding is done based on the IP header, and there is no way for a router in the middle of the provider's network to differentiate between traffic belonging to different VPNs.
- The default route cannot be used in the customer VPNs, since there is no way to differentiate between default routes belonging to different customers.
- The scalability of the solution is limited. The state that must be maintained in all the provider's routers is equal to the number of customer destinations from all VPNs. In a model where provider routers must maintain state for all the VPN routes, the maximum number of customer routes that can be serviced is limited by the number of routes that the provider's core routers can support. The VPN service cannot be scaled beyond the forwarding table size of the core routers.

Clearly, a tunnelling scheme would be beneficial in this case, as it would shield the provider's core routers from having to carry the routing information for all customer routes.

7.5 BUILDING THE BGP/MPLS VPN SOLUTION

The BGP/MPLS VPN solution is built on the peer model described in the previous section. This should not come as a surprise, for two reasons. First, we have seen that PE-based VPNs have attractive properties such as simple routing from the customer's point of view and easy addition of new VPN sites. Second, we have seen that early PE-based solutions were limited by the fact that traffic travelled as IP in the core. Tunneling would eliminate this limitation, and MPLS can provide the necessary tunnels.

The BGP/MPLS VPN model was first published in informational RFC 2547 [RFC2547], documenting a solution developed at Cisco. Following the success of 2547 VPNs there was a desire from some service providers to make it into an IETF standard. A new working group was started in the IETF, called ppvpn (for provider-provisioned VPNs). One of the work items of the group was to standardize MPLS VPNs, and the internet draft that resulted from this work was named 2547bis [2547-bis]. In the industry today, BGP MPLS/VPNs are often called 2547bis for this reason.[1] The ppvpn Working Group undertook work in both the L2 and L3 spaces, and was later split into the l2vpn and the l3vpn Working Groups [L3VPN, L2VPN].

In the following sections, we will build the BGP/MPLS VPN solution step by step, hopefully shedding light on some of the design decisions taken. Before we can start, let us remember the goals the VPN solution is trying to achieve:

- Isolation of traffic between the different VPNs
- Connectivity between customer sites
- Use of private address spaces in each site

7.5.1 VPN routing and forwarding tables (VRFs)

Isolation of traffic between the different VPNs means that a customer in one VPN should not be able to send traffic to another VPN. Figure 7.1 shows two customer VPNs, belonging to

[1] At the time of this writing, the 2547bis draft had been approved for progression to RFC status, so probably by the time of the publishing it will have a new RFC number.

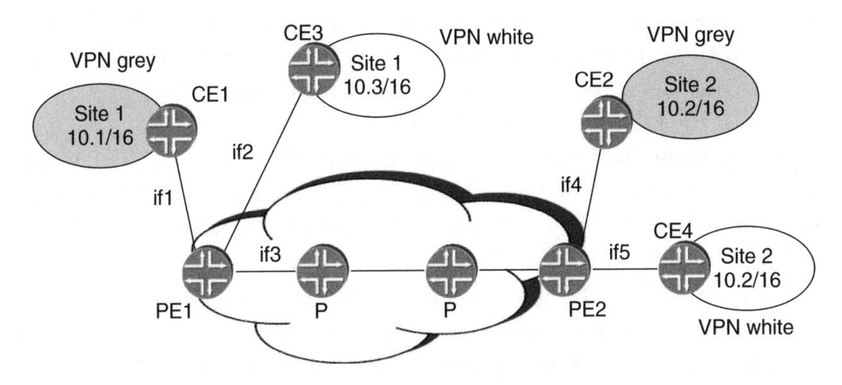

Figure 7.1 A simple network with two customers

customer 'white' and customer 'grey'. Each PE has sites from both VPNs attached to it. Assume that each PE is using a single routing/forwarding table. Use of a single table is problematic both in the case when the two VPNs use overlapping address spaces (as shown in the figure) and in the case where they use distinct address spaces. If the two VPNs use overlapping address spaces, then forwarding information cannot be installed for both in the common table, as there is no way of distinguishing the destinations in the two VPNs. If the VPNs use distinct address spaces, it is possible for a host in the VPN white site to send traffic to the VPN grey site, by simply sending IP traffic to the destination in VPN-grey; when the traffic arrives at the PE, the destination address is found in the routing table and the traffic is forwarded to VPN-grey.

Both these problems can be solved by attaching each customer site to its own virtual or physical PE device. Remember, though, from the description of the peer model that the PEs carry the burden of customer route distribution. Increasing the number of PEs with each new customer site is not scalable from either the routing or the network management point of view.

A more scalable solution is to use per-VPN routing and forwarding tables (VRFs), thus maintaining separate information for each VPN. These tables are in addition to the global routing/forwarding table used for non-VPN (Internet) traffic, and they contain routes for the customer's destinations both at the local site and at remote sites. How does the PE know which VRF to use when an IP packet arrives from the customer site? The solution is simple: associate each interface with a particular VRF through configuration. The term 'interface' here does not necessarily mean a physical interface; it could also be a logical one, such as ATM VCI/VPI, FR DLCI or Ethernet VLANs.

For example, in Figure 7.1, the interface if1 connecting PE1 to the CE1 is associated with the VRF for customer grey's VPN and the interface if2 connecting PE1 to CE3 is associated with the VRF for customer white's VPN. When an IP packet arrives over the CE1–PE1 interface if1, the destination of the packet is looked up in the VRF for customer grey and when it arrives on the CE3–PE1 interface if2, it is looked up in the VRF for customer white. When an IP packet arrives over an interface that is not associated with any VRF, the lookup is done in the global table. In later sections we will see how traffic arriving over core-facing interfaces such as if3 is handled.

The use of multiple forwarding tables at the PE is a necessary condition for allowing support of overlapping address spaces. However, multiple forwarding tables do not automatically ensure that traffic cannot be forwarded from one VPN to another. If the forwarding table for VPN-white were to contain information for destinations in VPN-grey, there would be nothing to prevent a host in VPN-white from sending traffic into VPN-grey. Thus, it is necessary to control the information that is installed in each VPN. This is accomplished by constraining the distribution of routing information, thus constraining the knowledge about reachable destinations.

7.5.2 Constrained route distribution

There are two approaches to constraining routing information per VPN. The first approach is to run separate copies of the routing protocol per VPN, very much like the overlay model, except that the routing peering is between PE routers rather than CE routers. This is not an attractive option from a management and scaling point of view, as the number of routing protocol contexts and the complexity of the routing peerings grow with the addition of each new VPN.

The second approach is to carry all VPN routes in a single routing protocol in the provider's network and constrain the distribution of VPN reachability information at the PEs. This is the method employed in the BGP/MPLS VPN solution, where BGP is the protocol carrying the VPN routes. Here are a few of the properties that make BGP the natural choice in VPN scenarios:

- Has support for route filtering using the community attribute; thus it can do constrained route distribution. Can attach arbitrary attributes to routes, so the community paradigm can be easily extended.
- Has support for a rich set of attributes, allowing control of the preferred routing exit point.
- Can carry a very large number of routes; thus it can support a large number of customer routes.
- Can exchange information between routers that are not directly connected; thus the routing exchanges can be kept between the PE routers only.
- Can carry label information associated with routes (we will see later on why this is important).
- Has support for multiple address families (we will see in the next section why this is required).
- Can operate across provider boundaries.

7.5.3 VPN-IPv4 addresses and the route distinguisher (RD)

We have seen that BGP has attractive properties as the routing protocol for carrying the VPN routes across the provider's network. However, a BGP speaker can only install and distribute one route to a given address prefix, which is problematic when carrying VPN addresses that are from private (and thus possibly overlapping) address spaces.

The solution is to make the private addresses unique by concatenating an identifying string called the route distinguisher (RD) to the IP prefix, in effect creating a new address family (the VPN-IPv4 address family). The BGP multiprotocol (MP) capability allows BGP to carry prefixes from multiple address families [RFC2858]. This is why sometimes, in the context of VPNs, BGP is referred to as MP-BGP. The address family (AFI) and subsequent address family (SAFI) used for encoding the VPN-IPv4 address family are 1 and

128 respectively. Sometimes when discussing VPN configuration, VPN-IPv4 is referred to as SAFI 128.

An interesting thing to note is that VPN-IP addresses only need to be known by routers in the provider's network, and only by those routers actually involved in exchanging routing information for VPN destinations. The customer is unaware of the existence of VPN-IP addresses. The translation between customer IP routes in a particular VPN and VPN-IP routes distributed between provider routers is performed by the PE routers. Before advertising a customer VPN route in BGP, the PE router attaches to it the appropriate RD for the VPN site, transforming it into a VPN-IP route. When receiving a VPN-IP route, the PE converts the route back to plain IP by removing the RD. The association between VPNs and RDs that must be applied to routes belonging to the VPNs is determined through configuration.

Since the RD's task is to make the VPN prefixes unique, it is important to ensure that the RDs themselves are unique. The structure of the RD is driven by the requirement that a service provider should be able to allocate unique RDs without the need for coordination. The RD is an 8-byte quantity consisting of three fields: a two-byte type field, an administrator field and an assigned number field. The type field determines how the other fields are to be interpreted. There are two options for the RD: a combination of a 2-byte AS number and a 4-byte locally assigned number and a combination of a 4-byte IP address and a 2-byte locally assigned number. Both the AS number and the IP address must be such that they ensure uniqueness of the generated numbers, if several providers are cooperating for providing the VPN service. For example, this can be done by using the AS number assigned to the particular network or an IP address from the public space assigned to the network.

In itself, the RD does not contain any information regarding the prefix to which it is attached. In particular, it does not convey information regarding the VPN to which the prefix belongs or the site from which the prefix originates. Since no meaning is associated with the RD, the association of RDs to routes is constrained by two factors: (a) the need for uniqueness and (b) the ease of configuration, management and tracking of RDs. One can imagine different RD allocation policies, with varying degrees of granularity for the RD scope. The most commonly used ones, as a compromise between achieving uniqueness and using a small number of RDs in the network, are using one RD per VPN per PE or using one RD per VPN. Some vendors recommend the use of one RD per VPN, though this is not technically necessary. This in turn creates the perception that the RD somehow helps identify the VPN, when in fact all it does is ensure uniqueness of VPN routes carried in BGP. Using a separate RD per VPN per PE can make troubleshooting easier if the RD is picked in such a way that it can unambiguously identify the PE that originated the route advertisement. It also makes it easy to handle the scenario of overlapping VPNs, where, for example, a particular site can be accessed from two VPNs,[2] since there is no confusion about how to build the RD for the routes in the common site.

To summarize, regardless of how RDs are allocated, their purpose is always the same: to make the VPN routes unique. This is necessary because all VPN routes are carried in the same routing protocol, and BGP can only distribute one route for a given prefix.

7.5.4 The route target (RT)

Let us now go back to the original problem: how to constrain the distribution of VPN routing information between the PEs, thus constraining the routing knowledge and defining which destinations are reachable from each VPN site. The requirement is broader than simply separating the routing information per VPN, for two reasons: (a) customers may require arbitrary and complex connectivity models between their sites and (b) support for overlapping VPNs means that the same route must be present in several VPN routing tables. In fact, what customers want is a flexible way to define policies that determine the connectivity between different sites. What is therefore needed is a way to do route filtering. For BGP/MPLS VPNs, this is done using BGP extended communities.

[2] An example for such a scenario is a case of two companies that partner with each other and therefore require common access to a resource such as a database. The VPN site where the database resides belongs to the VPN of both companies.

A BGP speaker can mark a route by attaching one or more communities to it [RFC1997]. The communities allow the receiver of the advertisement to filter the routes that it wishes to accept. One of the goals of the community attribute is to allow the service provider to allocate values that are locally significant, without the need for external coordination (a similar philosophy to the one we saw in the previous section on RD allocation). The BGP community attribute is a 32-bit string, where the first 16 bits represent the AS number allocated to the provider and the last 16 bits represent a locally assigned number. Since AS number assignments are unique, each provider can manage his or her own number space and define up to 2^{16} distinct values. This means that if a provider uses communities for route filtering in VPNs, he or she is limited to at most 2^{16} customers. Furthermore, the provider must make sure that the values used for VPNs and the values used for other policy functions do not clash. To overcome this limitation, extended communities were introduced [EXT-COMM]. Extended communities are just like communities, except that they use 32 bits for the locally assigned portion, thus allowing definition of 2^{32} distinct values. Because they provide a structured number space, the extended communities do not clash with communities used for providing other policy functions in the network. In the context of VPNs, the extended community used for route filtering is called the route target (RT). The RT is what accomplishes the constrained route distribution between PEs and ends up defining the connectivity available between the VPN sites.

Here are a few important properties of the RT:

1. One or more RTs can be attached to the same route.
2. Attaching an RT to a route can be done with arbitrary granularity: the same RT can be attached to all the routes in a particular site, or different RTs may be attached to each route. Determining which RT to use is defined through configuration. The use of policy language allows a flexible definition of matching criteria.
3. Up to 2^{32} RTs are available in each AS.

So how does the route distribution work? Let us take a look at the example in Figure 7.2, where sites from two VPNs (white and grey) are attached to the same pair of PEs. The requirement is to

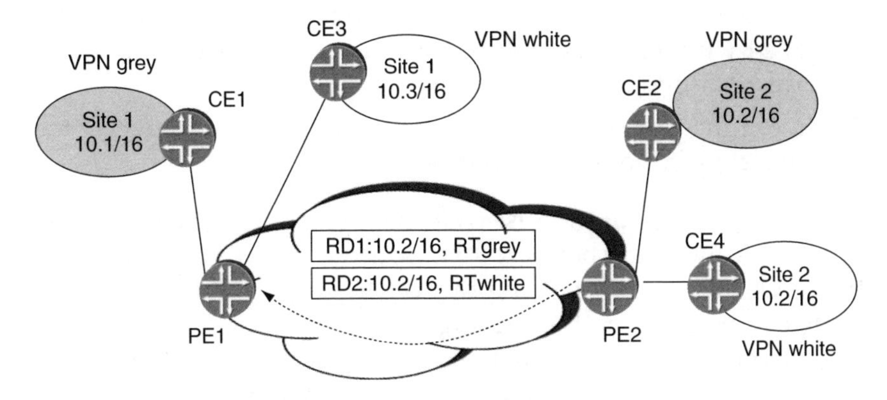

Figure 7.2 Using the RT and the RD

provide connectivity between the sites in each VPN. Here are the steps, for advertisements from PE2 towards PE1:

1. Assume that the PE2 knows which customer routes are reachable in each customer site that is attached to it. It may have received the customer VPN routes from the CE attached to it through a routing protocol or it may have knowledge of these routes through configuration.
2. The goal is to allow other sites in the VPN to forward traffic to these destinations. For this to happen, these routes must now reach the other VPN site; therefore they must be exported into BGP. We have seen in the previous section that the PE translates customer VPN routes into VPN-IPv4 routes before exporting them to BGP by attaching the RD. In addition to this, the route is also tagged with one or more RTs that are determined on a per-VRF basis using policies defined in the configuration. Since these RTs are applied at the time the route is exported, they are sometimes called export-RT. In the example, a single RT is attached to each route.
3. The routes are carried as VPN-IPv4 routes via BGP to the remote PE. The remote PE must decide in which VRF to install this routing information. The decision is done by matching the received routes against locally defined per-VRF import policies expressed in terms of RTs (import-RT). If the route is tagged with any of the RTs specified in the import-RT for a particular VRF, it is stripped of the RD and imported into the VRF and the

routing and forwarding state is installed. In the example, routes tagged with RT-white will be installed in the forwarding table, corresponding to the white VPN.

4. If a routing protocol is running between the PE and the CE, the routes may be advertised to the CE.

An interesting question is: what happens to the routes that do not match the RT? There are two options: discard them or keep them. On one hand, it is impractical to keep all advertisements received, because of the scaling limitations such an approach would put on the PE. On the other hand, if these routes are discarded, there is a problem relearning them when the need arises (e.g. following an addition of a new VPN site at a PE or a configuration change to the import-RT). The solution is to discard routes that do not match the RT, but have the ability to ask for them again when the need arises. This is accomplished through the route-refresh capability of BGP, described in [RFC2918]. This capability is negotiated at the session initialization time and allows a BGP speaker to request its peer to resend all the routes it previously advertised to it.[3]

To summarize, the constrained distribution of routing information is driven by import and export policies defined on a per-VRF basis. These policies are expressed in terms of RTs. A route that is tagged with several RTs is imported in a VRF if any of its RTs match any of the RTs defined in the import-RT of the VRF. Several RTs can be attached to the same route on export and different routes can be tagged with different sets of RTs. In order to ensure that a route is advertised from site 1 to site 2 in a given VPN, the route is tagged with one or more RTs at the time it is advertised by the PE servicing site 1, PE1. The RTs that are attached must be such that the import policy on the PE servicing site 2, PE2, matches it. Thus, the export-RT on PE1 must be such that it contains at least one RT that appears in the import-RT at PE2.

The import and export RTs are the central building block of VPNs, because they express the policies that determine the connectivity between customer sites. Here are a few examples of how the RT can be used to implement some of the common VPN connectivity models.

[3] The route-refresh capability is not VPN-specific; it was originally added to BGP to support nondisruptive routing policy changes.

Full mesh

All sites can communicate with all others directly, creating a 'full-mesh' topology. A single RT is used for both the import and the export policies at all the VPN sites; thus all sites end up installing a state for routes from all other sites.

Hub and spoke

Sites can communicate with one another indirectly through a designated site (the hub), creating a 'hub-and-spoke' topology. This is useful in the case where it is desired to pass all intersite traffic through a firewall, and the firewall resides in a particular site. This topology can be implemented through the use of two route targets, RT-spoke for the spoke sites and RT-hub for the hub site, as shown in Figure 7.3. On export, routes from the spoke sites are tagged with RT-spoke and routes from the hub site are tagged with RT-hub. The import policy for the hub site is to accept routes tagged with RT-spoke, thus learning the route information for all spoke sites. In addition, the hub site readvertises all routes it learned from spoke sites, or a default route, tagged with RT-hub.

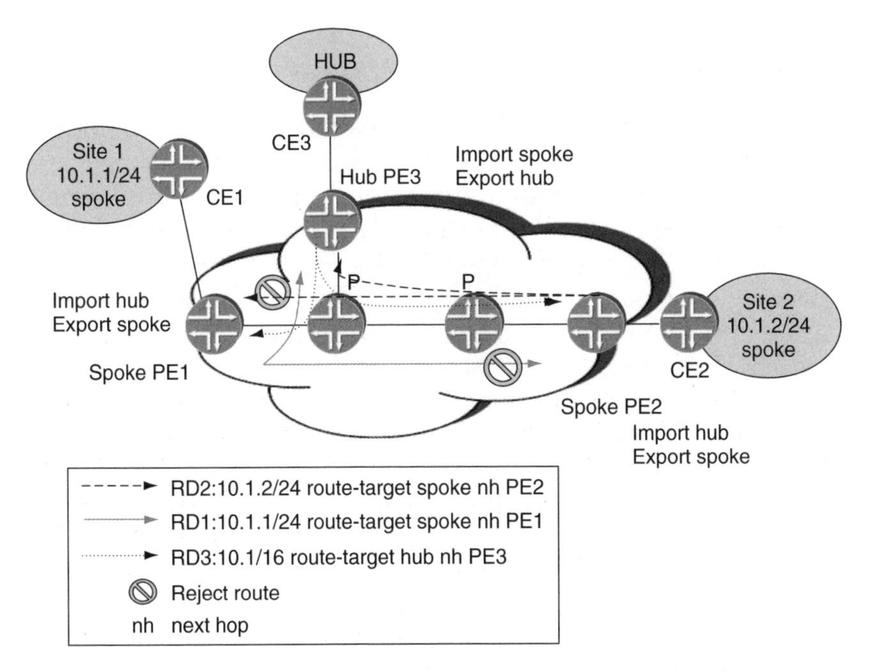

Figure 7.3 Hub and spoke

The import policy for the spoke site is to accept RT-hub only, thus learning the reachability information for all the other spoke sites through the hub. For example, the advertisement for 10.1.1/24, originated by PE1, is rejected at the spoke site serviced by PE2. However, the 10.1/16 advertisement, originated at the hub and tagged with RT-hub, is accepted. The result is that spoke sites do have connectivity to other spoke sites, but the traffic passes through the hub.

Overlapping VPNs

Designated sites from different VPNs can access a common site. For example, a common resource (such as a database or ordering system) is used by two companies. One option is to tag the route for the common resource with the RTs of both VPNs. The problem with this approach is that it allows connectivity to the shared resource from anywhere in the two companies. To provide more selective access to the shared resource, the solution is to tag it with a special route target on export and to import routes with this RT only in those sites that have access to the common resource.

Management VPNs

The goal of a management VPN is to allow the VPN provider to access all the CEs that it services, for network management purposes. A common way to accomplish this is to create a new VPN providing connectivity to all the CEs. The assumption is that because the CEs are managed by the provider, they all have unique addresses. The VRF corresponding to the management VPN has an import policy with a single route target, RT-management. In each of the customer VPNs, the routes corresponding to the PE–CE interfaces are exported with the RT-management route target, in addition to any other route targets necessary in the context of the particular VPN. Thus, only the routes for the PE–CE interfaces are imported in the management VPN. This is a very good example of how only a subset of the routes in a site can be tagged with a particular route target in order to achieve connectivity to only a subset of the destinations.

The route target is a flexible and granular way of constraining the distribution of routing information among sites. With the routing information in place, the next step is to see how traffic is forwarded between sites.

7.5.5 The solution so far – what is missing?

VPN routes are distributed as VPN-IP prefixes between the PEs using BGP. The next hop of such a route is the address of the advertising PE. Thus, in order to reach the destination, traffic must first be forwarded to the advertising PE. The traffic must be tunneled between PEs, for two reasons: (a) the P routers have no information on the VPN routes and (b) the BGP information is for VPN-IP addresses, which are not routable.

In Figure 7.4, let us take a look at what happens when CE1 in site 1 of the grey VPN sends traffic to host 10.2.1.1 in site 2 of the VPN. The IP packet arrives at PE1 over the CE1–PE1 interface. As a result, PE1 looks up the destination in the VRF for customer grey and determines that the packet must be forwarded to PE2. Let us assume that there is an MPLS tunnel between PE1 and PE2 and that the packet is sent over this tunnel. Upon arrival at PE2, the question is: which VPN does this packet belong to (or over which customer-facing interface should the traffic be forwarded)?

The problem is similar to the one we saw when introducing the VRF concept: when traffic arrives at the PE from the customer site,

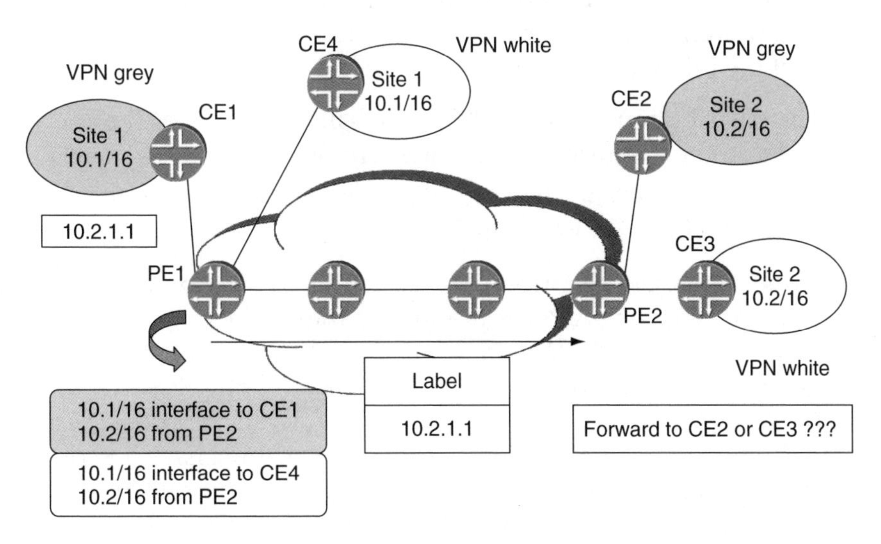

Figure 7.4 Forwarding traffic – what is missing

in which VRF should the lookup be done? In that case, the solution was to pick the VRF based on the interface over which the packet arrived from the customer site. Could a similar approach be used for demultiplexing traffic arriving from a remote PE?

In order to use the same paradigm as the 'incoming interface' approach, one must maintain several tunnels between the PEs, one for each VPN. At the ingress PE (the PE where the VPN traffic enters the provider network), the packets are mapped to the tunnel corresponding to the VPN to which they belong. At the egress PE (the PE where the VPN traffic leaves the provider network), the packets are forwarded to the correct VPN based on the tunnel on which they arrived. However, does this mean that the per-VPN state must be maintained in the core of the provider's network? The answer is 'no'. In the next section we will see how the per-VPN tunnels are created and associated with each VPN and how they are carried transparently over the provider's core.

7.5.6 VPN label

All that is needed in order to create a VPN tunnel with MPLS is to associate a label (the VPN label) with a VPN route. At the forwarding time, the VPN traffic is labelled with the VPN label at the ingress PE and sent to the egress PE. Based on the VPN label, the egress PE can demultiplex the traffic to the correct VPN.

Setting up and maintaining separate tunnels per VPN can only scale if the following two conditions apply:

1. The distribution of the VPN tunnel information is automatic and does not require manual intervention.
2. The P routers do not have to maintain a separate state for each one of the PE–PE VPN tunnels.

The first condition is satisfied by using BGP to distribute the VPN label along with the VPN route information, as explained in the section discussing BGP as a label distribution protocol in the Foundations chapter (Chapter 1).

The second condition is ensured by the label stacking properties of MPLS, which allow the creation of a hierarchy of tunnels, as described in the section discussing hierarchy in Chapter 1 (section 1.3.1). This is accomplished by stacking two labels: the VPN tunnel label (as the inner label at the bottom of the stack) and the PE–PE tunnel label (as the top label or outer label). Forwarding is always done based on the top label only, so the P routers need not maintain any state regarding the VPN tunnels. The VPN tunnel label is used for controlling forwarding at the PE.

To summarize, a VPN label is advertised along with each VPN-IP route exchanged using BGP. The next-hop of the VPN-IP route is the advertising PE. A PE–PE MPLS tunnel provides connectivity to the BGP next-hop of the VPN route and the VPN label controls forwarding at the egress PE. Three questions arise from this model:

1. What is the policy for VPN label allocation?
2. How is forwarding done at the egress PE?
3. What are the requirements from the BGP next-hop of a labelled VPN-IP route?

Let us try to answer each of these separately below.

1. What is the policy for VPN label allocation?

The purpose of the VPN label is to demultiplex VPN traffic arriving at the PE. From this point of view, any allocation policy, ranging from a separate label per route to a single label per site, fulfils the functional requirement. For most scenarios, one label per site provides the required functionality with the minimum amount of state. At the other extreme, one label per route can provide per/destination statistics and good visibility into the CE–CE traffic matrix. The cost is the extra state maintained, which can make troubleshooting difficult, as we have argued in the Foundations chapter in the context of LDP. Therefore, unless the functionality is needed, an approach that creates less state is preferable.

2. How is forwarding done at the egress PE?

From an implementation point of view, forwarding at the PE can be one of the following two options:

1. An MPLS lookup on the VPN label to determine the appropriate VRF, followed by an IP lookup in that VRF.
2. An MPLS lookup based on the VPN label, in which case the label provides an outgoing interface.

In the first case, the VPN label is used to identify the correct table for the lookup; in the second case, it is used to identify the correct outgoing interface. The end result is always the same: an IP packet is sent towards the correct VPN site. Most vendors support both types of lookup and let the user choose between the two through configuration. The reason is because most forwarding engines only look at the portion of the packet that contains the information for the type of forwarding they perform. For example, if one wanted to set the DiffServ behaviour of the traffic based on the IP header, the lookup would need to be done as IP rather than MPLS. The same holds true for any other features where the IP header information is required, such as applying firewall filters or doing accounting based on the IP header.

3. What are the requirements for the PE–PE connectivity?

The BGP next-hop of a labelled VPN-IP prefix distributed by BGP is the address of the PE that originated the advertisement (the egress PE). The intention is to forward the VPN traffic as MPLS, labelled with the VPN label. Thus, the requirement is to have a tunnel to the egress PE, which is capable of forwarding MPLS traffic. The process of finding an MPLS path to the BGP next-hop is often referred to as 'resolving' the BGP route. If the tunnel exists, the route is considered 'resolved' or 'active', meaning that traffic can be forwarded to its destination and that the route can be readvertised into other routing protocols. When the tunnel goes away, the route becomes 'unresolved' or 'inactive' and cannot be used for forwarding. If previously readvertised into a different routing protocol, the route must be withdrawn. From an implementation point of view it is important to note that the process of resolving and unresolving routes should be event-driven rather than based on a timer that scans the state of the LSPs. This is because the time it takes to update the routing state affects the convergence time and ultimately the time during which traffic may be blackholed.

MPLS tunnels are the most intuitive way to forward labelled traffic towards a destination. However, they are not the only option. The IETF defined extensions for carrying MPLS traffic in IPSec, GRE [RFC4023] and L2TPv3 tunnels [MPLS-L2TPV3], thus allowing providers to offer VPN services even over networks that do not support MPLS. This is particularly useful during network migration to MPLS or in cases where providers do not want to deploy MPLS in the core of their network. The development of the extensions to carry MPLS in other tunneling mechanisms is proof of the widespread acceptance and success of the BGP/MPLS VPN solution.

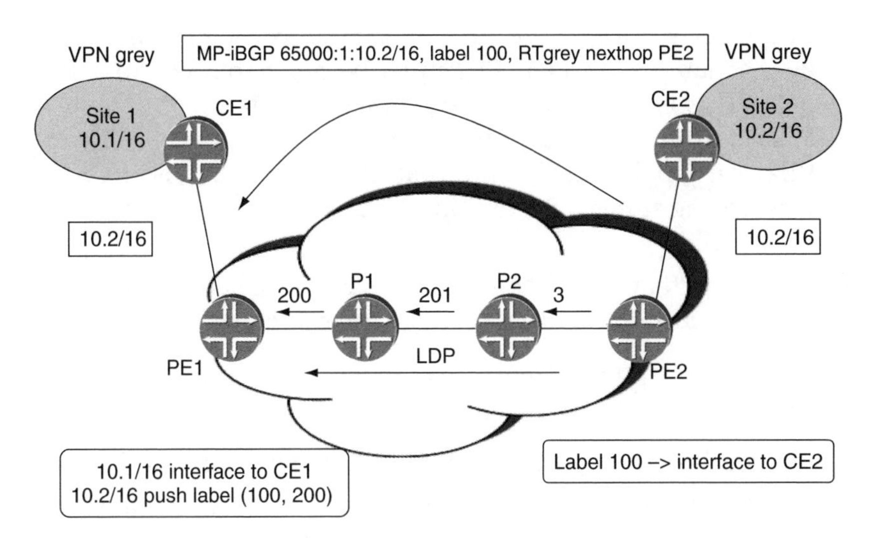

Figure 7.5 The routing and forwarding state created by VPN routes

Let us take a look at both the routing and the forwarding state created by the VPN routes for the network in Figure 7.5. For simplicity, we will look at the routing exchanges in one direction only, from PE2 towards PE1, in order to see the forwarding state that is used when forwarding traffic from PE1 towards PE2 (similar exchanges happen in the opposite direction as well):

1. Assuming that a dynamic routing protocol is running between PE2 and CE2 (attached to it), PE2 receives a route advertisement for prefix 10.2.0.0/16 from CE2. If no routing protocol is running, PE2 has this routing information configured as a static route. In Figure 7.5 BGP is assumed to be on the CE–PE link.
2. PE2 attaches an RD, e.g. 65 000:1, to the route and allocates a VPN label for this prefix, e.g. label 100. PE2 then exports the labelled route into MP-BGP, with the appropriate RT for customer grey's VPN.
3. PE2 creates the forwarding state binding the VPN label (100) to the outgoing interface leading to CE2.
4. PE1 receives the MP-BGP advertisement for the VPN-IP route 65000:1:10.2.0.0/16 with BGP next-hop of PE2 and label 100.
5. Based on the RT, PE1 determines this route must be installed in the VRF for customer grey. PE1 strips off the RD and installs a route to 10.2.0.0/16, with label 100 and next-hop PE2. In order to be able to forward traffic to this destination and make the route active, the 'next-hop PE2' information must be translated

to the forwarding state. It is not enough for PE2 to simply be reachable from PE1; PE2 must be reachable via an MPLS path. This is necessary because labelled traffic will be sent to PE2. Assuming that an MPLS tunnel exists between PE1 and PE2, set up by LDP, and that the LDP label is 200, the forwarding state is: destination 10.2.0.0/16, push a label stack of 100, 200.

6. Assuming that a routing protocol is running between PE1 and CE1 (attached to it), PE1 advertises the route towards CE1.

Figure 7.6 shows how the state created is used at the forwarding time:

1. PE1 receives an IP packet from CE1 with destination 10.2.1.1.
2. Based on the interface on which the packet is received, the route lookup is done in customer grey's VRF. The two-label stack (100, 200) is pushed on the packet and the packet is sent towards PE2.
3. The packet arrives at PE2 with a single-label stack of 100 (the LDP label was popped one hop before PE2, at P2 because of penultimate hop-popping).
4. PE2 has forwarding information binding label 100 to the outgoing interface to CE2 and the packet is stripped of its label and forwarded towards CE2.

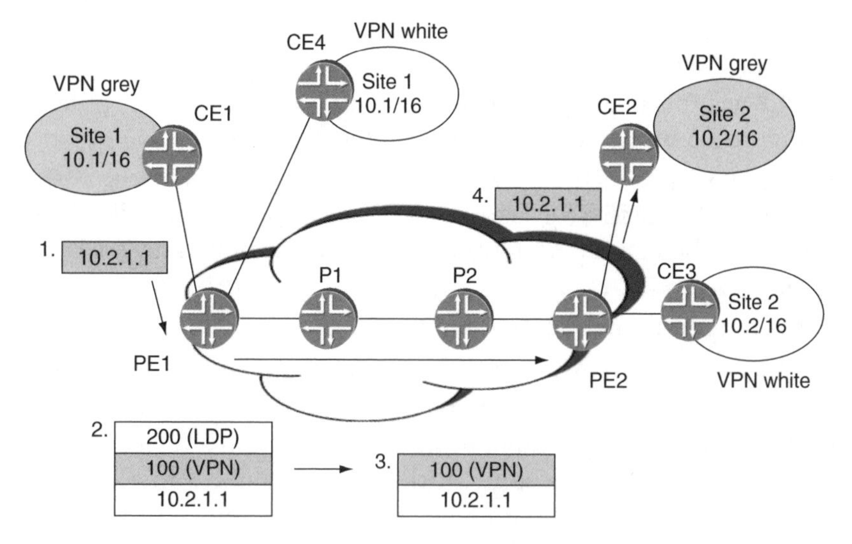

Figure 7.6 Forwarding traffic in a VPN

To summarize, the VPN label allows the PE routers to demultiplex VPN traffic arriving at the PE. BGP provides an automated way to distribute the VPN label by attaching it to the VPN-IP routes. The VPN tunnel information is hidden from the P routers and multiple VPN tunnels are carried inside a single PE–PE tunnel.

7.6 BENEFITS OF THE BGP/MPLS VPN SOLUTION

BGP/MPLS VPNs allow the customer to offload routing between the sites to the provider and enable the service provider to offer value-added services to its customers, such as firewall and authentication. The BGP/MPLS VPN approach allows the provider to leverage the infrastructure to service multiple VPN customers, rather than managing a virtual backbone for each customer. The PE–PE MPLS tunnels are used to carry traffic for multiple VPNs and multiple applications. By hiding the VPN information from the core of the network, the complexity is kept at the PE routers and the service can grow by adding more PE routers when needed.

The property of MPLS that is most powerful in the context of BGP/MPLS VPNs is tunneling. Tunneling using MPLS enables:

1. Building a hierarchy of routing knowledge. Tunneling makes it possible to forward traffic to addresses that are not known in the middle of the network, thus shielding P routers from any VPN knowledge.
2. Identifying traffic as belonging to a particular VPN at the egress point from the provider's network.
3. Providing straightforward and low-cost protection against packet spoofing.

The BGP/MPLS VPN solution builds on existing protocols and technology, only extending the protocols where necessary. It is a great example of the inventing versus reusing paradigm discussed in earlier chapters. The principles discussed in this chapter form the foundation of other MPLS applications, such as L2 VPNs and VPLS. However, before exploring them, let us finish the discussion on L3 VPN by looking at a few more advanced topics.

7.7 REFERENCES

[2547-bis]	E. Rosen and Y. Rekhter, 'BGP/MPLS IP VPNs', draft-ietf-l3vpn-rfc2547bis-03.txt, soon to become an RFC (currently in the RFC editor queue)
[EXT-COMM]	S. Sangli, D. Tappan and Y. Rekhter, 'BGP extended communities attribute', draft-ietf-idr-bgp-ext-communities-09.txt (work in progress)
[L2VPN]	http://ietf.org/html.charters/l2vpn-charter.html
[L3VPN]	http://ietf.org/html.charters/l3vpn-charter.html
[MPLS-L2TPV3]	M. Townsley, T. Seely, J. Young, 'Encapsulation of MPLS over Layer 2 Tunneling Protocol Version 3', draft-ietf-mpls-over-l2tpv3-00.txt (work in progress)
[RFC1997]	R. Chandra, P. Traina and T. Li, *BGP Communities Attribute*, RFC1997, August 1996
[RFC2547]	E. Rosen and Y. Rekhter, *BGP/MPLS VPNs*, RFC2547, March 1999
[RFC2858]	T. Bates, R. Chandra, D. Katz and Y. Rekhter, *Multiprotocol Extensions for BGP-4*, RFC 2858, June 2000
[RFC2918]	E. Chen, *Route Refresh Capability for BGP-4*, RFC2918, September 2000
[RFC4023]	T. Worster, Y. Rekhter and E. Rosen, *Encapsulating MPLS in IP or Generic Routing Encapsulation (GRE)*, RFC4023, March 2005

7.8 FURTHER READING

[MPLS-TECH]	B. Davie and Y. Rekhter, *MPLS Technology and Applications*, Morgan Kaufmann, 2000
[RFC4110]	R. Callon and M. Suzuki, *A Framework for Layer 3 Provider-Provisioned Virtual Private Networks (PPVPNs)*, RFC 4110, July 2005

8

Advanced Topics in Layer 3 BGP/MPLS Virtual Private Networks

8.1 INTRODUCTION

The previous chapter laid out the foundations of BGP/MPLS L3 VPN. This chapter explores some of the advanced topics that arise in the context of L3 VPNs such as scalability, resource planning, convergence, security and multicast support. All of these require a network-wide view and analysis. Therefore, it is necessary to first discuss two more important components of the VPN solution: PE–CE routing and route reflectors.

8.2 ROUTING BETWEEN CE AND PE

A key concept in the MPLS/VPN solution is that customer routes are kept in a VPN Routing and Forwarding (VRF). The VRF is populated with routes learned from the local CE and routes learned from remote CEs as VPN routes. In the previous sections we saw how customer routes are propagated as VPN-IPv4 routes across the provider's network from PE to PE and added to the

MPLS-Enabled Applications: Emerging Developments and New Technologies Ina Minei and Julian Lucek

appropriate VRF. In this section we will take a closer look at how routes are learned from the local CE.

There are several options for a PE to find out about routes from the CE attached to it: static routing, RIPv2, OSPF and BGP.[1] Regardless of how the PE finds out about the routes, it must install them in the VRF associated with the interface to the CE. Thus, a routing protocol must install routes learned over a CE–PE interface in the VRF associated with that interface. From an implementation point of view, this is accomplished by creating separate contexts for the routing protocols per VRF.

So far we have seen that the basic requirement is to have VRF aware routing. The next question is whether to use static or dynamic routing, and which routing protocol to use. One important thing to note is that this decision is local to each PE. In the same VPN, different methods may be employed at different sites, just as different sites may use different routing protocols within the site. This property is particularly useful in a situation where different sites in the VPN are managed by different administrations and may be running different routing protocols within the site (e.g. the sites of a company following an acquisition).

Let us take a look at some of the factors influencing the choice of the PE–CE routing method:

- Limitations of the CE device. For many VPN deployments it is required to have a large number of CE devices with limited functionality. This is true, for example, of a company with small branch offices, where many sites must access a central server. In this case, complex routing capabilities are not required, but due to the large number of CE devices, price is important. Simple, cheap devices often support just static routing or static routing and RIP. This is one of the reasons why static routing is one of the most popular CE–PE protocols in use today.
- The routing protocol running in the customer site. Running a different protocol on the CE–PE link than in the rest of the site means two things: (a) routes must be redistributed from one protocol to another and (b) the site administrator must deal with two routing protocols instead of one. This is why many times

[1] At the time of this writing, a proposal to standardize IS–IS as a CE–PE protocol was under discussion in the IETF. The requirement to allow IS–IS as a CE–PE protocol has not been addressed so far, as most enterprise deployments do not use IS–IS as the IGP.

customers prefer to run the same routing protocol on the CE–PE link as the one that is already running within the site.

- The degree of trust the provider has with regards to the customer's routing information. When trust is low, the provider may choose to use static routing towards the customer, to shield himself from dynamic routing interactions with the customer.
- The degree of control that a protocol gives to the provider. BGP allows route filtering based on policy and therefore gives the provider control over what routes to accept from the customer. Because it is relatively easy for the provider to be protected from a misbehaving customer, BGP is a popular PE–CE protocol.

Let us examine in more detail some of the issues that come up in the context of PE–CE routing:

1. *The use of static routing*. On one hand, static routing is simple and supported by any device. On the other hand, the hosts in the customer site have no knowledge of the reachability of destinations in other sites. If an LSP goes down in the core, the VPN routes resolving over that LSP become inactive at the PE. However, this information is not propagated to the customer site, causing a silent failure from the customer's point of view. This is a problem if the customer site is attached to two or more PEs for redundancy (multihomed) and could have picked a different link to exit the site. For this reason, dynamic routing is a requirement when using multihoming.
2. *Running the same protocol on the PE–CE link as in the customer site*. In principle, there is no reason not to use proprietary protocols, such as EIGRP, assuming that both the CE and the PE support them. However, one thing to bear in mind is that the CE and PE are in different domains and often under different administrations. Running any protocol whose algorithm is highly collaborative will make troubleshooting difficult in such an environment, as the customer and the provider site may be managed by different entities.
3. *Protection against a misbehaving CE*. The provider must protect itself from a misbehaving CE advertising a large number of routes. Using static routing avoids this problem altogether. When dynamic routing is used, the options are: (a) configure an upper limit on the number of routes the provider accepts from the customer or (b) use filtering to only accept a prenegotiated

subset of prefixes. The concept of limiting the number of routes can be extended to the entire VRF by setting a limit on the number of routes allowed in the VRF, regardless of whether they are received on the local PE–CE routing session or whether they are received from remote PEs.

4. *Using OSPF as the CE–PE protocol.* OSPF uses link state advertisements (LSA) to communicate topology information and therefore has the following properties:

   (a) LSAs cannot be filtered. Therefore, all LSAs arriving on the CE–PE link must be received and processed at the PE. This implies that there must be a degree of trust between PE and CE and that the CE will not flood the PE with large numbers of advertisements. This trust must exist because the protocol itself has fewer means to provide this control to the provider (e.g. when compared to a protocol like BGP).
   (b) The LSA type must be maintained as the OSPF routes are advertised between sites. Remember that the customer routing information is carried from one PE to another as VPN-IPv4 routes. Therefore, it is necessary to maintain all the information needed to create the correct type of LSA at the remote site, while the route is carried from one PE to another as a VPN-IPv4 route. This is done by using extended community attributes in BGP. The details of the operation are in [VPN-OSPF] and [OSPF-2547-DNBIT].

   Another issue can arise if a route received from the MPLS/BGP backbone is sent back into the backbone. The problem can be avoided by using OSPF route tagging. Although the use of OSPF may seem complex, it is an attractive option for customers who already run OSPF in their networks and do not want to deploy one more protocol.

5. *Using BGP as the CE–PE protocol.* When using BGP as the PE–CE protocol, there is an EBGP (external BGP) session between the provider and the customer (assuming that the provider and customers are in different ASs). BGP is attractive in this context because of the high degree of control it can give the provider over what routes to accept from the customer. BGP also has good scaling properties, because there are no periodic advertisements of any kind and only incremental updates are sent. For these reasons, BGP is a popular choice as a PE–CE protocol.

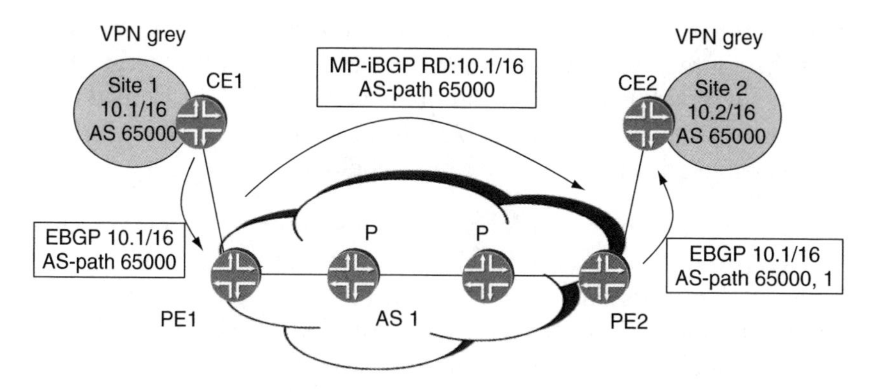

Figure 8.1 Using BGP as the PE–CE protocol may cause problems when the customer sites are in the same AS

An interesting situation arises in this case due to EBGP's built-in loop-prevention mechanism, which ensures that routes are not advertised back into the AS that first originated them. Loop prevention is implemented by appending the AS number of each AS which advertises the route in the AS path attribute attached to the advertisement and requiring a BGP speaker to reject an advertisement if it finds its own AS in the AS path (this is sometimes called an AS path loop). Figure 8.1 shows how this requirement affects route distribution in a VPN context. BGP routes arrive at PE1 from CE1 with the customer's AS number (65 000) in the AS path attribute. The provider's AS number (1) is attached to the AS path as the route is propagated from PE2 to CE2. If the remote customer site serviced by CE2 uses the same AS number as the customer site serviced by CE1, then the route is dropped, since a loop is detected in the AS path (the customer's AS number already appears in the AS path). This problem can be solved in one of the following ways:

- Use different AS numbers in each site. This solution may work if private AS numbers are used by the customer sites, but is not feasible if the customer sites belong to one of the ASs from the assigned number space.
- Configure the CE so that it does not enforce the requirement regarding loops in the AS path. This introduces the danger of routing loops caused by other advertisements with AS path loops in them.

- Remap the customer's AS number to the provider's AS number as the route is advertised as a VPN route (this is often referred to as AS override). This solution only works if the customer is not readvertising its routes via BGP to other networks from one of its sites, as explained in more detail below.

Let us take a look at a large VPN customer who peers with a service provider in one of its sites and advertises its routes to the Internet. The customer routes learned from remote sites and advertised as VPN routes over the provider's backbone will have the provider's AS number in the AS path. However, the customer wants to advertise its routes with its AS number. It makes no sense to include the VPN provider's AS number in such advertisements.

An elegant solution to this problem is described in [IBGP-PE-CE]. The idea is that BGP attributes received from the customer are stored in a new transitive BGP attribute that functions like a stack. The provider's BGP attributes are used within the provider's network. When the route is advertised from the remote PE to the remote CE, the stack is popped in order to discard the provider's attributes and the original customer attributes are restored. In this way, the customer BGP attributes are carried transparently over the VPN provider's backbone and the service provider's attributes, such as the AS number, do not appear in the routes received by the customer.

To summarize, in order to obtain information on CE routes, separate instances or separate contexts of the routing protocols are required per VRF. The choice of the routing method depends on several factors, among them the CE capabilities, the routing protocol running within the site and the degree of trust that the provider has in the customer's routing information.

8.3 ROUTE REFLECTORS AND VPNs

Let us now turn to the last remaining component of the VPN solution: route reflectors. The functional requirements and the tradeoffs regarding the use of route reflectors (RRs) are different in a VPN scenario than in a pure IP (Internet routing) scenario. In both cases the RR provides the following benefits from the configuration management point of view:

- Reduction in the number of BGP peerings. A BGP speaker (e.g. a PE) only needs to peer with the RR rather than with all other

BGP speakers. Thus, each speaker maintains a constant number of peerings, regardless of the number of BGP speakers in the network.

- Ease of configuration. Adding a new BGP speaker only requires setting up a BGP session to the RR, rather than multiple sessions from and to the new speaker.

The differences in the RR use in plain IP service provider networks and VPN networks stem from differences in the routing information carried and how it is advertised, in particular:

1. The routes carried by the PE routers.
2. The number of paths to a particular destination.

In a pure IP scenario, the routing information carried in BGP is the full Internet routing table. Traffic is forwarded as IP and therefore:

1. All routers need to have information for all the destinations.
2. Multiple paths exist for the same destination, because providers typically have several external peerings over which they learn Internet routes.

In this setup, the RR is often used to perform path selection, with the following consequences:

1. Reduction in the number of BGP paths in the network. The RR performs path selection and only advertises to its clients the best path.
2. Reduction in the number of updates generated/processed by each speaker. The RR clients have to process a single update rather than one for each one of the sessions advertising reachability to a particular destination.
3. Requirement for resources on the RR: memory for all BGP paths, CPUs for handling path selection and update processing and forwarding resources to keep the state for all the prefixes. (The RR does path selection based on local criteria; therefore traffic must flow through the RR.)

In contrast, in a VPN scenario:

1. PE routers only need to maintain the routing/forwarding state for the VPNs to which they are connected.
2. Multiple paths for the same destination are not prevalent. Even if the same destination is advertised from several VPN sites,

unless the same RD is used in all sites, path selection cannot take place (since the prefixes are treated as distinct).

Here are some observations regarding the use of RRs in a VPN network:

1. It is desirable for PEs to only receive BGP updates for the VPN routes that they will accept and install, rather than all VPN routes. This is in order to conserve CPUs on both the PE and the RR.
2. CPU requirements on the RR. The RR handles the update processing and replication for all route changes in all VPNs that have routes on the RR.
3. There is no requirement to maintain the forwarding state on the RR for all the VPN routes. Assuming distinct RDs, the RR is not performing path selection, so there is no need for traffic to be forwarded through the RR. Many vendors implement this capability in order to conserve forwarding resources on the RR. Figure 8.2 shows a VPN network where the RR is not in the forwarding path. PE1 and PE2 peer with the RR in order to learn the VPN routes, but VPN traffic is forwarded between PE1 and PE2 over the LSP taking the path PE1–P1–P2–PE2. For this to happen, the BGP next-hop must be propagated unchanged by the RR (thus, the advertisement 10.2/16 should arrive at PE1 with next-hop PE2).

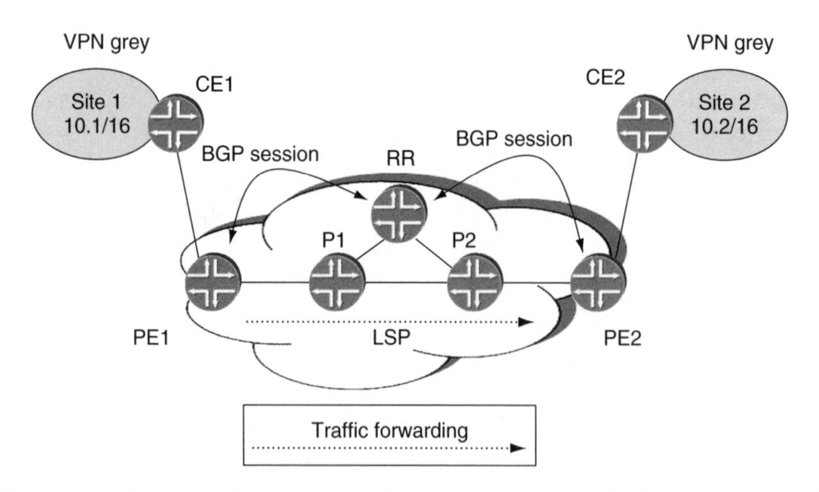

Figure 8.2 RR in a VPN setup, in the case where no traffic is forwarded through the RR

However, the RR can become a potential scaling bottleneck, as the one element in the network required to carry the state for all VPN routes. One way to avoid this problem is to partition the VPN routes among several reflectors. A PE router peers only with the route reflector that carries routes from the VPNs in which it is interested. However, this alone is not enough. What if a PE router is required to peer with all the reflectors, because it has customers whose routes reside on each one of the route reflectors? The PE would then receive updates for all the VPN routes in the network. What is needed is a way for the PE to inform the RR which routes it is interested in. In this way the routes are filtered at the RR rather than being propagated to the PE and filtered at the PE.

The solution is for the PE to advertise to the RR the set of RTs for which it is interested in receiving updates (typically, these are the set of RTs used in all the import policies in all VPNs on the PE). As a result, the RR only advertises to the PE the routes that are tagged with these RTs, resulting in less work for both the RR, which generates less updates, and the PE, which processes less updates.[2]

Two mechanisms are available for achieving this goal: outbound route filtering (ORF) [ORF] and route-target filtering [RT-CONSTR]. The difference between the two is the scope of the filter advertisements. With ORF, the scope is limited to two directly connected neighbors. For route-target filtering, the filtering information is propagated in the same way as routing information and can cross multiple hops and AS boundaries. ORF was introduced in BGP as a way to minimize the number of updates that are sent between two peers, long before VPNs became popular. However, in the context of VPNs it is useful to be able to propagate the filtering information automatically, which is how route-target filtering came into being.

Let us take a look at an example where route-target filtering is particularly useful. Figure 8.3 shows a US-based network with PEs on both the east coast and the west coast. The network has two route reflectors, RR1 and RR2, one on each coast (for simplicity, let us ignore the extra reflectors used for redundancy). PEs on each coast peer with the local route reflector (in the figure, the solid

[2] One important thing to note is that such a mechanism is not limited to interactions between the route reflector and its clients, and can be used for any BGP peerings.

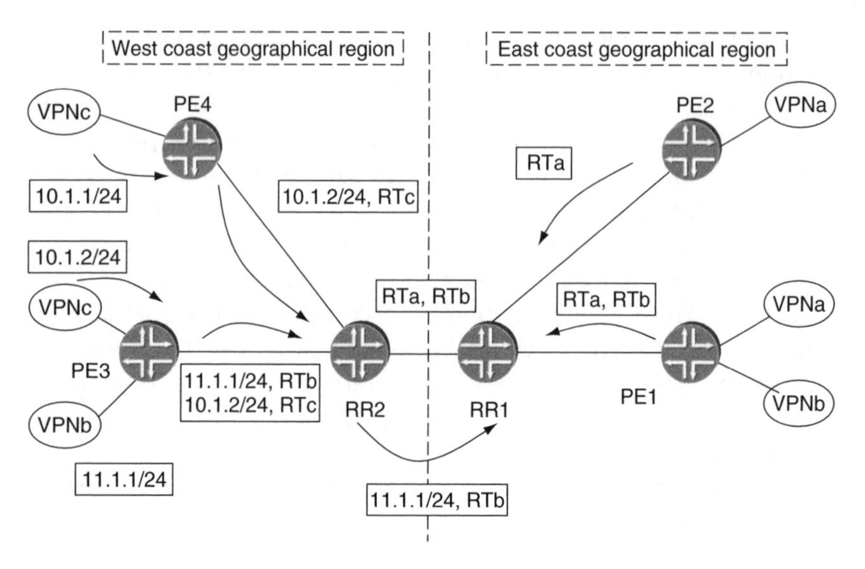

Figure 8.3 Route target filtering

lines represent connectivity, not necessarily direct links). It is reasonable to assume that most of the VPNs have sites within the same geographical area and that a small number of VPNs have sites on both coasts. PEs use route-target filtering to inform the RR which RTs they are interested in receiving updates for. For simplicity, Figure 8.3 shows the RT distribution in one direction only, from PE1 and PE2 to RR1 and the route propagation in the opposite direction, from PE3 and PE4 towards RR2 (for simplicity, the RDs are not shown in the route advertisements). The RRs peer with each other and propagate the filtering information they received from their clients. As a result, instead of exchanging all the routes, the RRs only exchange VPN routes for the VPNs that are present on both coasts. In the example, only the routes tagged with RTb are advertised by RR2 to RR1, instead of all the routes advertised by PE3 and PE4 to RR2. RR1 does not receive routes belonging to VPNc from RR2, because RR1 does not have any client PEs that are interested in those routes. This would not be possible with ORF, where the scope of the filtering information is limited to one hop, between the RR and the client.

In this section we have seen how route reflectors fit in a VPN network. Let us now take a look at the first topic requiring a network-wide view: scalability.

8.4 SCALABILITY DISCUSSION

Scalability is a crucial aspect of any network service. Without scalability, the service becomes the victim of its own success: as the popularity of a service grows, it becomes either technically difficult or economically unsound to provide the service to both new and exiting users. When evaluating the scaling properties of BGP/MPLS VPNs one must examine both these aspects and answer the following questions:

- From a solution point of view, is there a technical upper limit on the maximum number of VPN customers and sites per customer that can be serviced in a given network? In other words, is there a scalability bottleneck in the solution, or can the service be scaled by adding more routers to the network?
- How does increasing the number of customers impact the cost of the network? In other words, how is the scalability of each of the devices impacted by the growth in the number of customers and routes? At which point must new equipment be purchased?

8.4.1 Potential scaling bottlenecks

The first question is whether there is a scalability bottleneck in the solution. The dimensions in which a VPN service is expected to grow are the number of VPNs, the number of customer sites and the number of customer routes. Growing the VPN service may have implications on:

1. The provisioning burden on the provider.
2. The load on routing protocols.
3. The amount of VPN-related information that must be maintained on the provider's routers.
4. The amount of bandwidth needed to carry the VPN traffic.

What we will show in the following paragraphs is that such limitations do not exist, or can be avoided by splitting the load across several components in the network.

The provisioning burden on the provider

BGP/MPLS VPNs are based on the peer model. Thus, adding a new site to an existing VPN requires configuration of the CE router at the new site and of the PE router to which the CE router is attached. The operator is not required to modify the configuration of PE routers connected to other sites or touching the configuration of other VPNs. The PEs learn automatically about the routes in the new site. This process is called 'autodiscovery'. Adding a new VPN to the network does not require changing the configuration of existing VPNs or having knowledge of configuration parameters used for other VPNs.[3]

Another aspect of provisioning concerns the addition of a PE router to the network. BGP MPLS/VPNs require full meshes of BGP sessions (when no RRs are used) and of PE-to-PE MPLS tunnels between all PEs in the network. Therefore, in principle, when adding a new PE to the network all the existing PEs must be reconfigured to peer with the new PE. However, the provisioning of full meshes is not always a problem. For BGP, the full-mesh problem can be avoided using route reflectors. For MPLS, the full-mesh problem does not exist if the signalling protocol is LDP, as explained in Chapter 1. A full mesh of RSVP tunnels is a concern from a provisioning point of view, but new mechanisms such as the RSVP automesh (discussed in Chapter 12 discussing management of MPLS networks) help to address the issue.

The load on the routing protocols

Let us examine separately the protocols that run between PE and CE and the protocols that run between PEs. The CE has just one routing peering, regardless of the number of sites in the VPN (assuming that the CE is single-homed). The PE must maintain a peering with potentially each one of the CEs attached to it. As explained earlier in this chapter, this means an individual instance of the routing protocol. When the limit of protocol instances is reached on the PE, a new PE can be added to take over some of the customers. This scaling limitation on the PE affects the cost of growing the network, as we will see in the following section.

Let us now look at the protocol running between PE routers. The provider's PE routers use BGP to distribute the VPN routing information. There are several factors that allow BGP to successfully distribute this ever-growing number of VPN routes:

[3] The exception to this rule is the allocation of the RD, but even in that case, assuming a combination of IP address and allocated number, the free RD can be picked by looking at the configuration on the individual PEs.

1. BGP is designed to handle a large number of routes.
2. A PE only needs to maintain information about routes from VPNs that have sites attached to it.
3. Route reflectors reduce the number of BGP sessions a PE must maintain.
4. The route-target filtering scheme described earlier in this chapter can be used to limit the routes sent/received by a PE to only the relevant subset.

The amount of VPN-related information that must be maintained on the provider's routers

The state maintained on a given router depends on its role in the network:

- P routers do not participate in the routing exchanges for VPN routes, and the VPN traffic is forwarded across them over PE-to-PE MPLS tunnels. Thus, P routers are shielded from any VPN knowledge. The only state that gets created on these routers due to the VPN service is the state related to the PE-to-PE tunnels. This state is a function of the number of PEs and is not related to the number of VPNs or VPN routes. This statement is not necessarily correct for route reflectors, which we will discuss separately.
- PE routers do maintain the state for VPN routes, but only for the VPNs that they service. They also maintain the state for VPN sites, but only for those sites attached to them.
- Route reflectors maintain the state for VPN routes, but we have seen how this problem can be addressed via partitioning in the section dealing with route reflectors.

Thus, no router in the network is required to maintain the state for all VPN routes or sites.

The amount of bandwidth needed to carry the VPN traffic

As the number of customers and sites grows, the bandwidth required in the core for carrying the VPN traffic increases. This can be addressed by adding more capacity to the core. When the PE-to-PE tunnels are signalled using RSVP, traffic engineering may be used to obtain better utilization of the existing resources.

To summarize, we have examined the different areas that may be impacted by growing the VPN service and have found that since the BGP/MPLS VPN approach allows distributing the VPN load across several routers in the network, no single router needs to maintain the state for all the VPNs in the network. By splitting the load across multiple routers, BGP/MPLS VPNs can continue to scale as the service grows. Adding new routers implies an additional cost, so the question becomes: how does increasing the number of customers impact on the cost of the network? The following section shows how to determine whether a particular platform is adequate for its role in a VPN deployment and at which point more equipment must be installed. Readers not interested in this type of analysis can proceed directly to section 8.5, dealing with convergence times.

8.4.2 The cost of growing the VPN network

In order to see how the cost of the network changes with the addition of new customers, one has to analyze at which point it is necessary to add a new router to the network. The requirement to add a new router depends on the scaling limitations of the router. Physical properties such as CPU speed and memory, as well as logical properties such as the convergence time of the routing software, define the capabilities of a router. For a provider building an MPLS/VPN network, it is necessary to express these capabilities in VPN terms.

8.4.2.1 Scalability of PE devices

The number of (logical) interfaces that a PE can support is the first important PE scaling number, because it places an upper bound on the number of VRFs that the PE should be able to support. However, the number of VRFs supported may be smaller than the number of interfaces supported. For this reason, vendors express the scaling properties of potential PE devices by quoting the number of VRFs and number of routes per VRF that the device supports.

For example, a vendor may say that product X supports up to 1000 VRFs with 100 routes each, with RIP as the PE–CE routing protocol. This means that when device X is used as a PE, up to 1000 different VPN sites may be connected to it, and each site may advertise up to 100 routes, assuming that RIP is used for PE–CE route exchanges. When other routing protocols are used, the numbers may change. This should not come as a surprise, since different routing protocols have different resource requirements: a chatty protocol such as RIP requires more resources than plain static routes. It is important to keep in mind that, even for the same vendor, the VPN scaling numbers may depend on the PE–CE routing protocol used. When comparing numbers provided by several vendors, it is important to verify that they assume the same routing protocols on the PE–CE link. In addition, the physical properties of the platform and the software version used also impact the VPN scaling numbers a vendor advertises.

Apart from the number of VRFs and routes that a PE router supports, providers are also interested in the time it takes to bring up a router from an offline state. This time is expressed in minutes from boot-up time until the router is in a fully functional state, routing has converged, the forwarding state has been installed and the CPU utilization has gone down to normal. This time is important, since it determines the service impact the customer sites will experience every time the router is taken offline for maintenance or for software and hardware upgrades. This service impact in turn translates to the SLAs that can be guaranteed to the customer.

An often overlooked scaling dimension of a PE router is the number of routing protocol instances that it can support. If the network design calls for the use of a particular CE–PE routing protocol, the maximum number of instances supported becomes a scaling bottleneck.

The scaling properties of the PE routers in a VPN network determine the load that can be placed on them. This load is a function of the network design, in particular of the number of PE routers used. The fundamental question that a provider is faced with when building a new VPN service is whether the PEs match the load they will carry. In order to answer this question, the requirements on the PE routers must be derived from the requirements on the VPN network. Let us see how this can be done.

Requirements for a network providing a VPN service are expressed in terms of:

1. Number of customers. This is the total number of VPNs served by this network.

2. Number of interfaces (ports) per VPN. This is the average number of customer circuits or CE devices (assuming a single circuit per CE device) that belong to a given VPN.
3. Number of routes per CE. This is the average number of routes injected by each VPN CE device.[4]

The network design determines two parameters:

1. The number of PEs in the network. This is either a small number of large PEs or a large number of small PEs. Each approach has its tradeoffs in terms of network manageability and the impact of a PE failure, as well as the scale of the PE-to-PE BGP and MPLS meshes that must be maintained. In reality it is likely that a network would contain both large PEs and small PEs. Large PEs would be used for locations with high customer concentration, while small PEs would be used for locations with small customer concentration.
2. The number of interfaces in each VRF. This is a function of two factors: (a) the proximity of the PE device to the customer entry point in the network and (b) the Layer 2 technology used. For example, if Frame Relay is used, the provider can haul circuits belonging to the same VPN to a common PE. By increasing the number of interfaces in each VRF, the provider reduces the total number of VRFs required throughout the network to service a particular customer.

What we will try to do next is to see how these two network design parameters, along with the input of the network requirements, determine the load the PEs must carry. Two issues must be kept in mind when doing such an analysis:

1. The computations are based on averages. Therefore these averages should be maintained within acceptable margins of error.
2. The temptation is to design for the worst case, which results in over-provisioning of the network and increasing its cost. Here is an example of such a design for the worst case. Due to historical

[4] VPN routing information can be modeled either as a total number of routes per VPN or as the number of routes per CE device. We have chosen the latter for two reasons: (a) it correlates better with PE resource consumption and (b) it is easier to sanity check. One of the most common VPN scenarios is for the CE to interconnect with a branch office that contains a set of networks behind it. In this case there are several routes per CE and the number can be easily validated against the 'routes per CE' parameter.

reasons, some European countries have a high concentration of economical resources in one or two major cities. For instance, it is safe to assume that almost every VPN in France has circuits in Paris. Thus, assuming a naive analysis, the PE router in Paris would have to maintain the state for all VPNs. At this point, two distinctions must be made: first, the high concentration of customers in the Paris area means more PEs are used to service them and, second, whatever requirements are extracted for a PE at such a busy spot may not be applicable to other areas in the network.

Let us take an example network and see how the VPN service requirements and the design decisions determine the load on the PE routers. The requirements are: support 10 000 VPNs, with an average of 50 circuits (interfaces) per customer and 200 routes per customer site. The network design assumes 100 PEs and two interfaces in each VRF.

The first question is how many VRFs can be expected on each PE? When doing this computation, one could fall into one of two pitfalls:

1. Assume all PEs have sites from all VPNs (similar to the Paris example we saw earlier) and require support of 10000 VRFs per PE.
2. Compute the number of VRFs naively by dividing the number of VPNs by the number of PEs, and require 10000/100=100 VRFs per PE.

In both cases, the problem is that the number of customer sites in each VPN is completely left out from the computation, so the results do not reflect reality.

The number of customer interfaces affects the total number of VRFs in the network. In the example network, each customer has 50 interfaces and the design decision assumes two interfaces per VRF. Therefore, the customer's 50 interfaces translate to 25 separate VRFs, each on a different PE. In this case, a VRF translates to a single customer site (VRF=customer_interfaces/interfaces_in_VRF). The number of VPNs in the network must be multiplied by the number of sites in each VPN to obtain the total number of VRFs, yielding 10000×25=250000 VRFs in the network (VRF_total=VPN_total×customer_VRFs).

Assuming an average distribution of VRFs over PEs, each PE must service 250000/100=2500 VRFs (PE_VRFs=VRF_total/PE_number). Compare this number to the two numbers computed when disregarding the port information: it is four times smaller than the worst case computation and an order of magnitude bigger than the naive computation which implicitly assumed one site per VPN.

By inserting these formulas in a spreadsheet, one can see how different design decisions impact the requirements on the PEs. For example, increasing the number of interfaces in each VRF to five instead of two yields a total of 100000 VRFs. Thus, only 1000 VRFs need to be maintained per PE. Doing such an analysis can help evaluate different network designs, e.g. when deciding if it is cheaper to haul as many customer circuits as possible to the same PE or to add an extra PE at a particular geographical location, or when comparing the price/performance of different PE devices for a particular design. The same analysis can be applied when increasing the number of customers in the network in order to determine whether new PE routers must be installed, and how the growth in customer/routes translates to money spent on new PEs.

8.4.2.2 Scalability of route reflectors

We have seen in previous sections that route reflectors have the need for both large memory and a fast CPU. Let us take a look at some of the aspects of evaluating a device for deployment as a route reflector.

One of the most popular metrics service providers rely on is the initial convergence time. Assuming that all PEs are peering with two route reflectors for redundancy, this is the time during which there is a single point of failure in the network. Vendors often include the initial convergence time in the scaling numbers they provide to customers. The initial convergence time is given in minutes and is a function of the number of peers and the number of routes. When comparing convergence times provided by different vendors, it is important to distinguish whether the time is measured until the propagation of all routes has completed, or until the propagation has completed and the CPU utilization has returned to normal.

A more interesting analysis for a potential route reflector is to look at the anticipated load, expressed as the number of updates per second, and at the speed of the update processing on the reflector. The load is a function of the number of PEs peering with the reflector, the number of different VPNs on each PE and the number of VPN route flaps per unit of time. The speed of the update processing is measured as the time between receiving an update from one PE and propagating the update to all relevant PEs. Note that this time is not only affected by the CPU speed but also by software design decisions such as timer-based rather than event-based processing of updates.

Let us revisit the example network from the previous section: 10 000 VPNs, each with sites connected to 25 different PEs. Assuming one route change per minute per VPN, 10 000 updates are received every minute and propagated to an average of 24 PEs (assuming that route-target filtering is applied). This yields the following requirements: process 10 000 updates and generate 240 000 updates every minute. The above analysis is an excellent example of the benefits of deploying route-target filtering. Without route-target filtering, the updates are propagated to all PEs and the number of updates that the reflector must generate increases almost four times, to 990 000. From the same example it is easy to see that decreasing the number of sites by hauling more customer circuits to the same PE reduces the load on the route reflector.

An analysis such as the one above can be applied when evaluating a route reflector deployment to see whether the design can support the estimated load or whether it is necessary to split the VPN routes among several reflectors. Whatever the outcome, it is important to bear in mind that not using route reflectors at all is also an option. A full mesh of PEs with route-target filtering may be a cheaper option than deploying multiple-route reflectors. By understanding the requirements placed on the different components in a particular network design and the software and hardware features available it is possible to pick the best tradeoffs.

8.5 CONVERGENCE TIMES IN A VPN NETWORK

It is not enough to build a scalable VPN network. One must also make sure that the network meets the customers' expectation: to have the same convergence times as the alternative of buying circuits and running an IGP-based network over these circuits. When discussing convergence for VPN networks, there are two distinct scenarios: (a) a route delete/add in a customer site and (b) a failure in the provider's network, affecting connectivity between customer sites. Let us examine these separately below.

8.5.1 Convergence time for a customer route change

The propagation of a route add/delete in a customer site includes the following steps:

- Propagation of the route information between the CE and the local PE.
- Propagation of the route information between the local PE and the remote PEs. This step includes:
  - the update generation on the PE;
  - the processing and propagation of the update at the route reflector (if used);
  - the import of the route into the correct VRF on the remote PEs.
- Propagation of the route information between the remote PE and the CEs attached to it.

It is important to understand that in order to provide comparable service to an IGP-based network, the above steps must be performed event-driven, rather than timer-driven. Some implementations use periodic timers to scan for changes and process them (also known as scan timers). Such implementations are open to a maximum delay of the sum of the scan timer intervals, in addition to any processing and propagation delays that are incurred. Typically scan timers are in the orders of seconds, which can add up to tens of seconds of end-to-end delay. Therefore, when evaluating PE devices for a VPN network, it is important to examine the software behaviour in this respect in order to gauge the network-wide convergence time.

8.5.2 Convergence time for a failure in the provider's network

In the absence of mechanisms such as fast reroute, a link or router failure in the provider's core can affect the PE–PE LSPs and affect VPN connectivity. The convergence time in this case is defined as the time until the CE routers find out about the fact that a set of destinations has become unreachable. This time is made up of:

- The time it takes to detect of LSP failure and propagate the information to the PE.
- The time it takes to translate the LSP failure into a route withdrawal at the PE.

The LSP failure detection time is largely dependent on the label distribution protocol used. In the Foundations chapter (Chapter 1) we saw a scenario where LDP used in the independent control mode would yield a silent LSP failure that would never be detected. For a protocol such as RSVP, some types of failure would only be detected after the cleanup timer had expired, typically on the order of minutes. Liveness detection mechanisms (such as the ones described in Chapter 12 discussing management of MPLS networks) can help detect LSP failures within a bounded amount of time. The second part of the convergence time depends on whether the remote PE reacts to the LSP going down in an event-driven or in a timer-driven way. As explained in previous sections, for a timer-driven approach, the scan time interval adds to the total convergence time.

To summarize, VPNs are expected to give customers similar service to that of an IGP-based network. However, route propagation and failure detection may take far longer in a VPN. By understanding the interactions that take place, providers can get a better idea of the requirements from the software they deploy, as well as of the tools and the protocol choices they make.

8.6 SECURITY ISSUES

L3VPNs must provide the same security assurances as the alternative of connecting dispersed sites with circuits at Layer 2. The first security concern is the separation of traffic between VPNs. We have already seen that the L3 VPN solution has built-in mechanisms for isolation of addressing plans, routing and forwarding. However, since L3 VPNs operate over a shared infrastructure, additional concerns arise:

- Can traffic from one VPN 'cross over' into another VPN?
- Can a security attack on one VPN affect another VPN?
- Can a security attack against the service provider's infrastructure affect the VPN service?

Let us examine these separately below and see how the problem can occur and how it can be avoided.

Can traffic from one VPN 'cross over' into another VPN?

One of the most frequent configuration errors is to plug in the CE interface into the incorrect port on the PE. Thus, instead of belonging to VPN A the new site belongs to VPN B (recall that membership in a VRF is based on the interface).[5] Following such a misconfiguration it becomes possible to send traffic from one VPN to another, especially if the same addressing plan is used in both VPNs. If a routing protocol is running between CE and PE, the problem can be easily avoided by enabling authentication for the routing protocol exchanges. In the case of a misconfiguration, the routing session does not establish and routes are not exchanged between PE and CE.[6] In Chapter 12 discussing management of MPLS networks we will see a mechanism for detecting this misconfiguration even if no routing protocol is used on the PE–CE link.

Because the PE–CE link extends outside the provider's network and may cross shared-access media it is a natural target for attackers. Securing the routing protocol through authentication prevents an attacker from introducing incorrect routing information. The same can be achieved by setting up routing protocol policy on the PE. For example, the PE may limit the routes it accepts from the CE to the subnets used in the customer VPN. Setting up firewall filters on the CE–PE interface is another popular way to defend against attacks over this potentially insecure interface. For example, the filter would reject any traffic whose source address is not from the address space used in the customer VPN.

Can a security attack on one VPN affect another VPN?

Assuming that one VPN is compromised, other VPNs may be affected if the attack is such that it affects the PEs servicing the VPN. This is the case, for example, if the affected customer floods the PEs with traffic or with route advertisements. The way to protect against

[5] A similar configuration error can also happen for an L2 VPN.

[6] This works as long as different passwords are used for each session. If a default password is employed throughout, authentication will not detect the misconfiguration.

such a scenario is to limit the PE resources that a customer can consume. We have already seen that the PE can protect itself against a misbehaving CE by limiting the number of routes it accepts from the CE. Similarly, firewall filters on the PE–CE interface can rate-limit the amount of traffic that the CE can send, thus limiting the damage that an attacked VPN can inflict on other VPNs.

Can a security attack against the service provider's infrastructure affect the VPN service?

All VPN traffic is carried across the provider's core over the same infrastructure. Therefore, an Internet attack on the provider's core can impact the availability of the VPN service. To protect this infrastructure, providers conceal the core using two techniques: hiding the internal structure of the core and filtering packets entering the core. The goal is to make it hard for an attacker to send traffic to the core routers. Hiding the internal structure can be accomplished by: (a) using a separate address space inside the core (and not advertising these addresses outside the network) and (b) manipulating the normal TTL propagation rules to make the backbone look like a single hop.

The shared infrastructure over which L3 VPNs are built pose additional security challenges. We have seen just a few of the issues that can arise either from innocent misconfigurations or from malicious attacks. The responsibility for preventing such attacks cannot be placed on the provider alone or on the customer alone. Instead, it is shared between the provider and the customer.

8.7 QOS IN A VPN SCENARIO

From a QoS point of view, a BGP/MPLS VPN has to provide at least the same guarantees as a private network. Because it is sold as a premium service with QoS guarantees, the customer expectations from the QoS performance of a BGP/MPLS VPN are high.

As seen in the Traffic Engineering chapter (Chapter 2), QoS guarantees can be readily translated into bandwidth requirements. The question is, how are these requirements expressed? When discussing bandwidth requirements, two conceptual models exist:

1. Pipe model. A pipe with certain performance guarantees exists between two specific VPN sites. Therefore, the customer must

know the traffic matrix and must translate it into a set of pipes that meet these requirements. For example, for a VPN where branch offices connect to a central site, the amount of traffic between each branch office and the central site must be known. This approach is difficult to implement in practice because the traffic matrix is typically not known in advance. Furthermore, changes to connectivity (such as two branch offices starting to exchange traffic) require changing the pipe definitions.

2. Hose model. The bandwidth requirements are expressed as the aggregate bandwidth going out of and coming into a site. (This is similar to a hose because traffic is 'sprayed' from one point to multiple points.) It is much easier to define the bandwidth requirements in this case because the estimate is for the aggregate traffic rather than individual flows and the amount of traffic in/out of the site depends on its size and importance. Furthermore, such a model can easily accommodate changes to connectivity.

Once these requirements are defined, how are they implemented? In an MPLS network, a pipe can be implemented as an LSP and a hose can be implemented as a collection of pipes (a set of LSPs).

Does this mean, however, that separate LSPs must be set up for each and every VPN? The question is further complicated by the fact that within the same VPN several levels of service are offered, e.g. voice and data, requiring different QoS behaviors. Do the resources for each of the service levels come from the resources allocated to the particular VPN or from the resources allocated to the particular service level across all VPNs?

The answer depends on the goals that the provider is trying to achieve and is a tradeoff between the amount of state that must be maintained in the core of the network and the degree of control that the provider has over the different allocations. At one extreme, a set of PE–PE LSPs is shared by traffic from all service levels, belonging to all VPNs. At the other extreme, separate LSPs are set up for each service in each VPN. The number of LSPs that are set up is proportional to the number of PEs in the first case and to the number of PEs, VPNs and services within each VPN in the second case. Typically, providers deploy PE-to-PE tunnels that are shared across VPNs, because of the attractive scaling properties of this approach, but sometimes per/VPN tunnels can be set up, to ensure compliance with particular customer requirements.

How does the traffic receive its QoS guarantees as it is forwarded through these tunnels? Most providers offer QoS in IP/MPLS networks using DiffServ. In this model, customer traffic is classified at the local PE. When sending the traffic towards the remote PE over the PE–PE LSP, the EXP bits are set to ensure the correct per-hop behavior in the provider's core. However, one cannot rely on DiffServ alone to provide QoS; it is also necessary to have enough bandwidth along the path taken by the PE–PE LSP to ensure the correct per-hop behavior. This can be done by either overprovisioning the core or by setting up PE–PE LSPs with bandwidth reservations. In both cases, the assumption is that the traffic sent by the customer stays within the estimates used for the inter-PE LSP. Therefore, many providers police this traffic to ensure compliance.

In the chapter discussing DiffServ Aware Traffic Engineering (Chapter 4), we have seen that LSPs with bandwidth reservations cannot solve the problem of limiting the amount of voice traffic on links (a necessary condition to ensure bounds on jitter). To solve this problem, DiffServ-TE LSPs can be used to reserve bandwidth on a per-class basis. For example, instead of a single PE–PE LSP, two LSPs can be set up, one for voice and one for data traffic. Customer traffic is mapped to the correct PE–PE LSP based on the DiffServ classification.

8.8 MULTICAST IN A VPN

The discussion so far has focused on providing connectivity for unicast destinations in a VPN. How is multicast traffic handled? To answer this question, let us first determine what are the customer requirements for multicast traffic. From the customer's point of view, the requirement is simple: be able to forward multicast traffic between senders and receivers in different sites using the same procedures as if they belonged to a single physical network and using private address spaces. The current VPN multicast solution satisfies this requirement, but does so at a high cost to the provider. In this section we will describe the original solution to the multicast problem, as well as ongoing work in the IETF to overcome some of its limitations. This section assumes basic familiarity with PIM, P2MP LSPs and VPN concepts.

8.8.1 The original multicast solution

The original solution for multicast in BGP/MPLS VPNs is described in [DRAFT-ROSEN]. Although targeted to multicast traffic in a BGP/MPLS VPN setup, the solution departs from the VPN model described so far. Nevertheless, some of the elements of BGP/MPLS VPNs are reused: the multicast solution is still based on dedicated virtual PE routers (the VRF concept of BGP/MPLS VPNs) and customer traffic is still tunneled through the provider core between the PEs (the VPN tunnel concept of BGP/MPLS VPNs). Despite these high-level similarities, the multicast solution is very different from the unicast one. Exactly how different we will see in the remainder of this section.

To do so, let us start by listing the requirements for handling customer multicast traffic:

1. The relevant multicast trees must be established within the customer VPNs between the sources and receivers, even when these reside in different sites of the VPN. Assuming PIM is running within the customer VPN, this means that the relevant PIM state must be communicated between the sites.
2. Multicast traffic must be forwarded over the provider's core from the source to the receivers.

Figure 8.4 shows the conceptual model of traffic forwarding in a VPN. Two VPNs, VPN A and VPN B, are shown. A multicast source in site 1 of VPN A is sending traffic to receivers in sites 2 and 3 of the VPN. The traffic must be delivered from PE1 to PE2 and PE3 and from there to the appropriate CEs, CE4 and CE3. Similarly, a source in site 1 of VPN B is sending traffic to receivers in sites 2 and 3 of VPN B. The LANs in the centre of the diagram are not real, but as far as the PEs are concerned, they appear to be on a virtual LAN, and therefore a packet sourced at PE1 will reach both PE2 and PE3. Note that that this virtual LAN only connects the PEs servicing a particular VPN.

Similar to the unicast case, which is based on a peer VPN model, the distribution of the PIM state between the customer sites is also based on a peer VPN model. Rather than maintaining PIM adjacencies between the customer's CE routers, CE routers only require a PIM adjacency with their local PE router. This is similar to the unicast case, where if a routing protocol is in use, the

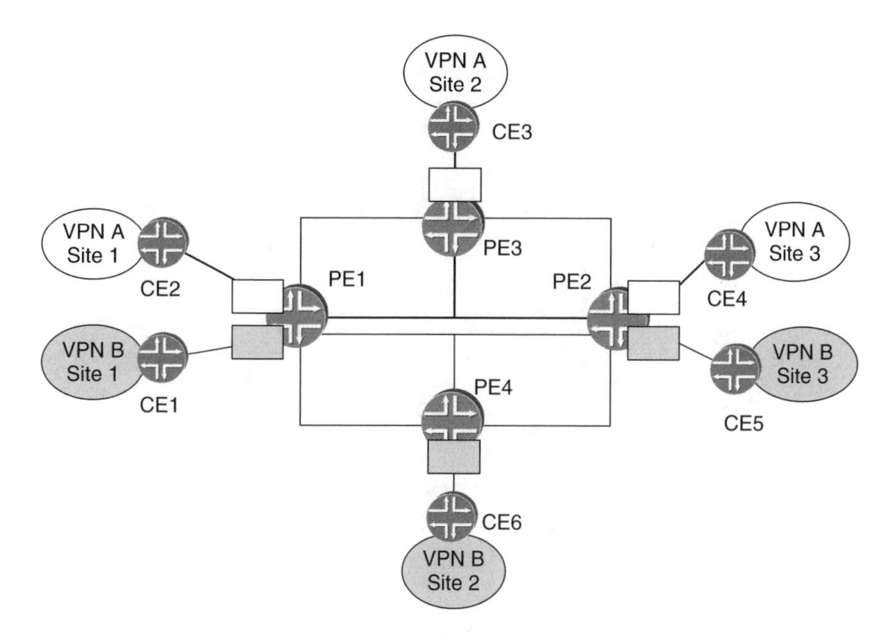

Figure 8.4 Forwarding multicast traffic in a VPN

customer's CE routers maintain an adjacency only with the local PE router. To propagate the PIM information between the PE routers and onwards to other VPN sites, PIM adjacencies are set up between the PEs that have sites in the same VPN. Because the PIM information exchanged is relevant only in the customer's VPN context, it is necessary to set up these PIM adjacencies on a per-VRF basis. These per-VRF PIM adjacencies ensure that the necessary multicast trees can be set up within each customer VPN. The need for per-VRF PIM adjacencies is in contrast to the unicast case, where a single BGP session exchanges routes belonging to all VPNs. The PIM sessions set up are shown in Figure 8.5. These VPN-specific PIM instances are referred to as PIM C-instance (where C stands for customer).

Similar to the unicast case, forwarding in the provider's core is done using tunnels between the PEs. For multicast, these inter-PE tunnels must carry traffic from one PE (the one servicing the customer site with the source) to multiple PEs (the ones servicing the customer sites containing receivers), as previously shown in Figure 8.4. This tunnel is called the multicast distribution tree (MDT). Conceptually, the MDT creates the abstraction of a LAN to which all the PEs belonging to a particular VPN are attached. This

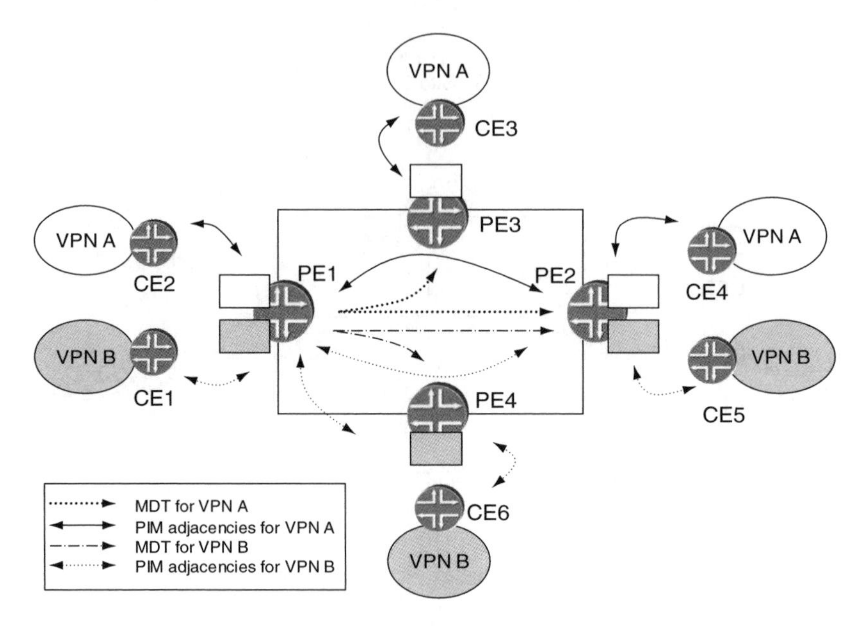

Figure 8.5 PIM sessions for a VPN

property is very important for the C-instance PIM sessions between the PEs, which can consider each other as directly connected neighbours over this LAN. The details of how the MDT is built and how it is used for forwarding cause the solution for VPN multicast to differ significantly from the BGP/MPLS VPN model.

Let us therefore take a look at some of the design decisions that the MDT forces on the solution:

1. Separate MDT per VPN. Conceptually, the MDT provides the same function for the multicast traffic as do VPN tunnels for the unicast traffic. Therefore, it is very intuitive that a separate MDT is required per VPN, connecting all the PEs with sites in the given VPN.[7]
2. Using PIM to build the MDT. The MDT is a point to multipoint tree and the PIM protocol is well suited for building such trees. The instance of PIM that is used to build the MDT runs in the provider backbone and for this reason is called a PIM P-instance (where P stands for provider). Note that this is a different

[7] We will see that more than one MDT can be used per VPN when discussing data-MDTs later in this section.

instance of PIM than the one used within the VRF to set up the intersite trees. Therefore, two levels of PIM are used, the C-instance for building the trees in each customer's VPN and the P-instance for building the MDTs in the core (note that a single P-instance builds all the MDTs).

The choice of PIM as the signalling protocol, although intuitive in the MDT discussion, forces a series of subsequent decisions. PIM uses the multicast address to identify the members in the point-to-multipoint tree it builds. Using PIM to signal the MDT means that the MDT is identified by a multicast address, with the following consequences:

1. A multicast destination address must be assigned to each VPN and manually configured on each PE. This address is called the default MDT group address or P-group address (P stands for provider).
2. Traffic must arrive at the egress PE with a destination address equal to the MDT group address. This address allows the PE to determine which VPN the packet belongs to, the same functionality achieved by the use of a VPN label in the unicast case. In the multicast case, traffic is forwarded through the core using either GRE tunnels with the MDT group address, or IP-in-IP tunnels with this destination.
3. P routers must participate in PIM for the setup of the MDT. Because multicast traffic is forwarded to the MDT multicast address through the provider core, the P routers in the core must be aware of this address, so they must participate in the PIM exchanges for building the MDT.

The multicast solution described above differs from the BGP/ MPLS VPN solution used for unicast traffic, both in the control and the data planes:

- The control protocol used to convey customer routing information between PEs is PIM. In the unicast case, BGP is used to carry the customer routes across the provider's core.
- The control protocol for building inter-PE tunnels with VPN context is PIM. In the unicast case, BGP is used to carry the VPN label which provides the PE-to-PE tunnels at the VPN level.
- Multicast traffic is forwarded through the provider core encapsulated in tunnels set up with either GRE or IP-in-IP. In the

unicast case, traffic is encapsulated in tunnels usually set up with MPLS (using either LDP or RSVP).

Let us take a look at the scalability properties of the solution so far.

Large number of PIM adjacencies

Per-VRF PIM instances are required for carrying customer PIM information across the provider core. Thus, for each L3 VPN requiring multicast service that a given PE is involved with, the PE forms a PIM C-instance adjacency with each of the other PEs servicing sites belonging to the VPN. Note that a separate adjacency is required for each VPN. Assume that a given PE services N VPNs running multicast. If each of these N VPNs has sites on M other PEs, then the PE must form ($M \times N$) PIM adjacencies. Depending on the values of M and N, this can result in a large control plane overhead, because the maintenance of each PIM session requires the periodic exchange of PIM hello messages. It is interesting to note the contrast with the unicast routing case, where in the worst case a PE has a single routing protocol adjacency (a BGP session) with each remote PE, or when route reflection is in use, perhaps two to four adjacencies with route reflectors.

Per-VPN state is maintained on the P routers

P routers participate in the PIM P-instance exchanges necessary to set up the MDT across the core. Because there is at least one MDT per VPN, the amount of state maintained by the P router is equal to at least the number of VPNs that have multicast support. Recall that in the unicast case the P routers did not need to maintain any per-VPN routing state.

Manual configuration of the MDT group address

The tunnel multicast destination address tells the receiving PEs which VPN the packet belongs to. For this purpose, the mapping between the multicast destination address and the VPN must be manually configured on each PE.

Potentially large number of trees in the core

Each MDT can carry traffic from multiple multicast groups, as long as those groups belong to the same VPN. The detail of how many trees are formed depends on the implementation. Assuming the tree carries traffic for all multicast groups within a VPN, there are two options:

1. For each VPN, there is a single multicast tree rooted at a rendezvous point (RP). If there are N VPNs in the network, then there are N multicast trees within the SP part of the network.
2. For each VPN, there is a multicast tree rooted at each PE. Hence if there are N VPNs in the network present on each of M PEs, there are $(M \times N)$ multicast trees present in the SP part of the network.

Note that in the discussion so far we have always assumed that each multicast tree extends to all the PEs that service a site of the given VPN. This can be wasteful of bandwidth because each PE in a VPN receives all the multicast traffic of that VPN, even if it has no CEs attached that are interested in the multicast groups in question.

One solution to this is the data multicast distribution tree ('data-MDT') scheme in which a multicast group having a high volume of traffic has its own dedicated multicast tree that extends only to those PEs attached to customer sites that contain active receivers for the multicast group in question. However, these additional trees increase the amount of multicast state in the core even more. Not only do the core routers carry the per-VPN state, they also carry the state for those individual multicast groups within a VPN that have their own dedicated multicast tree.

Having seen the basic operation and the limitations of the existing multicast solution, let us now take a look at the ongoing efforts in the IETF to improve it. Some of the topics discussed in the next section are documented in [VPN-MCAST] and [MCAST-ENCAPS].

8.8.2 Improving the existing multicast solution

A fundamental problem with any scheme for carrying VPN multicast traffic over the service provider network is that it is not possible to simultaneously satisfy the following:

1. Maintain the minimum state in the core of the SP network.
2. Achieve bandwidth efficiency, by not sending traffic for a multicast group to PEs that do not need to receive it and by not using ingress replication.

These two goals are conflicting because goal 2 implies having knowledge about the individual multicast groups in the core of the network. However, it is useful to be able to provide the capability to service providers to choose where between the two extremes they wish to operate. In order to address this, and to address the scaling issues described in the previous section, work is underway in the IETF in several areas.

Reduction in the PIM state

In order to deal with the scaling issue introduced by the potentially large number of PIM adjacencies required between PE routers, the following improvements are proposed:

1. Reduce the amount of PIM overhead by minimizing the exchange of PIM hellos.
2. Reduce the number of periodic join messages by adding a refresh reduction scheme to the PIM protocol.
3. Use BGP to reduce the number of PIM adjacencies. Instead of using per-VPN PIM adjacencies between PEs to convey join/prune messages arriving from the customer's CE routers, use BGP. This is accomplished by defining a new BGP subsequent address family (SAFI). Here is how it works. When receiving a PIM join message from an attached CE, the PE converts it to a BGP update containing the multicast group and, if the join is an (S, G) join, the multicast source. The PIM join is advertised to the remote PEs as a BGP update. The attraction of this scheme is that the same BGP sessions and same BGP infrastructure used for the unicast routing exchanges can be used to carry this multicast information. For example, if route reflectors are in place, they can also be used for signalling multicast reachability as well as unicast reachability.

Reduction in the number of multicast distribution trees

The goal is to allow the service provider some flexibility in trading off bandwidth efficiency against the amount of multicast state in the core of the network. Several alternatives are proposed, listed below in ascending order of the amount of multicast states required within the core of the SP network:

1. No multicast trees in the core. Rather than using multicast trees for distribution, each PE uses point-to-point tunnels between pairs of PEs and replicates the multicast traffic on each of these tunnels.
2. A single shared multicast distribution tree. All VPNs share the same distribution tree, resulting in a single tree in the core. This reduces the amount of state in the core of the network, at the expense of potentially wasting bandwidth, by sending traffic to PEs that do not need to receive it. For example, if a network supports two VPNs, VPN A and VPN B, and routers PE1, PE2 and PE5 service sites belonging to VPN A and PE routers PE1, PE2, PE3 and PE5 service sites belonging to VPN B, then multicast traffic arriving at PE1 from a site in VPN A would be sent to PE2, PE3, PE4 and PE5. This would be the case even though PE4 definitely does not need to receive the traffic, not having any site of VPN A attached to it.
3. Multiple shared multicast distribution trees. An alternative to using a single distribution tree is to have multiple distribution trees, each carrying a different subset of VPNs. One criterion for sharing trees could be a similar geographical distribution, which implies an overlap in the PEs that service the VPNs in question. Yet another alternative would be to have multiple distribution trees, each carrying a subset of multicast groups. Therefore, from a given VPN, some multicast groups could be carried on one tree, other multicast groups on another tree and so on, regardless of which VPN the groups belong to. An example of where this is useful is if a service provider has POPs in key major cities and other smaller cities. If several VPNs have some multicast groups that only have members in the major cities, it may be advantageous to have a multicast tree dedicated to serving those particular multicast groups.

Recall from the previous section that the multicast distribution tree in the core was used to identify which VPN traffic belongs to. In the cases where multicast traffic from multiple VPNs share the same distribution tree, a method is needed for each receiving PE to identify the VPN to which a packet arriving on a multicast distribution tree belongs. This is achieved by using an MPLS label. Who assigns this label?

If the label is allocated by the egress routers, then the egress routers need to agree on which label value to use, because all the egress routers must receive identical replicas of the packet, including the label value. Having a scheme for negotiating the value between all egresses is cumbersome. Instead, the role of picking the identifying MPLS label is given to the ingress router. This method is called upstream label allocation, because it is the router upstream of the direction of the traffic flow that allocates the label (and the traffic flows downstream towards the egress), and is documented in [UPSTREAM-LABEL].

Having the ingress allocate the label is not without its challenges. Assume that two VPNs, VPN A and VPN B, exist in the network. For VPN A, the source is in a site attached to PE1 and the receivers to PE3 and PE4. For VPN B, the source is serviced by PE2 and the receivers by PE3 and PE5. Assume that by chance both PE1 and PE2 pick label L1 as the identifying label. When traffic arrives at PE3 over the shared multicast distribution tree, with label L1 there is no way to distinguish whether the traffic belongs to VPN A or to VPN B. To avoid this problem, a mechanism must be set in place to uniquely identify the labels within the context of the neighbor who allocated it. Such mechanisms are currently under discussion in the IETF.

Use of P2MP LSPs

The reason for the multicast distribution trees in the core is to forward traffic from one source to multiple destinations. As an alternative to being PIM-based, the multicast distribution trees can be replaced by P2MP LSPs, discussed in the Point-to-Multipoint LSPs chapter (Chapter 6). As with the PIM-based trees, P2MP LSPs can either be rooted at each PE or at a central point. In the latter case, a point-to-point tunnel is used to carry traffic from each PE to the central point.

These alternatives are illustrated in Figure 8.6. Figure 8.6(a) shows a P2MP LSP rooted at PE1 which extends to all the other PEs. Each of the other PEs also have a P2MP LSP extending to the other PEs. These P2MP LSPs are not shown in the diagram. Figure 8.6(b) shows the centralized scheme in which a P2MP LSP is rooted at P3 and extends to all the PEs in the network. To send traffic into the P2MP LSP, PE1 has a unicast tunnel (the dotted line in the diagram) extending from PE1 to P3. Similarly, each of the other PEs has a unicast tunnel to P3, also not shown in the diagram.

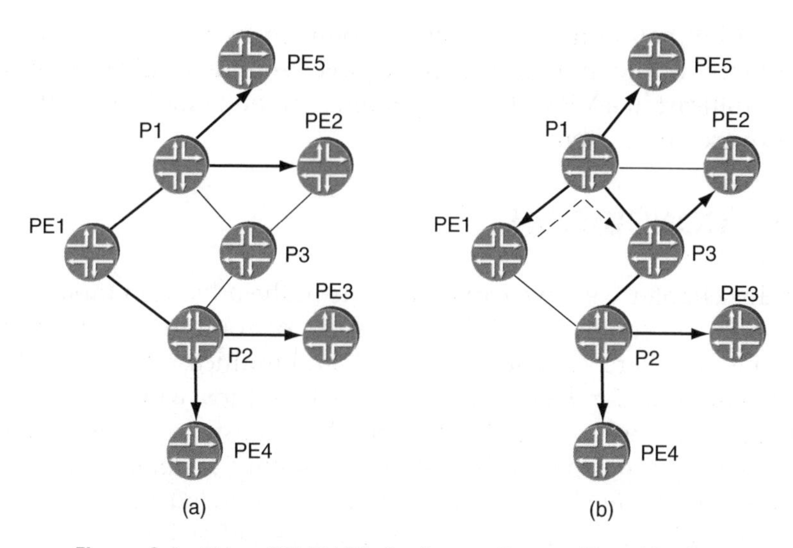

Figure 8.6 Using P2MP LSPs for forwarding multicast traffic

By using P2MP LSPs, the same protocol, namely RSVP or LDP, can be used to signal MPLS tunnels for the purpose of carrying multicast traffic as for unicast traffic. Also, if BGP is used to convey membership information of multicast groups as described at the beginning of this section, then the same signalling protocol is being used for multicast and unicast reachability. This results in a reduction in the number of protocols being used within the SP core.

Using RSVP as the signalling protocol for the P2MP LSPs brings several advantages:

- Ability to do bandwidth reservations for the P2MP LSPs.
- Precise control over the path followed by each of the sub-LSPs within the P2MP LSP.
- Flexible optimization options in the core. For example, the service provider can choose to optimize bandwidth consumption by creating a minimum cost tree or can optimize latency by creating a shortest-path tree.
- Traffic protection using MPLS fast reroute. The ability to protect multicast traffic using MPLS fast reroute is especially useful if the L3 VPN customers already enjoy such protection for their unicast traffic and expect equivalent traffic protection in the multicast case.

A VPN solution would not be complete without support for multicast traffic. In this section we have seen the original solution for multicast in a VPN, its scaling limitations and potential solutions for overcoming them.

8.9 CONCLUSION

In this chapter we have explored some of the advanced topics that arise in the context of L3 VPN: the use of route reflectors, VPN scaling numbers, security issues, QoS and multicast. However, the discussion so far has been restricted to setups where the VPN spanned a single AS and where the VPN customers were enterprises (implying that each customer has only a small number of routes). In the next chapter we will look at more advanced setups, where VPN customers may themselves be either Internet service providers or VPN providers and where a single VPN may span across several ASs.

8.10 REFERENCES

[DRAFT-ROSEN] E. Rosen, Y. Cai and E. Wijnands, 'Multicast in MPLS/BGP IP VPNs', draft-rosen-vpn-mcast-08.txt, replaced by [VPN-MCAST]

[IBGP-PE-CE] P. Marques, R. Raszuk and L. Martini, 'RFC2547bis networks using internal BGP as the PE-CE Protocol', draft-marques-l3vpn-ibgp-00.txt (work in progress)

[MCAST-ENCAPS] T. Eckert, E. Rosen, R. Aggarwal and Y. Rekhter, 'MPLS multicast encapsulati ons', draft-rosen-mpls-multicast-encaps-00.txt

[ORF] E. Chen and S. Sangli, 'Address prefix based outbound route filter for BGP-4' draft-ietf-idr-bgp-prefix-orf-01.txt (work in progress)

[OSPF-2547-DNBIT] E. Rosen, P. Psenak and P. Pillay-Esnault, 'Using an LSA options bit to prevent looping in BGP/MPLS IP VPNs', draft-ietf-ospf-2547-dnbit- 04.txt (work in progress)

[RT-CONSTR] P. Marques *et al.*, 'Constrained VPN route distribution', draft-ietf-l3vpn-rt-constrain-02.txt (work in progress)[8]

[UPSTREAM-LABEL] R. Aggarwal, Y. Rekhter and E. Rosen, 'MPLS upstream label assignment and context specific label space', draft-raggarwa-mpls-upstream-label-00.txt (work in progress)

[VPN-MCAST] E. Rosen and R. Aggarwal, 'Multicast in MPLS/BGP IP VPNs', draft-ietf-l3 vpn-2547bis-mcast-00.txt (work in progress)

[VPN-OSPF] E. Rosen, P. Psenak and P. Pillay-Esnault, 'OSPF as the Provider/Customer Edge Protocol for BGP/MPLS IP VPNs', draft-ietf-l3vpn-ospf-2547–04.txt (work in progress)

8.11 FURTHER READING

[2547-bis] E. Rosen and Y. Rekhter, 'BGP/MPLS IP VPNs', draft-ietf-l3vpn-rfc2547bis-03.txt, soon to become an RFC (currently in the RFC editor queue)

[ADV-MPLS] V. Alwayn, *Advanced MPLS Design and Implementation*, Cisco Press, 2001

[L3VPN] http://ietf.org/html.charters/l3vpn-charter.html

[MPLS-TECH] B. Davie and Y. Rekhter, *MPLS Technology and Applications*, Morgan Kaufmann, 2000

[MPLS-VPN] I. Peplnjak and J. Guichard, *MPLS and VPN Architectures*, Cisco Press, 2000

[RFC2918] E. Chen, *Route Refresh Capability for BGP-4*, RFC2918, September 2000

[8] At the time of this writing, this draft had been approved for progression to RFC status, so by the time this text is published, the document may be an RFC.

[RFC4111] L. Fang, *Security Framework for Provider-Provisioned Virtual Private Networks (PPVPNs)*, RFC 4111, July 2005

[VPLS-MCAST] R. Aggarwal, Y. Kamite and L. Fang, 'Multicast in VPLS', draft-raggarwa-12 vpn-vpls-mcast-00.txt (work in progress)

[VPN-SECURITY] M. Behringer, 'Analysis of the security of BGP/MPLS IP VPNs', draft-behringer-mpls-secutiry-10.txt, soon to become an RFC (currently in the RFC editor queue)

9

Hierarchical and Inter-AS VPNs

9.1 INTRODUCTION

In the previous two chapters we have seen the basic operation of BGP/MPLS L3 VPNs, where a service provider offers a VPN service to an enterprise customer as replacement for the mesh of circuits connecting geographically dispersed locations. The problem of connecting geographically dispersed locations is not unique to enterprise customers. Carriers may have similar problems, especially following an acquisition of a new network from a different carrier. In this case, connectivity is needed to the new network and could be (and sometimes is) accomplished by buying L2 circuits. However, just like the enterprise case, it is possible to also connect the remote locations via an L3 VPN service.

In the following sections we will see scenarios where the VPN customers are themselves Internet service providers (ISPs) or VPN providers and they obtain backbone service from a VPN provider who acts as a 'carrier of carriers' [2547-bis]. An important thing to note in the context of a 'carriers' carrier' scenario is the fact that all sites of a customer who is a carrier belong to the same AS. We will discuss inter-AS solutions in a separate section afterwards, because the concepts introduced in the carriers' carrier discussion will facilitate the understanding of the inter-AS case.

MPLS-Enabled Applications: Emerging Developments and New Technologies Ina Minei and Julian Lucek

9.2 CARRIERS' CARRIER – SERVICE PROVIDERS AS VPN CUSTOMERS

The biggest challenge in providing VPN service to customers who are themselves providers is the sheer number of routes that may be advertised from each site: the entire routing table when the customer is an Internet provider and the total number of VPN routes in the customer's network when the customer is a VPN provider. A large number of customer routes from each such customer places a significant burden on the (carriers' carrier) network, both in terms of the memory needed to store the routes at the PEs and in terms of the resources necessary to send and receive advertisements every time any of these routes flap.

In order to be able to scale the solution and support a large number of carrier customers it is necessary to shield the VPN provider (carriers' carrier) from the carrier-customer's routes. The idea is to split the load between the customer and the provider as follows:

1. The carrier-customer handles the route advertisements between the sites, via IBGP (internal BGP) sessions between routers in the different sites. The routes exchanged in these sessions are called 'external routes' because they are not part of the carrier-customer network. They may either be Internet routes or they may be VPN routes belonging to the carrier-customer's own VPN customers. For example, Figure 9.1 shows an ISP as a VPN customer. In this context, prefix 20.1/16 learned over the EBGP peering is considered to be an external route.
2. The carriers' carrier VPN provider builds the connectivity necessary to establish the BGP sessions between the customer sites. In particular, the loopback addresses of the BGP peers in the carrier-customer's sites are carried as VPN routes. These addresses are the BGP next-hops for the external routes exchanged in the customer's iBGP sessions and are called 'internal routes' because they are part of the carrier-customer's own network. For example, in Figure 9.1, ASBR1's loopback address is considered an internal route, because it is part of the carrier-customer's network.

This approach is different from the one taken with enterprise customers, where the advertisement of routes from site to site is the responsibility of the provider and where routing peerings

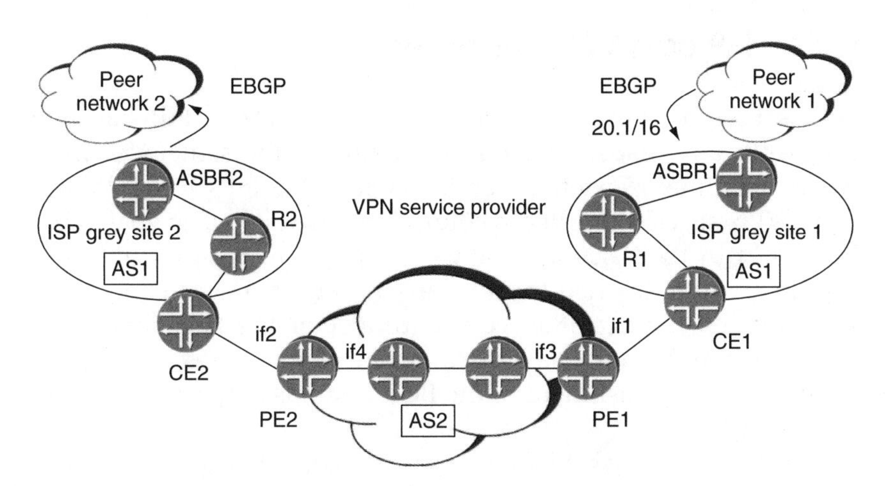

Figure 9.1 An Internet service provider as a VPN customer

between customer sites are not necessary. Note, however, that this fundamentally different approach is catering to a fundamentally different type of customer: one for which routing and route distribution are his or her daily business and who would be running BGP sessions between routers anyway.

To summarize, in a 'carriers' carrier' environment, the backbone VPN provider has two tasks: (a) to facilitate the establishment of the BGP sessions between customer sites by advertising the internal routes and (b) to permit forwarding of traffic to the external destinations learned via these BGP sessions. In the following sections we will see how this is accomplished, using the following principles:

1. Use tunneling to forward traffic across nodes that do not have routing information for the packet's final destination.
2. Use a label to identify the VPN on the PE.
3. Use the next-hop in the BGP advertisement to provide coupling between the VPN routes and the routing to the remote PE.

Let us now look at the two main carriers' carrier (CsC, also known as carrier of carriers or CoC) scenarios. We will first look at the case where the customer carrier is an ISP that only carries Internet routes and does not offer the L3 VPN service to its end customers. We will then look at the case where the customer carrier offers L3 VPN service to its end-customers.

9.2.1 ISP as a VPN customer

Let us take a look at the ISP in Figure 9.1. The ISP has two geographically dispersed sites (belonging to the same AS), from where it maintains external peerings with other ISPs. The ISP buys an L3 VPN service in order to connect the two sites. The goal of the ISP is to run IBGP sessions between routers in the two sites and exchange routes learned from the external peers in each one of the sites. In this context, it is very natural that the backbone VPN provider should not need to carry the ISP's routes, but rather facilitate the establishment of the BGP sessions between routers in the two sites.

In order to establish a BGP session, there must be connectivity to the BGP peer's loopback addresses. This can be easily accomplished by advertising the loopback addresses as VPN routes, using the same mechanisms we have seen in the L3 VPN Foundations chapter (Chapter 7). Once these addresses are reachable from both sites, the BGP session can establish. The conceptual model of the route exchanges is depicted in Figure 9.2 for an IBGP session between ASBR1 and ASBR2. The routes that are exchanged as VPN routes are the loopback addresses of the routers between which the IBGP session will be established, ASBR1 and ASBR2. The Internet routes from the customer sites are exchanged over

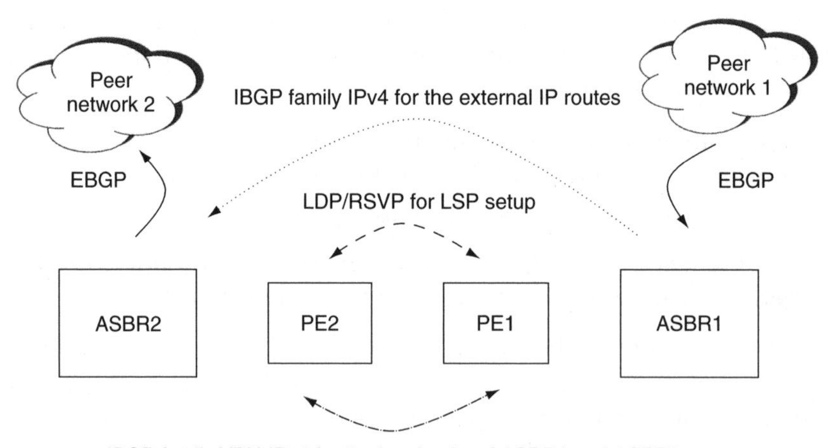

Figure 9.2 Conceptual model of the route exchanges for ISP as a VPN customer

this IBGP session between ASBR1 and ASBR2 and thus are not part of the VPN. The BGP next-hop of such an Internet route is the address of the remote end of the BGP session over which the route was learned.

Figure 9.3 shows the route advertisements that take place. The figure only shows the information relevant to the discussion, in particular:

- The advertisements are shown in one direction only, from site 1 towards site 2.
- The figure focuses on a single route learned from an external peering, 20.1/16, learned at ASBR1.
- For simplicity, a single BGP session is shown between the routers in the two sites, namely the session between ASBR1 and ASBR2.
- The full mesh of IBGP sessions between all the routers in each site is not shown. In particular, remember that although not shown in the picture, IBGP sessions exist on the CEs.

Here are the route advertisements that take place when setting up the BGP session between the ASBRs and exchanging the external routers (we will see later on that these are not the only route advertisements):

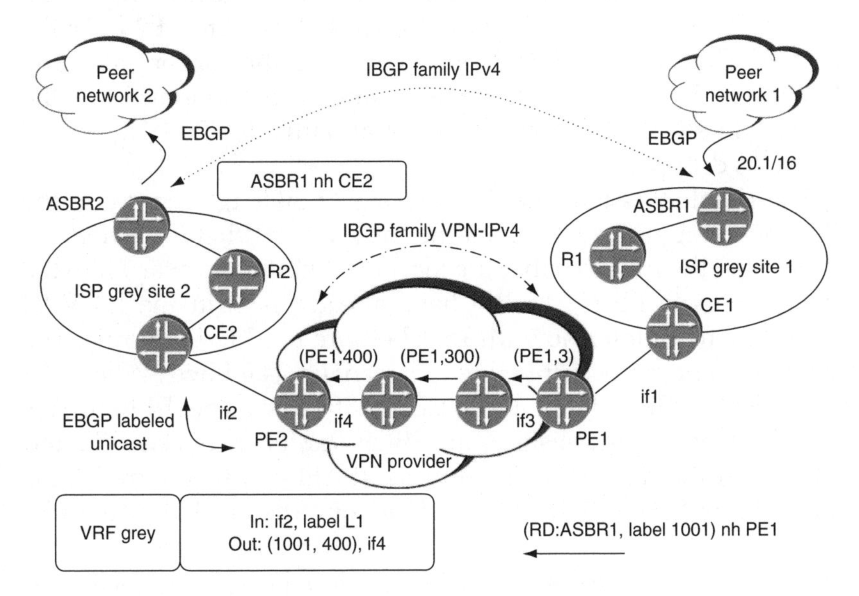

Figure 9.3 The route advertisements for ISP as a VPN customer

1. A label switched path is set up between PE1 and PE2, using either LDP or RSVP.
2. The loopback of ASBR1 is advertised as a VPN route in the same way as we have seen in Chapter 7. As a result:
   - At the PEs, in the VRF associated with this ISP customer, there is a route for ASBR1's loopback.
   - The route for ASBR1's loopback is advertised to CE2 and CE2 advertises it with itself as the next-hop to all routers in site 2.
3. An IBGP session is established between ASBR1 and ASBR2[1] and routes learned from external peers are advertised between sites. In particular, ASBR2 learns the route 20.1/16, with BGP nexthop ASBR1.
4. The route is advertised via IBGP to all the routers in site 2; thus all routers in site 2 have a route for destination 20.1.1.1 with next-hop ASBR1. The route to ASBR1 was advertised as a VPN route and was advertised to all routers in site 2 by CE2. (Thus, the BGP next-hop of the route for ASBR1 is CE2.)

Let us take a look at the solution so far and investigate a problem that happens when attempting to forward traffic from ASBR2 to destination 20.1.1.1. All routers in site 2 have knowledge of this destination, with next-hop ASBR1, which was learned from CE2. The traffic arrives at CE2 and is forwarded as IP to PE2 over the CE2–PE2 interface. PE2 performs a lookup in the appropriate VRF. However, the VRF only contains routes for the loopbacks of the routers in site 1 and does not have an entry for 20.1.1.1, so the packet is dropped.

The problem is that the PE only has knowledge regarding the BGP next-hop of the route, not the route itself. What is needed is a way to tag the traffic so that the local PE (PE2) can forward it to the correct remote PE (PE1). We have already seen in the L3 VPN Foundations chapter how MPLS labels are used to tag traffic. The same concept can be applied in this scenario as follows. When PE2 advertises ASBR1's loopback to CE2 it attaches a label L1 to it. This can be done by establishing an EBGP session between PE2 and CE2, with the family labeled-unicast (SAFI 4, also referred to as labeled-inet [RFC3107]). When advertising the labeled route, PE2

[1] This assumes that the route for ASBR2 is known in site 1. The figure does not show the route exchange in this direction.

installs the forwarding state, swapping the label L1 to the VPN label that it received from PE1. When CE2 receives the advertisement for the labeled route for ASBR1, it installs the forwarding state which pushes label L1 to the traffic destined for ASBR1 before forwarding it to PE2. Figure 9.4 shows the conceptual model of forwarding traffic from CE2. At CE2, label L1 is pushed on incoming IP traffic which has ASBR1 as the next-hop. At PE2, label L1 is swapped for the VPN label advertised by the remote PE (PE1) for the VPN route for ASBR1. The label for the next-hop of this VPN route (PE1) is pushed on top of the VPN label, just as for the normal VPN scenario. Traffic arrives at PE1 with the VPN label and can be forwarded towards CE1.

To summarize, in order to extend the LSP between the CEs, BGP is used on the PE–CE link and advertises a label along with the prefix. Conceptually, what is done is to use BGP to extend the PE–PE MPLS tunnel all the way to the CE, thus shielding the PEs from knowledge about external routes. The LSP is made up of different segments, in this case a BGP-advertised label on the PE–CE segment and the VPN tunnel stacked on top of the LDP/RSVP transport tunnel on the PE–PE segment. The segments are 'glued' together by installing the forwarding state, swapping the label

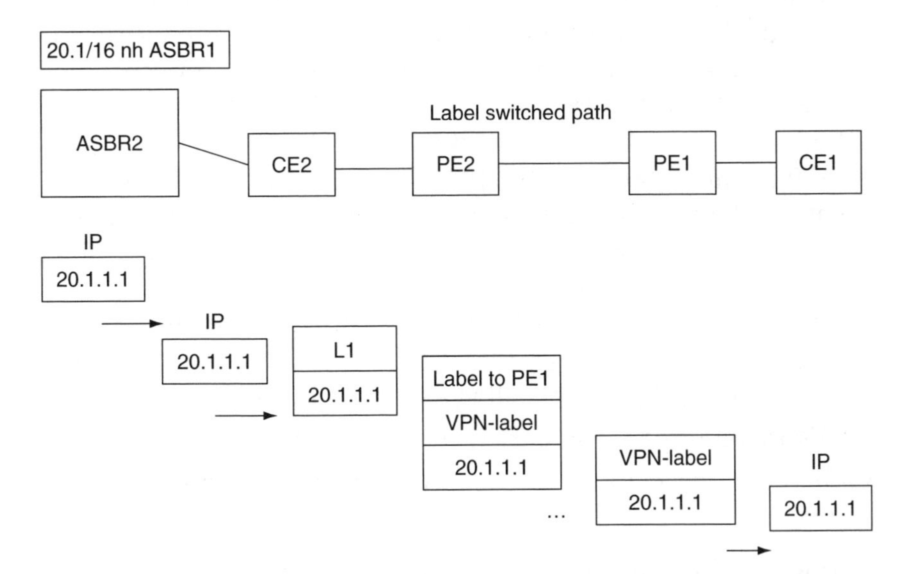

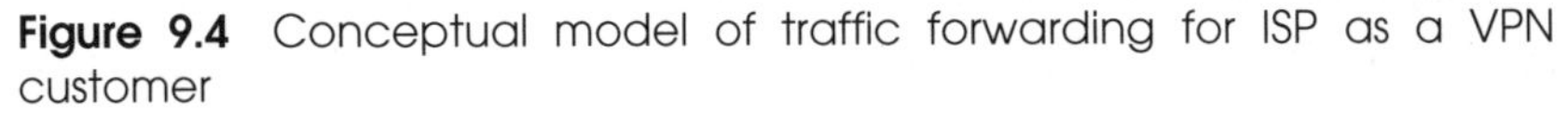
Figure 9.4 Conceptual model of traffic forwarding for ISP as a VPN customer

from one segment to the label from another segment. This is an important concept that will be applied throughout the remaining sections.

A potential security issue arises in this scenario if the label advertised in BGP by the PE for a particular route can be guessed (spoofed) by a different customer attached to the same PE. This is only a concern if the forwarding state on the PE (swapping between the assigned label and the VPN label) is not stored on a per-VPN basis. Some implementations solve this problem by maintaining separate MPLS tables per VPN; others may do so by having the PE reject labeled traffic arriving on any interface except the one over which the label was advertised.

Let us summarize the key properties of the solution:

- The ISP's Internet routes are advertised via an IBGP session between routers in the ISP's sites. These routes are external routes, outside the ISP's own network.
- All routers within each customer site must keep the routing state for the external routes and a full mesh of IBGP sessions is required within each site.
- The VPN provider carries as customer VPN routes only routes that are internal to the ISP's network (in the discussion so far, these were loopback addresses, but in practice both loopbacks and interface addresses are exchanged). As a result, the VRF on the PE maintains a small number of routes for the VPN corresponding to this ISP. The routes in the VRF are the BGP next-hops for the routes exchanged over the customer's IBGP sessions.
- Traffic cannot travel as pure IP between CE and PE since the PEs have no knowledge of the customer routes. The BGP next-hop is the glue that ties the customer routes and the forwarding information on the PE.
- MPLS tunnels are necessary between the CEs. The tunnels are made up of several segments, CE–PE, PE–PE and PE–CE, and are glued together by installing a swap state for the labels.
- CEs are required to support MPLS and MP-BGP.
- There is a need to protect the PE from the possibility of label spoofing.

This type of setup is more commonly seen when the backbone provider and the ISP are different divisions of the same company.

Another option for providing the same service is to connect the ISP sites at layer 2, rather than implementing the layer 3 solution described above. The arguments for and against each of these approaches are the same as the ones put forth in the comparison of overlay and peer VPN models in the chapter discussing basic L3 VPN functionality (Chapter 7).

9.2.2 VPN service provider as a VPN customer – hierarchical VPN

The second type of carrier-customer supported by the 'carriers' carrier' scenario is a VPN provider. For readability purposes, in order to distinguish between the VPN provider who acts as a customer and the one who acts as a provider, we will refer to the providers in Figure 9.5 by the names given to them in the figure. Figure 9.5 shows a VPN provider, provider 'grey' with two geographically dispersed sites, sites 1 and 2, both belonging to the same AS. Provider grey services two VPN customers, customer red and customer blue, with sites attached to PEs in both sites 1 and 2. The goal of provider grey is to run IBGP sessions between its PEs and advertise the VPN routes for customer red and

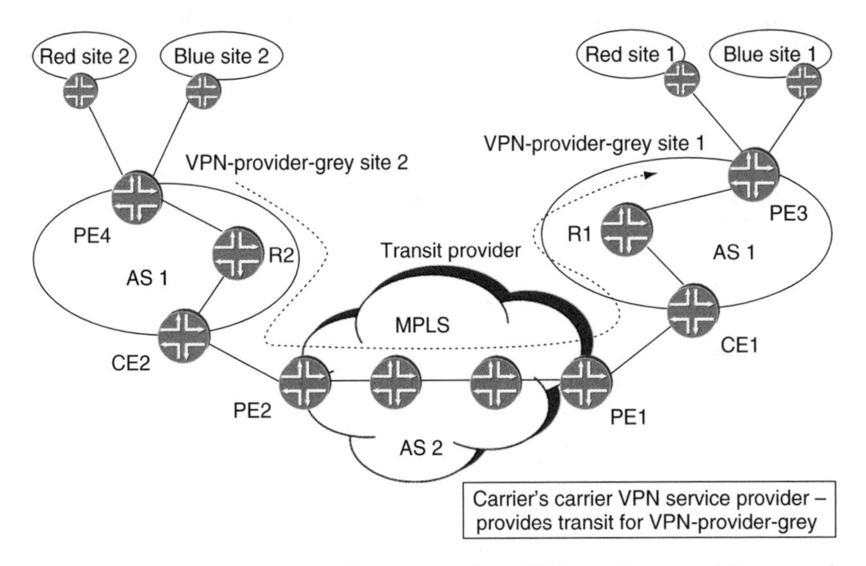

Figure 9.5 A hierarchical VPN where the VPN service provider grey is himself a VPN customer

customer blue. For this purpose, provider grey buys a VPN service from a 'carriers' carrier' VPN provider. In this context, it is natural that this carrier should not need to carry the routes of provider grey's customers, but rather to facilitate the establishment of the BGP sessions that exchange these routes. The conceptual model of the route exchanges is depicted in Figure 9.6.

This scenario is very similar to the one we saw in the previous section, with the difference that the routes exchanged over the IBGP sessions between the customer's PEs are VPN-IP routes rather than IP routes. In order to forward the (labeled) VPN traffic, a label-switched path is required between the customer's PEs. In the previous section we have seen how such a path can be built between the customer's CE routers by running a labeled-inet (SAFI 4) EBGP session between the provider's PE and the customer's CE routers. To extend the label-switched path to the customer's PE routers one can take the same approach and run a labeled-inet IBGP session between the customer's CE and PE routers. Figure 9.7 shows a conceptual model of the advertisement for PE3's loopback between provider grey's sites.

The reachability information for PE3's loopback is advertised as a VPN route from site 1 to site 2, as shown in Figure 9.7. At the VPN provider's remote PE (PE2), the route is advertised as a labeled route to CE2, with label L1, just as in the previous section, and the forwarding state is installed to swap label L1 to the VPN label. At CE2, the route is advertised with label L2 and the forwarding state

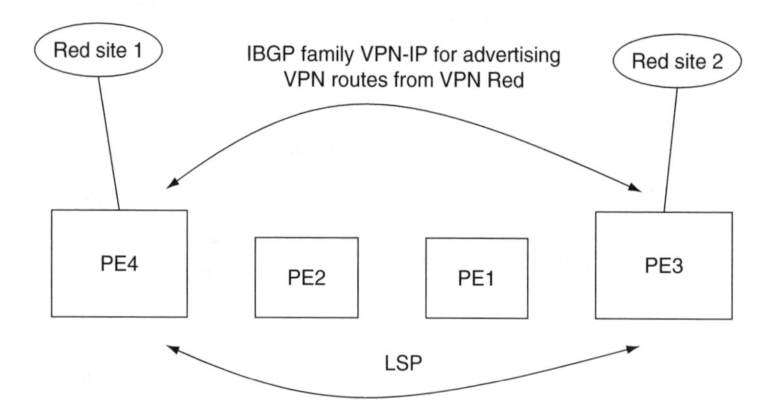

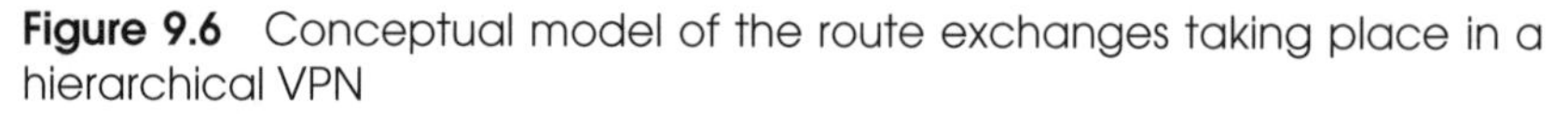

Figure 9.6 Conceptual model of the route exchanges taking place in a hierarchical VPN

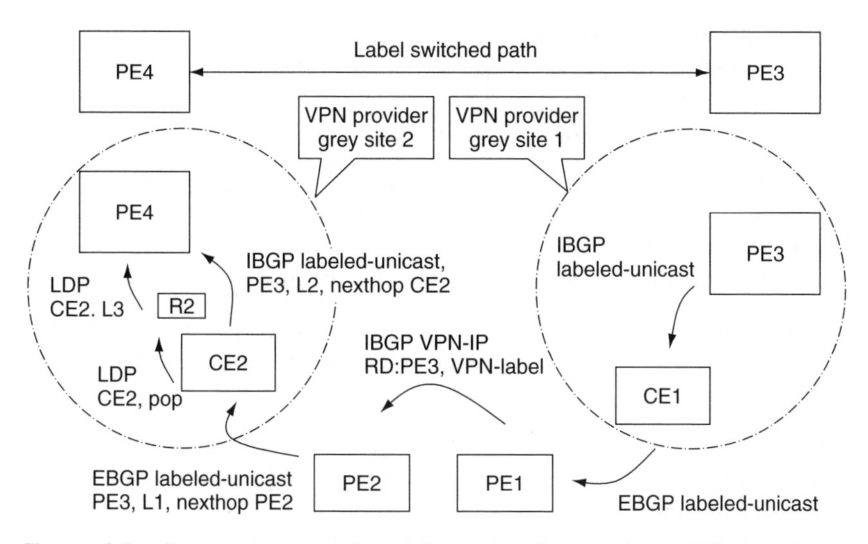

Figure 9.7 Conceptual model of the advertisement of PE3's loopback between the provider grey's sites

is installed to swap L2 to L1. At PE4, the route for PE3 is received with label L2 and next-hop CE2. The forwarding state is installed to push label L2 to traffic with BGP next-hop of PE3 and forward it towards CE2. Since this is labeled traffic, it is required that CE2 be reachable via an LSP. This LSP may be built with LDP, RSVP or even BGP. Assuming that LDP is running in site 2, label L3 exists at PE4, advertised by R2 for CE2's loopback. Thus, to forward traffic to PE3 from PE4 two labels must be imposed: the top label is the label for CE2's loopback, L3, and the bottom label is the label identifying the destination PE3, L2.

When traffic arrives from customer red's VPN attached to PE4 it must be forwarded towards PE3, with the following labels imposed: the bottom label is the VPN label that was advertised via the BGP session between PE3 and PE4 (not shown in Figure 9.7) and the top labels are the labels that carry the traffic to PE3, labels L2 and L3. Thus, a three-label stack is pushed at PE4.

So far we have not discussed the BGP sessions shown in Figure 9.7 for site 1. In principle, from a technical point of view, there is no need to advertise PE3's loopback as a labeled route in BGP in the local site (site 1), and therefore the BGP advertisement for PE3's loopback can be an unlabeled IPv4 route. However, note that for the sake of simplicity in Figure 9.7 the routing exchanges are shown in one direction only, from site 1 towards site 2. However, a symmetrical exchange is happening in the opposite direction from site 2 towards site 1. Thus, a BGP session for the family labeled-inet (SAFI 4) is necessary between PE3 and CE1 for carrying the label for PE4's loopback. In order to maintain a single BGP session rather than two, many vendors recommend advertising a null label for the route in the local site.

At this point let us stop and make the following observations:

- It is required to run MPLS within the customer's sites. In the example above, it was required to have a label-switched path between the customer CE and PE devices (CE2 to PE4).
- It is possible to isolate the knowledge regarding the addresses in the remote site to the routers at the edge of the site (routers CE2 and PE4 in Figure 9.7). Therefore, the IGP in site 2 need not carry information about the addresses of routers in site 1.

Note that the solution presented above solves the problem of distributing a label for PE3's loopback by using two label distribution protocols within the remote site, site 2: one protocol (BGP) for PE3's loopback and another one (LDP) for CE2's loopback. An alternate approach is to use LDP only. The idea is to configure LDP to advertise a label for an FEC corresponding to PE3's loopback. Remember from the Foundations chapter (Chapter 1) that in order for the label-switched path to establish, the FEC must also be present in the IGP. Thus, at CE2, PE3's loopback is injected in both LDP and the IGP. As a result, a single label is pushed at PE4 in order to reach PE3. When VPN traffic belonging to customer red is forwarded, it receives a two-label stack, the label to reach PE3 and the VPN label appropriate to the red VPN. Note that LDP is only distributing labels *within* each site of AS1. BGP is still used between AS1 and AS2.

Using LDP instead of BGP for label distribution has several advantages:

1. If LDP is already running, there is no need for an additional protocol.
2. Fewer labels are imposed at the time the packet is forwarded; in this case, two labels are required instead of three. This used to be an important consideration in older hardware implementations, which had limitations handling deep label stacks.

The main disadvantages of using LDP is that it requires redistribution of routes into both LDP and the IGP, with the following consequences:

1. Requires redistribution of the routes for the loopbacks of the remote PEs (BGP peers) into the IGP. These routes are advertised via BGP and providers are wary of redistributions from BGP into the IGP, since a mistake in the redistribution policy can inject a large number of routes in the IGP and cause IGP meltdown.
2. The IGP must carry prefixes from a different site, which could impact scaling of the IGP. When the two sites are in different ASs (discussed later in this chapter), the provider requires more control over the routing information injected into one AS from another. This control is readily available with the BGP solution.

The differences from the ISP-as-a-customer scenario in the previous section are:

- MPLS is used within the sites.
- Only the routers imposing the label stack are required to have knowledge of the external routes. VPN routes are exchanged for the BGP next-hops of the external routes. Thus, the exchanges of the VPN routes can happen over sessions between the routers that actually do label imposition.

9.3 MULTI-AS BACKBONES

The previous section showed how a VPN provider can be a VPN customer, and its sites are in the same AS. In this section we will take a look at what happens when the sites are in different ASs. This can be the case when a provider spans several ASs (e.g. following an acquisition) or when two providers cooperate in order to provide a VPN service to a common customer. In the latter case, it is necessary for the providers to agree on the conditions and compensation involved and to determine the management responsibilities. To distinguish the two cases, they are referred to as inter-AS and interprovider respectively.

The problem with multi-AS scenarios is that the routers in the two sites cannot establish an IBGP session to exchange external routes. Instead, an EBGP session must be run. [2547-bis] describes three ways to solve the multi-AS scenario. These methods are often referred to by their respective section number in [2547-bis], as options A, B and C. An important thing to bear in mind when reading this section is that multi-AS scenarios are not targeted in particular at carrier-customers and provide a general solution for VPNs crossing several ASs.

9.3.1 Option A: VRF-to-VRF connections at the ASBR

The simplest method to exchange the VPN routes across an AS boundary is to attach the two ASBRs directly via multiple subinterfaces (e.g. VLANs, or Virtual LANs) and run EBGP between them. Each ASBR associates one of the subinterfaces with the VRF for a VPN requiring inter-AS service and uses EBGP to advertise customer IP routes within each VRF, as shown in Figure 9.8.

This solution requires no MPLS at the border between the two ASs. As far as ASBR2 is concerned, ASBR1 appears to be equivalent to a regular CE site. Similarly, ASBR2 appears to be a regular CE site to ASBR1. The separation of routing and forwarding information is accomplished by using separate VRFs per VPN at the ASBR, with the following consequences:

- The ASBRs are configured with VRFs for each of the VPNs crossing the AS boundaries. There is a separate subinterface associated with each of the VRFs.
- There is per-VPN state on the ASBRs and the provider is required to manage the subinterface assignment for the different VPNs.

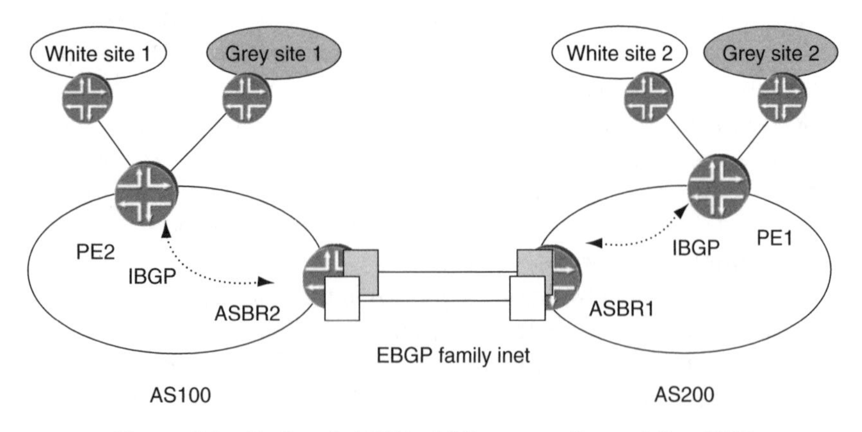

Figure 9.8 Option A: VRF-to-VRF connections at the ASBR

- The ASBRs must exchange all the VPN routes from all VPNs crossing the AS boundary.
- Multiple EBGP sessions are maintained (one per VPN).

Despite these less than desirable scaling properties, the solution works and is deployed for situations when the number of VPN customers and the number of VPN routes is small. There are several benefits to the solution. Option A is simple to understand and deploy and is contained by the routers providing the VPN service. Furthermore, it simplifies interworking among providers, because the interconnection between providers is simply an interface. Therefore, this interface becomes the element of control on which policing, filtering and accounting can be done with per-VPN granularity.

9.3.2 Option B: EBGP redistribution of labeled VPN-IPv4 routes

The undesirable scaling properties of option A are caused by the fact that the VPN routes are exchanged as IP routes, so the per-VPN state must be maintained by the ASBRs. Furthermore, every time a new VPN is added that requires inter-AS service, the ASBR must be configured with the correct VRF information. Thus the addition of a new VPN is no longer limited to the configuration of the PE routers and involves ASBR configuration as well.

To avoid keeping the per-VPN state at the ASBR, VPN-IPv4 routes can be advertised instead. The option B solution uses a single EBGP session between the ASBRs regardless of the number of VPNs and advertises labeled VPN-IPv4 routes over it. The routes are exchanged between PE and ASBR via an IBGP session. The conceptual model is shown in Figure 9.9. In order to ensure that unauthorized access to the VPN is not possible, the EBGP session must be secure and VPN-IPv4 routes should not be accepted on any other session except the secure one.

At the end of the routing exchanges, the PEs in the different ASs have received the VPN routes for their customers with the appropriate VPN labels assigned by their peers. However, in order actually to be able to forward traffic between the two customer sites, a label-switched path must exist from one PE to the other, across the AS boundaries.

It is possible to build the necessary LSP by using BGP as a label distribution protocol on the inter-AS link, as we have seen

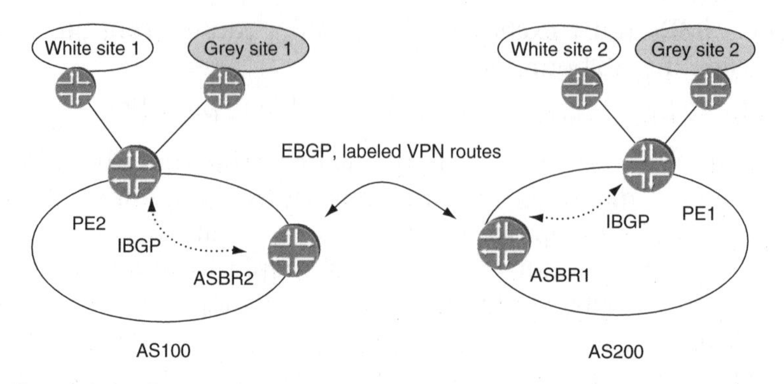

Figure 9.9 Option B: EBGP redistribution of labeled VPN-IPv4 routes

in the carriers' carrier scenario. Let us take the example network of Figure 9.9 and see how this is done for a VPN route advertised from PE1. The assumption is that MPLS is running in each one of the ASs, so label-switched paths exist between the ASBRs and the PEs. What is needed is a way to stitch between the LSP within each AS and the one-hop LSP over the inter-AS link, by installing the appropriate MPLS swap state. When ASBR1 receives a labeled VPN route from PE1, with the VPN label L1, it allocates a label L2 for it and advertises it to ASBR2 with itself (ASBR1) as the next-hop. At the same time, it installs the forwarding state that swaps between L2 and L1. ASBR2 advertises the VPN route with label L2 in its AS. When traffic is sent from PE2, for this VPN prefix, it is labeled with L2. At ASBR1, the label is swapped to label L1 advertised by PE1. Note that in this scenario, the VPN label changes – the label used by the remote PE is not the same as the one that was allocated by the local PE.

A security breach may arise if the label assigned by the ASBR is spoofed by an outsider. To avoid this situation, the ASBR should only forward labeled traffic arriving over an interface over which the label was actually advertised. Another potential security threat results from peering with an unauthorized source who wants to capture the traffic for a particular VPN. To avoid such unauthorized peering, the providers in the two ASs must negotiate and agree upon which routers are allowed to exchange VPN routes and which RTs they will use in the route advertisements.

Let us take a look at some of the properties of this solution:

- There is no need for per VPN configuration and per VPN interface assignments at the ASBRs.
- The ASBRs must keep the state for all VPN routes.
- It cannot easily do traffic filtering at the IP level for traffic crossing the AS boundary.
- A single EBGP session must be maintained between the ASs.
- It requires a PE–PE inter-AS LSP.

9.3.3 Option C: multihop EBGP redistribution of labeled VPN-IPv4 routes between the source and destination AS, with EBGP redistribution of labeled IPv4 routes from one AS to the neighboring AS

The solution in option B still requires that all VPN routes be maintained and advertised by the ASBR. This makes the solution unsuitable for cases where there are a lot of VPN routes. Both the problem and the solution are very similar to the hierarchical VPN scenario discussed in section 9.2.2: the customer uses a multihop EBGP session between its PE routers to carry external prefixes as labeled VPN-IPv4 routes and the provider provides connectivity to the PE loopbacks by advertising them as labeled IPv4 routes from one AS to another. In this way, the ASBRs do not carry any of the VPN routes. Thus this option is the most scalable of the three options discussed here.

Note that in order for this solution to work, the BGP next-hop of the VPN-IPv4 routes exchanged over the multihop EBGP session must not be changed. Figure 9.10 shows a conceptual model of the route exchanges, based on the network of Figure 9.5. As with the hierarchical VPN scenario, three labels are imposed on the traffic at ingress, unless the addresses of the PE routers are made known to the P routers in each domain, in which case a two-label stack is imposed.

In addition to the security concerns seen so far, a new problem arises in this scenario, if the approach of using the LDP rather than the labeled BGP is used (as explained in Section 9.2.2). Since the addresses of the PE routers are advertised, this means that the IGP

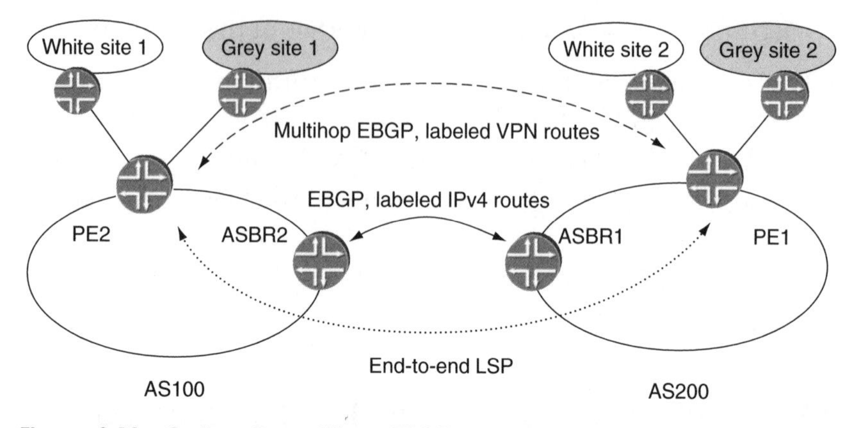

Figure 9.10 Option C: multihop EBGP redistribution of labeled VPN-IPv4 routes between the source and destination AS, with EBGP redistribution of labeled IPv4 routes from one AS to the neighboring AS

in one provider's network carries addresses from a different provider's network. Sometimes, this is viewed as a security concern because the addresses of the PEs in one AS are known to the routers in the core of another network, thus revealing the addressing structure of the remote AS. This makes option C undesirable for inter-provider (as opposed to plain inter-AS) setups. Furthermore, in contrast to option A, interworking among providers is seen by some as more difficult because the connection is an end-to-end 'fat pipe' without any VPN context. The LSP between PE3 and PE4 in Figure 9.10 carries traffic from all VPNs. Thus, there is no per VPN context and the ability to do policing, filtering and accounting per VPN at the ASBR is lost.

Table 9.1 compares some of the properties of the three options described above.

9.4 INTERPROVIDER QoS

From the customer's point of view, the QoS expectations are the same, regardless of whether the VPN service is implemented by a single provider or by multiple providers. When VPN traffic crosses several domains, it is necessary for each of the domains to enforce its own policy to ensure the desired QoS.

In the Advanced L3 VPN chapter (Chapter 8) we have seen how QoS can be provided for BGP/MPLS VPNs by marking customer

Table 9.1 Comparison of the three inter-AS solutions

	Option A	Option B	Option C
State at the ASBR	Per-VRF state; all VPN routes in all VPNs are maintained	No per-VRF state, but all VPN routes in all VPNs are maintained	Only the addresses of the remote PEs are maintained (the BGP next-hops of the VPN routes)
Per-VRF configuration at the ASBR	Yes	No	No
VPN label	Not used	May change at the ASBR (unless an inter-AS LSP is set up via RSVP)	Remains constant
Requirement to run MPLS across the AS boundary	No	Yes	Yes
EBGP session	Single hop, family IP, multiple sessions (one per VRF)	Single hop, family VPN-IPv4, single session	Multi-hop, family VPN-IPv4, single session.
Security concerns	None	Prevent label spoofing at the ASBR	Access to all PEs from the remote AS

traffic with the EXP bits, providing the correct per-hop behavior in the network. The marking is done at the PE and forwarding the traffic in the core is based on the EXP bits. The solution requires mapping DSCP to EXP consistently at the PE–CE boundary. In an interprovider setup, it cannot be expected that the same EXP bits will be used to represent the same per-hop behaviors in both networks. Therefore, it is necessary to remark the EXP bits of the VPN traffic as it crosses the inter-AS link.

The inter-AS link poses other challenges as well. The link is shared between two administrative domains, so it is more difficult to upgrade and can easily become a point of congestion. This problem can be made worse by the fact that often the same link is used to carry Internet traffic between the providers in addition to the VPN traffic. In order to ensure SLAs for the VPN traffic, it is necessary to set up and enforce policies for prioritizing and rate-limiting traffic at the AS boundary. This can be done either at the aggregate level across all VPN customers, or with per VPN visibility, as possible in option A.

However, it is not enough to enforce the customer SLA end-to-end. When a service is shared between two providers it is important to be able to measure, report and troubleshoot the performance consistently in both ASs. SLAs are usually tied to compensation, so in the case of SLA violation it is necessary to be able to determine which provider is at fault. Today, the issue of measuring and reporting is solved on a case-by-case basis using the tools that are available in each provider's network.

9.5 CONCLUSION

The L3 VPN solution supports VPNs spanning across several ASs, as well as VPN hierarchies. The key to achieving these in a scalable manner is the use of MPLS tunnels. The tunnels are built by stitching together several tunnel segments using either BGP or LDP.

This chapter concludes the discussion on Layer 3 BGP/MPLS VPNs. In the following two chapters, we discuss Layer 2 VPNs and VPLS and will see that much of the protocol machinery used for the Layer 3 VPN has been extended to Layer 2 VPNs and VPLS.

9.6 REFERENCES

[2547-bis] E. Rosen and Y. Rekhter, 'BGP/MPLS IP VPNs', draft-ietf-l3vpn-rfc2547bis-03.txt, soon to become an RFC (currently in the RFC editor queue)

[RFC3107] Y. Rekhter and E. Rosen, *Carrying Label Information in BGP-4*, RFC3107, May 2001

9.7 FURTHER READING

[ADV-MPLS] V. Alwayn, *Advanced MPLS Design and Implementation*, Cisco Press, 2001

[L3VPN] http://ietf.org/html.charters/l3vpn-charter.html

[MPLS-TECH] B. Davie and Y. Rekhter, *MPLS Technology and Applications*, Morgan Kaufmann, 2000

[MPLS-VPN] I. Peplnjak and J. Guichard, *MPLS and VPN Architectures*, Cisco Press, 2000

[VPN-TUTORIAL] I. Minei, 'BGP/MPLS Layer 3 VPNs', tutorial at Nanog 30, http://nanog.org/mtg-0402/minei.html

10

Layer 2 Transport over MPLS

10.1 INTRODUCTION

This chapter describes the rapidly growing area of Layer 2 transport over MPLS networks. This is a key component of a multiservice network as it allows service providers to migrate Frame Relay, ATM and leased-line customers to an MPLS network while maintaining similar service characteristics from the customer's point of view. It also enables new Layer 2 service offerings based on Ethernet access. In this chapter, we compare the two main schemes for achieving Layer 2 transport over MPLS, one based on LDP signaling and the other based on BGP signaling. We also discuss Circuit Cross Connect (CCC), which was the precursor to these schemes and is still in use in several service providers' networks. In addition, we describe a scheme by which Layer 2 traffic can be transported over MPLS across multiple service provider networks while maintaining the required QoS characteristics.

10.2 THE BUSINESS DRIVERS

Native Layer 2 services have existed for several years, based on Frame Relay or ATM. Often these services are used by an enterprise to

MPLS-Enabled Applications: Emerging Developments and New Technologies Ina Minei and Julian Lucek

build its corporate Layer 2 VPN by interconnecting its LANs over a wide area. Service providers can offer near global reach, either directly or through interconnection agreements with partners. The services are a valuable source of revenue to service providers, at the time of writing far outstripping revenues from IP services. In these networks, customer sites are interconnected at Layer 2, sometimes in a full mesh but more typically in a hub-and-spoke topology. The role of the service provider is to transport the ATM cells or Frame Relay frames over the wide area, at an agreed bit-rate for each circuit.

As well as being used to carry general LAN interconnection traffic, these services, especially in the ATM case, are sometimes used to carry traffic requiring more stringent SLAs from the network, e.g. with respect to delay variation, such as video traffic or Private Automatic Branch eXchange (PABX) interconnections.

In many cases, a service provider can migrate these services to an MPLS network while retaining the same connectivity, as far as the customer is concerned, and maintaining similar service characteristics. In these cases, the presentation to the customer is still over ATM or Frame Relay and a similar service-level agreement (SLA) is offered. For example, in the Frame Relay case, a CIR (committed information rate) is agreed for each circuit and SLA is defined for parameters such as packet loss, latency and delay variation.

Migrating these services to an MPLS network saves the service provider capital and operational expenses compared to running separate networks for Layer 2 connectivity and Layer 3 connectivity. Also one of the schemes discussed later in this chapter greatly reduces the operational burden of provisioning Layer 2 connections within the service provider part of the network, especially in cases where a high degree of meshing is used between customer sites, which leads to a further saving in operational costs.

Another growing application of Layer 2 transport over MPLS is Ethernet services, in which a customer's Ethernet frames are transported between the customer's sites over the service provider's MPLS network. The appeal to the end customer is that Ethernet is the standard Layer 2 protocol used within the enterprise and hence is familiar to the corporate IT staff. Using Ethernet to interconnect their sites over the wide area is a natural extension of the use of Ethernet within their premises. In many cases where customers have been using ATM or Frame Relay services for LAN interconnection, there is no fundamental reason why ATM or Frame Relay should be used as the interconnectivity method. Ethernet has the attraction that it is more flexible in terms of access rates – the service provider can offer, for example, a 100 Mbps Ethernet tail that is rate-limited to the level paid for by the customer. This allows for smoother upgrades in access speed than having, for example, to change from an E1/T1 access circuit to an E3/T3 access circuit. These factors, along with the fact that Ethernet-based equipment tends to be less expensive than ATM or Frame Relay equipment, by virtue of volume means that in some cases a customer might migrate from a native ATM or Frame Relay based service to an Ethernet service in order to reduce costs. In order to address this market, some service providers now offer point-to-point Ethernet services with similar SLA parameters to some of the Frame Relay or ATM services that they replace. However, at the time of writing, few service providers offer an Ethernet service equivalent to ATM CBR (constant bit rate). CBR services are uncontended, i.e. the bandwidth is guaranteed end to end, and accompanied by tight SLAs on delay variation. However, MPLS networks are capable of providing such a service through a combination of control plane techniques such as DiffServ Aware TE and packet scheduling mechanisms that prioritize the CBR traffic appropriately at each hop in the network.

Besides point-to-point Ethernet services, many service providers offer multipoint Ethernet services, known as the Virtual Private LAN Service (VPLS). VPLS is the subject of the next chapter of this book, while this chapter discusses point-to-point services.

Whether a customer can migrate to Ethernet depends on whether local Ethernet access is available, bearing in mind that ATM and in particular Frame Relay (FR) access networks have much higher geographical penetration in many territories, reaching the smaller cities, whereas Ethernet may only be available in larger cities. The incumbent service providers, who tend to own the large ATM or Frame Relay networks, may choose to retain the access part of those networks but migrate the core to an MPLS network, as illustrated in Figure 10.1.

Apart from these geographical penetration considerations, some customers may be using applications that require ATM in particular, e.g. video codecs or PABXs having an ATM interface, which would preclude them from migrating to an Ethernet service. As a consequence of the factors discussed above, service providers providing Layer 2 services over MPLS are likely to need to support a variety of access media types, including ATM, Frame Relay and Ethernet for the foreseeable future.

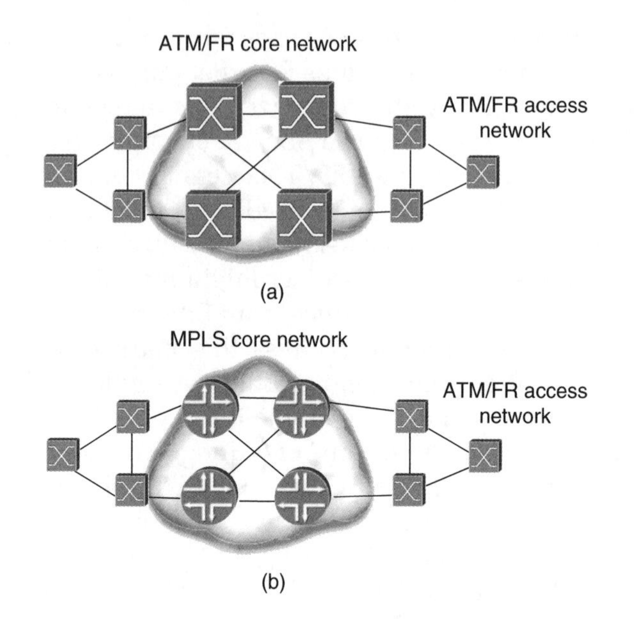

Figure 10.1 Migration of an ATM or Frame Relay core to an MPLS core: (a) prior to migration and (b) after the migration

10.3 COMPARISON OF LAYER 2 VPNs AND LAYER 3 VPNs

The introduction to the Layer 3 VPN chapter (Chapter 7) discussed the two main models that exist for VPN connectivity: the overlay model and the peer model. BGP/MPLS-based Layer 3 VPNs fall within the peer model. In contrast, when an enterprise builds a Layer 2 VPN, by buying Layer 2 transport services from the service provider they are building an overlay network. Hence the differences between Layer 2 and Layer 3 VPNs are as follows:

1. In the Layer 2 case, no routing interaction occurs between the customer and service provider. In the L3 VPN case, the CE and PE router can exchange routes.
2. In the Layer 2 case, the customer can run any type of Layer 3 protocol between sites. The SP network is simply transporting Layer 2 frames and hence is unaware of the Layer 3 protocol that is in use. Although IP is prevalent in many enterprise networks, non-IP protocols such as IPX or SNA are often in use.

This would preclude the use of a Layer 3 VPN to transport that type of traffic.

3. Multiple (logical) interfaces between each CE and the corresponding PE are required in the Layer 2 case, one per remote CE that each CE needs to connect to. For example, if the CE routers are fully meshed and there are 10 CE routers in total, each CE needs nine interfaces (e.g. DLCIs, VCs or VLANs, depending on the media type) to the PE, each leading to one of the remote CE routers. In the Layer 3 VPN case, one connection between each CE and the local PE is sufficient as the PE is responsible for routing the traffic towards the appropriate egress CE.

For some customers, L3 VPN is the better choice, for others L2 VPN, depending on what protocols need to be carried and the degree to which the customer wishes to do their own routing or to outsource it to the service provider. Hence, in order to address the widest possible market, many service providers offer both Layer 3 and Layer 2 services over their MPLS infrastructure. There exist PE routers that are capable of supporting both types of service simultaneously, in addition to VPLS, which is discussed in the VPLS chapter of this book (Chapter 11).

10.4 PRINCIPLES OF LAYER 2 TRANSPORT OVER MPLS

There are two main approaches to Layer 2 transport over MPLS: one involving LDP signaling [MRT-TRS, PWE3-CON] and the other based on BGP signaling [KOM-BGP]. In the forwarding plane, these approaches are the same, in terms of how Layer 2 frames are encapsulated for transport across the MPLS network. However, the two approaches differ significantly in the control plane. In later sections, we will discuss how each approach operates and then compare and contrast the two.

A single point-to-point Layer 2 connection provided over an MPLS network is sometimes called a pseudowire, to convey the principle that as far as possible the MPLS network should be invisible to the end customer, in such a way that the two CEs interconnected by the pseudowire appear to be directly connected back to back.

One of the problems with traditional Layer 2 VPNs is the administrative burden of adding a new site to an existing VPN, and the associated lead-times. If the sites are fully meshed, when a new site is introduced a new circuit must be provisioned between the new site and every other site in the network, and hence extra configuration at every site in the network is required. Indeed, often this administrative burden has forced customers to adopt a hub-and-spoke arrangement. We will show how the BGP approach greatly reduces the administrative overhead associated with traditional Layer 2 VPNs by making it much easier to add new sites to an existing mesh.

Examples of Layer 2 protocol types that can be carried over an MPLS network are as follows:

- *ATM*. Two main modes exist: a mode in which AAL5 PDUs are transported on the pseudowire and a mode in which ATM cells are transported on the pseudowire. In the latter case, the cells could belong to any AAL type, since the AAL PDUs are not reassembled by the MPLS network.
- *Ethernet*. The mapping of traffic into a pseudowire can be on a per-VLAN or on a per-port basis. In the per-VLAN case, if an Ethernet connection between the customer CE router and the service provider's PE router contains multiple VLANs, each VLAN can be mapped to a different pseudowire for transport to a different remote CE.
- *Frame Relay*. The mapping of traffic into a pseudowire can be on a per-port basis or on a per-DLCI basis. In the per-DLCI case, if a Frame Relay connection between the customer CE router and the service provider's PE router contains multiple DLCIs, each DLCI can be mapped to a different pseudowire for transport to a different remote CE.

The examples above are those that dominate in current deployments, for the reasons discussed in section 10.2 of this chapter. In addition, the transport of HDLC and PPP frames is also supported by some vendors. Also, there is interest in the transport of TDM circuits [TDM-PWE3] (e.g. E1 or T1) in pseudowires across MPLS networks, using the same control plane mechanisms as for Layer 2 transport.

Figure 10.2 illustrates an example network in which a service provider is using its MPLS network to provide a Layer 2 service to a customer. The three customer sites are fully meshed, so each site has a circuit corresponding to each remote site that it needs to connect to. In the example, the media type chosen by the customer is Ethernet

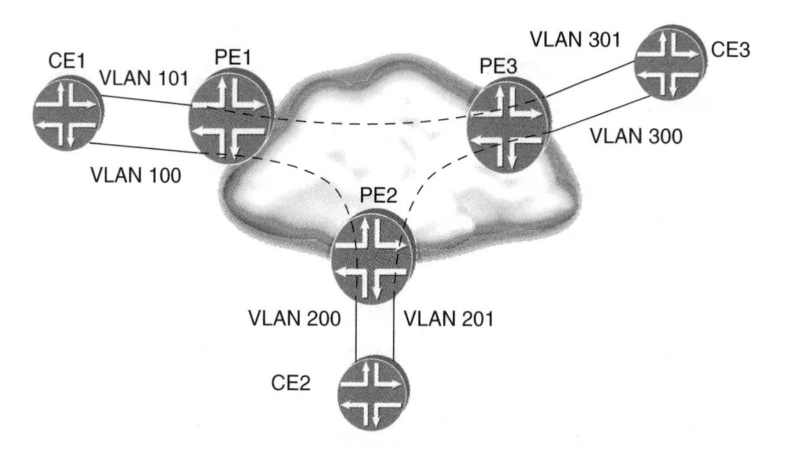

Figure 10.2 Example Layer 2 VPN showing connectivity between three customer sites

but the same principle applies if the customer were using ATM or Frame Relay instead. The PEs belong to the service provider and the CEs belong to the customer, so the boundary between the service provider and the customer is the set of VLAN access circuits. CE1 uses VLAN 100 to connect to CE2 and uses VLAN 101 to connect to CE3. As far as CE1 can tell, it is directly connected to CE2 and CE3 and is not 'aware' of the presence of the service provider network. The ingress PE must send the packet to the appropriate egress PE, from where the packet is forwarded on the appropriate circuit to the receiving CE. For example, if a packet arrives at PE1 from CE1 on VLAN 100, PE1 must forward the packet to PE2 and PE2 in turn must forward the packet to CE2 on VLAN 200. The following sections describe how this is achieved, in terms of how the packet is encapsulated for transport across the MPLS network, and the operation of the control plane.

10.5 FORWARDING PLANE

This section describes the operation of Layer 2 transport in the forwarding plane. Note that the encapsulation method is the same regardless of whether the BGP or LDP control plane scheme is in use. The detail of how Layer 2 packets are encapsulated for transport over an MPLS network is contained in an Internet draft entitled 'Encapsulation methods for transport of Layer 2 frames over IP and

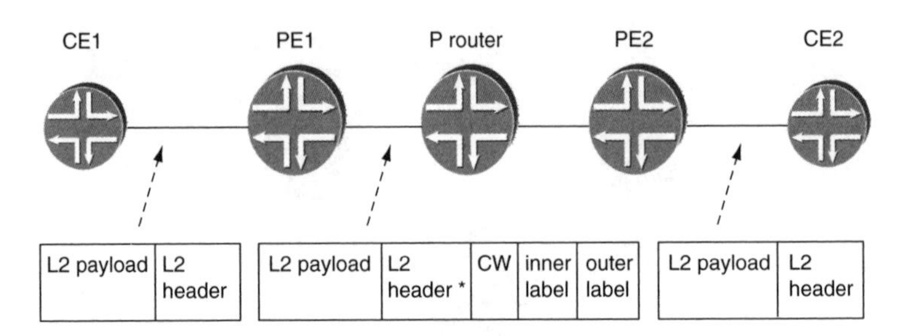

Figure 10.3 Forwarding plane operation of L2 transport over MPLS

MPLS networks' [MRT-ENC] and also in a series of documents, one per media type, produced by the Pseudowire Emulation Edge-to-Edge (PWE3) Working Group in the IETF. The PWE3 documents contain more encapsulation variants than [MRT-ENC] for some Layer 2 media, but most existing implementations are closer to [MRT-ENC].

Figure 10.3 shows a cross-section through the network shown in Figure 10.2, showing the transport of Layer 2 packets on a pseudowire between CE1 and CE2. When the Layer 2 frame arrives at PE1, PE1 carries out the following forwarding operations:

1. Parts of the L2 frame that do not need to be transported to the remote PE are removed. For example, in the Ethernet case the FCS (Frame Check Sequence) is removed.
2. In some cases, a 4-byte Control Word (CW) is prepended to the L2 frame. The Control Word can include a sequence number so that the egress PE can detect mis-sequencing of packets. Depending on the media type, the Control Word may also contain flags corresponding to control bits within the header of the native Layer 2 frame. This allows the value of those control bits to be conveyed across the core of the network to the egress PE without having to transport the entire native Layer 2 header.
3. PE1 looks up the value of the MPLS inner label that PE2 expects for the frame and prepends an MPLS header having that label value.
4. PE1 determines how to reach PE2. As with L3 VPN, the network operator has a choice of tunneling technologies in the core, including LDP and RSVP-signaled LSPs and GRE and IPSec tunnels. Indeed, the same tunnel can be shared by L3 VPN, L2 traffic and VPLS traffic. If an LDP or RSVP-signaled LSP is used, PE1 determines the MPLS label value required to reach PE2 and

stacks an MPLS header containing that label value on top of the inner MPLS header. In networks where MPLS transport is not used between PEs, PE1 determines the appropriate tunnel to reach PE2 (e.g. a GRE or IPSec tunnel).

5. PE2, on receiving the packet, examines the label value of the 'inner' MPLS header before popping it. From that, it determines that the underlying L2 frame must be sent on VLAN 200 to CE2. If the Control Word is present, PE2 may check the sequence number and take appropriate action should the packet be out of sequence. The processing of the sequence number by the egress PE is optional. Actions that a receiving PE can take on receiving an out-of-order packet are to drop the packet or to reorder the packets into the correct sequence. PE2 then regenerates the L2 frame, which may involve determining the values of control bits in the frame header by referencing the corresponding flags in the Control Word. In the example in the figure, the Ethernet frame arrived with a VLAN identifier (ID) of 100. However, CE2 expected a VLAN ID value of 200, so PE2 must rewrite the value of the VLAN ID accordingly. PE2 then forwards the frame to CE2.

Let us now examine the various Layer 2 encapsulations described in the IETF drafts. Note that not all implementations necessarily support all of the variants described below.

10.5.1 ATM cell

The PWE3 ATM encapsulation draft [PWE3-ATM] specifies two modes, the *N*-to-one mode and the one-to-one mode. In the *N*-to-one mode, cells from one or more Virtual Channel Connections (VCCs) or from one or more Virtual Path Connections (VPCs) are mapped to a single pseudowire. In this case, the VPI–VCI fields of each cell are preserved when the cell is transported across the core, so that the egress PE knows which VPI/VCI a particular cell belongs to. If desired, the *N*-to-one mode can be used to transport all the VPCs on a particular port to a remote port in the network.

In the one-to-one mode, cells from a single VCC or a single VPC are mapped to a single pseudowire. In the single VCC case, the VPI and VCI fields are not sent across the core as they can be regenerated at the egress PE. In the single VPC case, the VPI field is not sent across the core as it can be regenerated at the egress PE.

Whichever mode is used, the Control Word can be used to carry the value of the ATM Cell Loss Priority (CLP) bit from the ingress PE to the egress PE. The egress PE can copy the value of the bit into the regenerated ATM cell.

Most implementations today support the *N*-to-one mode rather than the one-to-one mode. The latter is regarded as optional by the PWE3 ATM draft [PWE3-ATM] and is not mentioned in [MRT-ENC]. Whichever mode is used, the Header Error Check (HEC) field of each ATM cell is not sent across the core. Instead, it is regenerated by the egress PE.

Either mode allows for multiple ATM cells to be sent in a single MPLS packet across the core. The number of cells that the user wishes to send is a tradeoff between bandwidth efficiency, delay variation and the number of cells that the user can afford to lose should an MPLS packet be lost.

10.5.2 ATM AAL5

In the case where the ATM data to be transported belong to AAL5, it is more bandwidth efficient to reassemble the AAL5 frame at the ingress PE and transport it across the core as a single entity than to transport unassembled cells. In this mode, there is a one-to-one mapping between ATM VCCs and pseudowires.

10.5.3 Frame Relay

In the 'one-to-one mode', a single Frame Relay DLCI is mapped to a single pseudowire [PWE3-FR]. The Frame Relay header and FCS are not transported. The Control Word, if used, contains a bit corresponding to each of the Frame Relay parameters in the list below:

- FECN (Forward Explicit Congestion Notification bit)
- BECN (Backward Explicit Congestion Notification bit)
- DE bit (Discard Eligibility bit)
- C/R (Command/Response bit)

The use of these bits is not mandatory, but if used, the ingress PE copies the value of each bit from the Frame Relay frame into the corresponding field in the Control Word, thus allowing the state of those parameters to be conveyed across the core of the network. The egress PE then copies the value of each bit into the Frame Relay frame that it sends to the CE. The Control Word also contains a 16-bit sequence number, although its use is not mandatory.

In addition to the mode described above, the PWE3 Frame Relay draft [PWE3-FR] also describes a port mode. In this mode, all of the DLCIs on a particular port are transported across the network in a single pseudowire to a particular remote port. This means that, unlike the one-to-one case described above, the Frame Relay address field must be transported across the core. Unlike the one-to-one mode, if the Control Word is used, the fields corresponding to the Frame Relay control bits described above are not used. The 16-bit sequence number can be used, although its use is not mandatory. Note that the port mode is regarded as optional by the PWE3 Frame Relay draft and is not mentioned in [MRT-ENC].

10.5.4 Ethernet

Two modes of Ethernet transport [PWE3-ETH] exist, one in which the mapping to pseudowires across the core is on a per-VLAN basis and another in which an entire Ethernet port, which may contain multiple VLANs, is mapped to a single pseudowire. The use of the Control Word is optional, but if used there is a 16-bit sequence number that can be used if required. The FCS is stripped off at the ingress PE and regenerated by the egress PE. The Control Word in the Ethernet case is generally regarded as less useful than in the ATM or Frame Relay cases.

10.6 CONTROL PLANE OPERATION

Let us see how the control plane for Layer 2 transport operates. We will examine the LDP-based scheme [MRT-TRS] and the BGP-based scheme [KOM-BGP]. Both approaches have the following characteristics in common:

1. A means for a PE, when forwarding traffic from a local CE via a remote PE to a remote CE, to know the value of the VPN label (inner label) that the remote PE expects.
2. A means for signaling characteristics of the pseudowire, such as media type and MTU. This provides a means to detect whether each end of a pseudowire are configured in a consistent manner or not.

3. An assumption that the pseudowire formed is bidirectional. Hence, if there is a problem with transport in one direction, forwarding is not allowed to occur in the opposite direction.
4. A means for a PE to indicate to remote PE(s) that there is a problem with connectivity, e.g. if the link to a CE goes down.

The two schemes differ significantly in the way in which a PE knows which remote PE(s) it needs to build pseudowires to. In the LDP-based scheme, this information must be manually configured on the PEs. The BGP scheme, in contrast, has in-built autodiscovery properties, so this manual configuration is not required.

10.6.1 LDP signaling scheme

A targeted LDP session is created between each pair of PEs in the network (or at least each pair of PEs between which L2 transport is required). On each PE the identity of the remote PE for each pseudowire is manually configured. The PE–PE LDP session is used to communicate the value of the 'inner label' or 'VPN label' that must be used for each pseudowire. In general, there may be multiple L2 pseudowires between a particular pair of PEs, each pertaining to a different customer, but only one LDP session, so an identifier, known as the VC ID, is used to distinguish between the connections being signaled. The same VC ID is configured on each of the two PEs taking part in each L2 pseudowire.

Referring again to Figure 10.2, on PE1 an association (through a command line interface, CLI, configuration) would be created between VLAN 100, a VC ID and an IP address of PE2 (typically the loopback address), which is the address used for the targeted LDP session. Similarly, on PE2, an association would be created between VLAN 200, a VC ID having the same value as on PE1, and the address of PE1. PE1 uses the LDP session to inform PE2 of the VPN label value that PE2 must use when forwarding packets to PE1 on the pseudowire in question; similarly PE2 informs PE1 of the label value that PE1 must use.

Let us look at the information that is communicated over the LDP session, in addition to the inner label value itself. A new FEC element has been defined [MRT-TRS] in order that LDP can signal the requisite information. This contains the following fields:

- VC ID.
- Control Word bit. This indicates whether a Control Word will be used.
- VC type. This indicates the encapsulation type (PPP, VLAN, etc).
- Interface parameters field. This contains information such as media MTU and, in the ATM cell transport case, is an indication of the maximum numbers of concatenated cells the PE can support.

There is no concept of a VPN as such in the LDP-based scheme. The pseudowires are created in a pair-wise manner without the network being 'aware' that a set of connections actually form a VPN from the customer's perspective. Note that this lack of VPN awareness means that if a customer requires that its CEs are fully meshed with pseudowires and an additional CE is added to the network, each PE involved (i.e. each PE having a CE of that customer attached) must be configured with a VC ID corresponding to the new connection. This can cause a large provisioning overhead if there are a large number of CEs in the existing mesh.

The fact that LDP is being used as the signaling mechanism for the pseudowires does not mean that LDP must be used as the signaling mechanism for the underlying transport tunnels used to carry the packets from the ingress PE to the egress PE. The transport tunnels could be RSVP-signaled or LDP-signaled LSPs or could be GRE or IPSec tunnels.

10.6.2 BGP-based signaling and autodiscovery scheme

The BGP-based approach [KOM-BGP] aims to give operational characteristics to the service provider that are familiar from a Layer 3 VPN. As with a Layer 3 VPN, BGP is used to convey VPN reachability information. With a Layer 3 VPN, service providers take it for granted that a new CE site can be added to an existing PE without having to add extra configuration to all the other PEs in the network. This is because the other PEs learn through BGP about the existence of the new site (or rather the routes associated with that site and which PE is attached to that site). This autodiscovery property has been carried through to Layer 2 VPNs in the BGP-based approach. This greatly reduces the operational burden of adding new CEs to an existing L2 VPN.

As a consequence of the autodiscovery property, rather than having to manually configure a pseudowire between each pair of CEs, the pseudowires are created automatically. As with the L3 VPN, a PE derives the inner label (VPN label) in order to reach a particular remote CE from information carried in the BGP advertisements.

One difference, however, between Layer 2 and Layer 3 VPNs is that in the L2 VPN case, the inner label (the VPN label) used to reach a particular CE depends on the CE that the packet originated from (so that the egress PE can determine which CE the packet came from and hence can forward the packet on the appropriate logical interface to the receiving CE). In principle, each PE could advertise a list of labels for each attached CE, each label on the list corresponding to one PE–CE logical interface (VLAN, DLCI, etc.). However, in fact a more compact method is used, in which each PE advertises through BGP sufficient information for remote PEs to *calculate* the label value to use. Without this scheme, either each PE would receive information that it is not interested in (inner label values that other PEs need to use) or the information sent to each PE would have to be tailored to that PE. Using BGP to carry the necessary information allows the reuse of much of the protocol machinery already developed for L3 VPNs, such as the use of route distinguishers and route targets. Also, if the service provider offers the L3 VPN service, as well as the L2 VPN service, the same BGP sessions and same route reflectors can be used to support both services.

Let us look in more detail at the mode of operation of the scheme. For each CE attached to a PE, a CE identifier (CE ID) is configured on the PE. This CE ID is unique within a given L2 VPN. Also, each of the circuits from the CE to the PE is associated with a particular remote CE ID. This association is either explicitly made through configuration or implicitly by mapping circuits to CE IDs in the order in which they appear in the configuration. In this way, when a packet arrives from the local CE on a particular circuit, the PE knows to which remote CE the packet should be forwarded. The PE obtains the knowledge about the location of a remote CE (in terms of the PE to which it is attached) from the BGP updates originated by other PEs. Each PE advertises the CE IDs of the CEs to which it is attached and also sufficient information for any other PE to calculate the pseudowire label required in order for the packet to be forwarded to the CE by the PE.

Let us look in more detail at the content of a BGP update message:

- Extended community (route target). As with L3 VPNs, this allows the receiving PE to identify which particular VPN the advertisement pertains to.
- L2-Info extended community. This community is automatically generated by the sending PE. Encoded into the community are the following pieces of information:
  - Control flags, e.g. a flag to indicate whether a control word is required or not
  - Encapsulation type (PPP, VLAN, etc.). This allows the receiving PE to check that the local and remote 'tails' of the L2 connection are of consistent media type.
  - MTU (so that the PE can check that the remote 'tail' circuit is configured with the same MTU as the local tail).
- Other BGP attributes such as the AS path, etc.
- The NLRI. This contains the following items:
  - Route Distinguisher. As with L3 VPNs, this allows 'routes' pertaining to different VPNs to be disambiguated.
  - CE ID.
  - Label base.
  - Label-block offset.
  - Circuit status vector (CSV) sub-TLV.

The label base and label-block offset are the information required for a remote PE to calculate the VPN label to use when sending traffic to the CE ID on that PE. Bear in mind that the value of the label by a remote PE depends on which CE it is forwarding traffic from. A PE allocates 'blocks' of labels. Each block is a contiguous set of label values. The PE does not explicitly advertise each label within the block. It simply advertises the value of the first label in the block (the label base) and the size of the block (the latter being the length field of the CSV sub-TLV).

In simple cases, there is only one label block whose size is sufficient that each remote CE has a label to use within the block. In such cases, the label value that a remote CE, having a CE ID of value *X*, must use to reach the CE in question is computed as follows:[1]

$$\text{Label value} = \text{label base} + X - 1$$

For example, let us refer again to Figure 10.2. Suppose that PE1 advertises a label base of 100 000. PE2 and PE3 receive the advertisement, either directly or via a route reflector. Assume that the CE ID of CE1 is 1, that of CE2 is 2 and that of CE3 is 3. When PE3 receives a packet on VLAN 301, it knows (through the configuration) that the packet must be sent to CE ID 1. It looks in its routing table and sees it has an entry of CE1 and sees that the label base is 100 000. The formula above yields a label value of 100 002, which should be used as the inner label. The BGP next-hop of the route is PE1, so it knows that the packet should have an outer label pertaining to the tunnel (RSVP- or LDP-signaled LSP) that leads to PE1.

Sometimes, there may be more than one label block, e.g. if the original label block was exhausted as more sites were added to the Layer 2 VPN. In this case, each block is advertised in a separate Network Layer Reachability Information (NLRI). The first label block corresponds to the CEs having the lowest IDs, the next label block to the next lowest and so on. Note that in the BGP NLRI, there is a 'label-block offset' parameter. This is equal to the CE ID that maps on to the first label in the block. For example, there might be two label blocks, each with a range of 8. The first would have a label-block offset of 1, so CE ID 1 would map to the label base. The second would have a label-block offset of 9, so CE ID 9 would map on to the label base of that block. Let us suppose that the label base of the second block is 100 020. A PE forwarding traffic from CE ID 12 would choose the fourth label in the second block, namely 100 023. The label blocks used in this example are illustrated in Figure 10.4.

It should be noted that the process by which a PE allocates label blocks and advertises them through BGP is fully automated. Hence there is no need for the network operator to be explicitly aware of what is happening.

The circuit status vector allows a PE to communicate to remote PEs the state of its connectivity. Each bit within the vector corresponds

[1] In the [KOM-BGP] IETF draft, the CE ID numbering scheme is one of the examples starts at 0. At the time of writing, in existing implementations the CE ID numbering scheme starts at 1. The text in this chapter follows the scheme used by existing implementations.

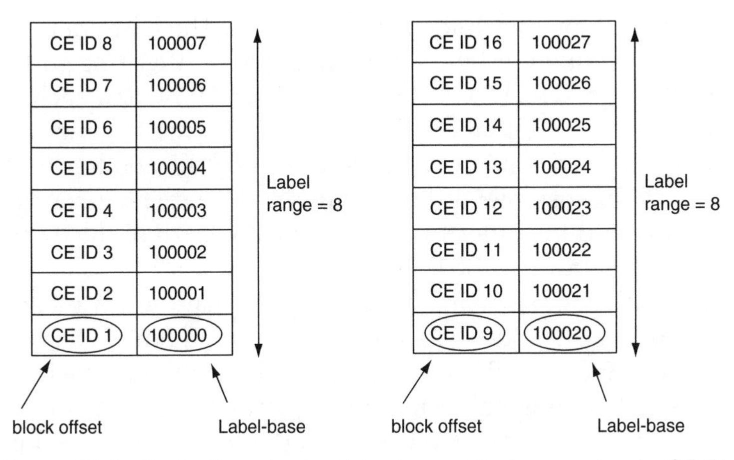

Figure 10.4 Illustration of label blocks and their mapping to CE IDs

to a different circuit between it and the local CE. A value of 0 indicates that the local circuit in question and the tunnel LSP to the remote PE is up, while a value of 1 indicates that either or both of them are down.

The BGP scheme makes it straightforward for the network operator to create any desired topology for an L2 VPN. For example, if a hub-and-spoke topology is required, this can be achieved by only provisioning on each PE that is attached to a spoke CE the remote CE ID of the hub CE. An alternative method to achieve a desired topology is through the manipulation of BGP extended communities. For example, in the hub-and-spoke case, a PE attached to a spoke site can be configured only to accept BGP NLRI originated by the PE attached to the hub CE, by virtue of the fact that the NLRI is configured with a particular extended community.

A key property of the BGP-based scheme is that a PE does not require configuration of the location of any remote CE, in terms of the identity of the remote PE to which it is attached. This information is learnt through BGP, the BGP next-hop telling a PE which remote PE 'owns' the CE having a particular CE ID. This autodiscovery property greatly simplifies the administration of an L2 VPN as new sites are added or moved. In addition, if preprovisioning is used, when a new CE site is added to an existing VPN only the local PE needs any additional configuration.

An example of preprovisioning is as follows: assume that a customer orders an L2 VPN having 10 sites (CEs), but predicts that over time they may wish to grow to 20 sites. The customer wishes the sites to be fully meshed. Rather than just configuring sufficient circuits (e.g. VLANs, DLCIs, etc.) between each PE and each CE to accommodate the initial 10 sites (i.e. nine circuits), the service provider can provision 19 circuits between each PE and the local CE. Then, when the customer orders a new site, the local PE is configured accordingly and the other PEs automatically learn of the existence of the new site through BGP.

Using BGP as the signaling mechanism also has the advantage that if an entire new PE is added to the SP network and route reflectors are in use, only the new PE and the route reflectors of which it is a client need an additional configuration. An additional advantage is that some developments made for L3 VPNs can also be used in the L2 VPN case without having to reinvent the wheel. For example, in L2 VPN interprovider scenarios, similar techniques can be used as described for the L3 VPN in the Hierachical and Inter-AS VPNs chapter of this book (Chapter 9). For example, if the interprovider option C method described in Section 9.3.3 is used in the L2 VPN case, the label operations are exactly as described in the L3 VPN chapter. In fact, the ASBRs are not actually aware of whether they are carrying L2 VPN traffic or L3 VPN traffic (or indeed VPLS traffic, as discussed in the next chapter). Another advantage of using BGP as the signaling mechanism is that the route-target filtering (RTF) scheme described in the Layer 3 VPN chapter (Chapter 8) can be used to control the flow of L2 VPN routing updates as well as L3 VPN updates.

10.6.3 Comparison of BGP and LDP approaches to Layer 2 transport over MPLS

Table 10.1 compares the BGP and LDP approaches to Layer 2 transport over MPLS. The key difference is that the BGP scheme has autodiscovery properties similar to those familiar from L3 VPNs. This makes provisioning straightforward, both in terms of building the initial mesh when a Layer 2 VPN is first deployed and when new CE sites are added to the mesh over time as, in both cases, the required pseudowires are automatically created rather than having to be individually configured. The LDP scheme, in contrast, does not have any VPN awareness and hence requires manual pairwise configuration of pseudowires between PEs.

Table 10.1 Comparison of LDP and BGP control plane schemes for Layer 2 transport

	LDP-based scheme	BGP-based scheme
Control plane sessions	Fully meshed	Can use route reflectors or confederations to avoid full mesh
Explicit VPN awareness	No	Yes
In-built autodiscovery	No	Yes
Configuration burden of setting up a full mesh of connections between *N* sites	$O(N^2)$	$O(N)$
Interdomain capability	Difficult to achieve	Yes, using schemes analogous to those for L3 VPN

Hence, the provisioning burden of creating a full mesh of pseudowires between N sites is of order N^2, because on each PE, for each local CE, a connection must be provisioned to each remote CE. Whenever a new CE site is added, every PE in the network requires additional configuration. As a consequence of these differences, while either scheme might be suitable for a small deployment that is unlikely to grow over time, for any sizable deployment the BGP scheme is more appropriate.

The use of the BGP scheme becomes even more attractive if the same PEs are involved in the Layer 3 VPN service as well as the Layer 2 VPN service. Is this case, the same BGP sessions can be used for both the L3 VPN NLRIs and the L2 VPN NLRIs, and the same BGP infrastructure can be used, e.g. route reflectors. This is less cumbersome than having to invoke and maintain separate protocols (BGP and LDP) for the control plane of the two services. Furthermore, if the service provider is offering the VPLS service (the subject of the next chapter in this book), this can also use that same infrastructure if BGP signaling is used for the VPLS service.

10.7 FAILURE NOTIFICATION MECHANISMS

An important requirement for Layer 2 transport schemes is to provide a mechanism for a PE to indicate to a local CE that there is a problem with the connection to one of the remote CEs. For example, there may not be connectivity to the remote PE due to problems in the service provider's network, or the link between the remote PE and remote CE could be down.

First of all, how does a PE become aware of such connectivity problems? As described in a previous section, the BGP scheme provides a circuit status vector so that a PE can advertise to remote PEs the state of its local PE–CE circuits and the state of its LSPs to remote PEs. Earlier versions of the LDP scheme stipulated that a PE should withdraw the VPN label that it advertises to a remote PE if there are problems of this nature. More recently, a Status TLV has been added to the LDP scheme that allows one PE to signal to the remote PE the status of its connectivity. The action taken by a PE if there is a connectivity problem depends on the Layer 2 media type.

In the ATM case, operations, administration and management (OAM) cells can be generated by the PE and sent to the local CE. This action tells the ATM CE equipment that there is a problem with the VC or VP in question, so the CE will stop sending traffic on that connection. In the case where pseudowires are provided on a per-VC basis, AIS F5 OAM cells can be sent, and in the case where pseudowires are provided on a per-VP basis, AIS F4 OAM cells can be sent.

In the Frame Relay case, local management interface (LMI) frames can be used in a similar way to the OAM cells in the ATM case.

Ethernet is more problematic as most deployed equipment does not have any OAM functionality, although Ethernet OAM is currently under study in the standards bodies. If the pseudowires are being provided on a per-port basis, some implementations provide a mechanism to bring an Ethernet port down, in order to make the attached CE aware that there is a problem and to prevent it forwarding traffic to the PE. In the case where pseudowires are provided on a per-VLAN basis, this strategy of course cannot be used.

10.8 LAYER 2 INTERWORKING

The Layer 2 schemes described so far require both ends of a Layer 2 transport connection, or all the tail circuits of a given Layer 2 VPN, to be of the same Layer 2 media type. This can be a constraint in situations where a customer uses more than one media type, perhaps as a consequence of the prevalence of different media types in the various regions in which the customer is located, or as a consequence of mergers and acquisitions or because the customer is in the middle of migrating from one type of access medium to another. One way of relaxing this constraint is Layer 2 interworking, also known as 'Layer 2.5 VPNs'. This allows different media types to be used as the tails of the Layer 2 connections, with the proviso that the packets being transported over the connection must be IP packets.

In this scheme, when a packet arrives at a PE from the local CE, the entire Layer 2 encapsulation is stripped off, exposing the underlying IP packet. Note this is unlike the treatment of the Layer 2 frames described in Section 10.5 of this chapter, in which certain parts of the Layer 2 header are retained for transport to the remote PE. The underlying IP packet has the VPN label applied plus any transport labels and is sent to the remote PE. The remote PE extracts the IP packet and applies the appropriate Layer 2 encapsulation corresponding to the local Layer 2 tail circuit.

Only IP packets can be transported using this scheme because there is no way for the receiving PE to know which Layer 3 protocol the packet belongs to (this information having been discarded when the Layer 2 encapsulation was removed from the packet by the ingress PE; e.g. in the Ethernet case the information is carried in the Ethertype field). Hence the receiving PE simply assumes that the packet is IP and sets the relevant field accordingly when it builds the Layer 2 header ready for forwarding to the local CE.

10.9 CIRCUIT CROSS CONNECT (CCC)

This section describes Circuit Cross Connect (CCC)[CCC]. CCC was the first method to be devised and implemented for carrying Layer 2 traffic over a MPLS network and was the precursor to the LDP and BGP schemes discussed so far in this chapter. It is still used by service providers, and indeed is having a renaissance as a method to couple Layer 2 traffic into point-to-multipoint LSPs.

The main difference between CCC and the other schemes described in this chapter is that CCC always uses an RSVP-signaled LSP as the transport tunnel between PEs. Each CCC connection has a dedicated RSVP-signaled LSP associated with it, so unlike the LDP and BGP schemes discussed previously in this chapter, the transport tunnel cannot be shared between multiple connections. This is fine for small deployments, but if a large number of connections are required between particular pairs of PEs in a network, the number of RSVP-signaled LSPs will be correspondingly large. As a consequence of having a dedicated LSP for each connection, the inner label (VPN label) that is used in the BGP and LDP schemes to identify the connection that a packet belongs to is not required in the CCC case. The Layer 2 media types supported by CCC are the same as for the BGP and LDP schemes.

By default, in most RSVP implementations the egress router declares an implicit null label for the last hop of the LSP, so penultimate hop-popping (PHP) occurs. However, in the case of CCC the egress PE needs to know on which LSP traffic is arriving so that the traffic can be mapped on to the appropriate local Layer 2 interface (bearing in mind that there is no inner label or VPN label). Hence PHP is not used in the CCC case and a non-null label is used for the last hop of the LSP.

A new RSVP object called the Properties Object, carried within the RSVP Path messages, was defined to carry information pertaining to the CCC connection. It contains a Circuit Status TLV, which allows the PE at each end of the connection to convey the status of its PE–CE link to the other PE. Hence if the PE–CE link at one end goes down, the PE at the other end becomes aware of this.

For a point-to-point CCC connection, the connection is bidirectional, so an RSVP-signaled LSP is required in each direction between the two PEs. Configuration-wise, on each of the two PEs, the user creates an association between the local PE–CE interface (VC, VLAN, DLCI, etc.) and the outgoing and incoming RSVP-signaled LSPs corresponding to that connection. If one PE learns from the other PE (via the Circuit Status TLV) that the PE–CE link at the remote end has gone down, it declares the CCC connection down and ceases forwarding traffic on that connection. Similarly, if the LSP in one direction goes down, the CCC connection is declared down in both directions. The various media-specific OAM actions taken by the PE are similar to those described in the previous section.

As well as point-to-point connections, CCC is also used to transport point-to-multipoint traffic over P2MP LSPs. In the case of point-to-multipoint CCC connections, the connection is unidirectional, from the ingress router of the RSVP-signaled P2MP LSP to each of the egress routers. At the ingress PE, the user creates an association between the local PE–CE interface on which the Layer 2 PDUs enter the router and the P2MP LSP. At each egress PE, the user creates an association between the P2MP LSP and the local PE–CE interface on which the Layer 2 PDUs leave the router. If one or more of the egress PEs goes down (or one or more egress PE–CE connections), the P2MP CCC connection remains up so that the remaining PEs can still receive the traffic. As discussed in the P2MP MPLS chapter of this book (Chapter 6), some broadcast video codecs generate traffic in the form of ATM cells, so a P2MP CCC connection is the ideal way to transport these to multiple destinations across an MPLS network, thus emulating a point-to-multipoint VC. The fact that there is a one-to-one mapping between a CCC connection and a P2MP LSP is a good fit for the broadcast video application, as typically a P2MP LSP is created expressly for the purpose of carrying one and only one video flow.

10.10 RSVP SIGNALING FOR INTERDOMAIN PSEUDOWIRES

In earlier sections of this chapter, we have discussed BGP-based and LDP-based signaling of pseudowires. In this section, we discuss the motivation for pseudowires created by RSVP-based signaling.

Many enterprises have a presence in all corners of the world where no one single service provider can reach. Hence with present-day native ATM and Frame Relay services, interconnection agreements exist between service providers that enable the global transport of customers' traffic across multiple providers. Although the creation of such interconnection arrangements between service providers is nontrivial, the result is that customers can benefit from global connectivity.

In order for a similar connectivity to exist for Layer 2 services over MPLS, mechanisms are required for the signaling of pseudowires across AS boundaries, including parameters such as bandwidth requirements. Also, control over the route is likely to be required, at least in terms of which service providers the connection passes through. In addition, protection properties such as MPLS fast reroute are desirable. Some service providers contain multiple ASs, for administrative reasons or as a result of mergers and acquisitions, so for them interdomain pseudowire schemes are required in order to offer end-to-end service when the end-points are in different ASs.

RSVP-TE is a natural fit for these requirements. As seen in the Traffic Engineering chapter (Chapter 2), RSVP has mechanisms for creating bandwidth reservations and for explicit routing. Hence work is underway in the IETF to extend RSVP-TE to facilitate the signaling of pseudowires [RSVP-PW] – this section summarizes the current proposal. Note that the fact that RSVP is being used for the signaling of the pseudowire does not necessarily mean that RSVP must also be used for the signaling of the underlying transport tunnel. The underlying transport tunnel within each domain could be an RSVP-signaled LSP or an LDP-signaled LSP or could be a GRE tunnel, and the various domains that the pseudowire passes through could be using different types of transport tunnels. However, if bandwidth guarantees are required on behalf of the pseudowire, RSVP-signaled LSPs are required as the transport tunnel in order to ensure that the bandwidth guarantees can be honored. In this case, admission control of RSVP-signaled pseudowires on to transport RSVP-TE LSPs would be performed.

The way in which interdomain operation is achieved is by the concatenation of pseudowires. The end-to-end connection thus formed is known as a multihop pseudowire. For example, let us look at Figure 10.5. The multihop pseudowire has as its end-points

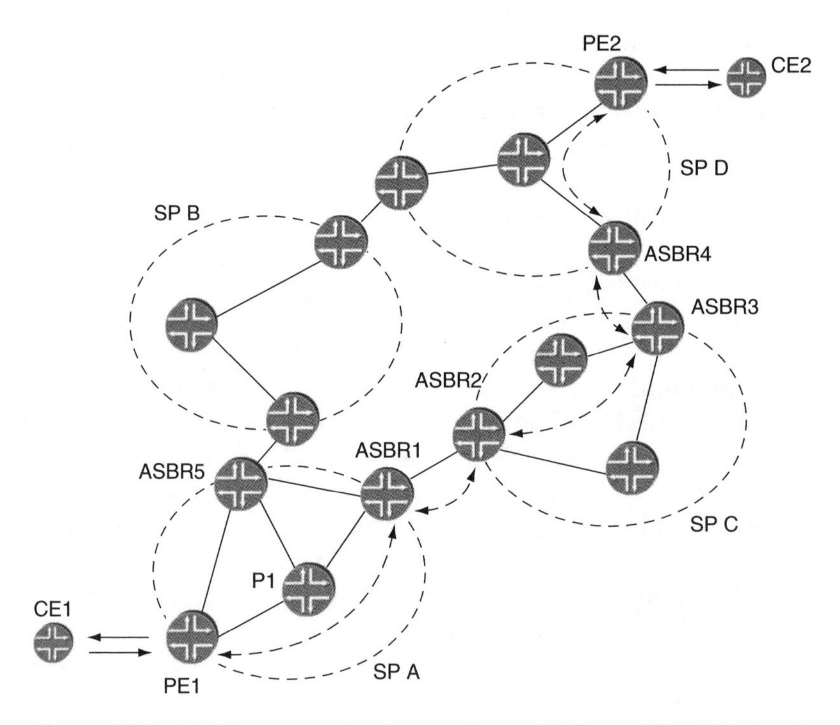

Figure 10.5 Multihop pseudowire carrying a 30 Mbps ATM CBR circuit

PE1, which is in the network of Service Provider A, and PE2, which is in the network of Service Provider D. In the example, the pseudowire corresponds to the customer's ATM CBR circuit having a bandwidth of 30 Mbps.

The individual pseudowire hops that constitute the multihop pseudowire are as follows:

- PE1 to ASBR1
- ASBR1 to ASBR2
- ASBR2 to ASBR3
- ASBR3 to ASBR4
- ASBR4 to PE2

As can be seen from the list above, a single pseudowire hop may comprise multiple router hops. For example, the pseudowire hop from PE1 to ASBR1 passes through P1. When the signaling process is complete, each hop on the multihop pseudowire knows the value of the inner label (VPN label) required to reach the next hop along the pseudowire. Thus PE1 knows the value of the inner label or VPN label, V, expected by ASBR1 for the pseudowire. In turn, ASBR1 knows the value of the VPN label, W, expected by ASBR2 for the pseudowire and so on. Hence, when ASBR1 receives a packet having an inner label V, it knows to swap the label for one having the value W and forward it to ASBR2.

Note that as for normal single-hop pseudowires, these labels are distinct from the labels used for the transport tunnels that exist within the domains. For example, when PE1 sends a packet on the pseudowire to ASBR1, it first pushes the inner label, V, expected by PE1 and then pushes the outer label corresponding to the transport tunnel to ASBR1 expected by P1.

Let us see how RSVP signaling of the multihop pseudowire works. The RSVP signaling uses the concept of nonadjacent neighbors. For example, the routers between PE1 and ASBR1, such as P1, are unaware of the existence of the pseudowire and so are not involved in the processing of the RSVP messages associated with the pseudowire. Therefore, for the purposes of pseudowire signaling, PE1 is a nonadjacent RSVP neighbor of ASBR1 and sends a directed RSVP Path message to ASBR1.

In the Foundations chapter of this book (Chapter 1), we saw that the LSPs created by RSVP signaling are unidirectional. The LSP is set up by Path messages traveling from the ingress towards the egress and Resv messages from the egress to the ingress, with the Resv messages allocating the label for each hop. In contrast, the signaling for multihop pseudowires is such that a single RSVP session is used to signal a bidirectional pseudowire. RSVP already has mechanisms to cater for this bidirectional approach – these are described in the Generalized MPLS (GMPLS) RSVP-TE extensions specifications [GMPLS-TE]. Path messages are used to allocate labels for traffic traveling along the pseudowire in one direction and Resv messages are used to allocate labels for traffic traveling in the opposite direction. For the example in the figure, if PE1 initiates the signaling of the pseudowire, PE1 sends an RSVP Path message to ASBR1 which contains the pseudowire label that ASBR1 must use when sending traffic to PE1; similarly, the Resv message sent by ASBR1 to PE1 contains the pseudowire label that PE1 must use when forwarding traffic on that pseudowire to ASBR1.

In order to be able to set up a multihop pseudowire, the following are required:

1. A means to identify the pseudowire that the RSVP messages relate to (bearing in mind that there may be multiple pseudowires between PE1 and ASBR1 and so on). One proposal to achieve this is to define a new TLV, known as the Pseudowire TLV.
2. A means to specify the hops to be traversed by the pseudowire. This information is conveyed in the Explicit Route Object (ERO) of the RSVP Path message. The ERO could contain the full set of hops, if that information is known to PE1, e.g. through manual configuration or offline computation. In that case, the ERO would have the form {PE1, ASBR1, ASBR2, ASBR3, ASBR4, PE2}. Alternatively, the ERO generated by PE1 may only contain a partial set of hops, which is added to by the intermediate hops that have the missing information. For example, if PE1 knows that the pseudowire in the figure needs to pass through ASBR1, but does not know the details of the hops required beyond there to reach PE2, then the ERO sent by PE1 would be {PE1, ASBR1, PE2}. ASBR1, ASBR2 and ASBR3 would then be responsible for adding in the missing hops into the ERO of the Path message. This process is known as ERO expansion. PE1 could know that the pseudowire needs to pass through ASBR1 by reference to the routing table, e.g. if ASBR1 is the BGP next-hop of the route to PE2. Alternatively, the exit ASBR from SP A's AS may have to

be manually configured on PE1, if the ASBR that would be chosen by routing is unsuitable. For example, if the ASBR that would be chosen by reference to the routing table is ASBR5, but SP B does not support interprovider pseudowires, the manual configuration would be necessary to ensure that ASBR1 is chosen as the next-hop of the pseudowire in order that the pseudowire passes through SP C's AS.

3. A means to signal the QoS requirements of the pseudowire. In the case of the example in the figure this would be the fact that a bandwidth reservation of 30 Mbps for the class type (CT) corresponding to ATM CBR is required. At described in the Traffic Engineering chapter and DiffServ Aware Traffic Engineering chapter of this book (Chapters 2 and 4 respectively), RSVP has the requisite mechanisms to signal QoS requirements.
4. For the setup of the pseudowire to be successful, sufficient bandwidth resources have to be available on a transport LSP corresponding to each hop of the pseudowire. Admission control of the pseudowire on to an LSP needs to be performed at each hop of the pseudowire. For example, PE1 has to ensure that 30 Mbps of bandwidth of the appropriate CT is available on an LSP from PE1 to ASBR1. Similarly, ASBR2 needs to ensure bandwidth is available on an LSP to ASBR3 and so on. The same process also occurs in the reverse direction, bearing in mind that the pseudowire is bidirectional.

10.11 OTHER APPLICATIONS OF LAYER 2 TRANSPORT

In this chapter so far, we have discussed how the Layer 2 transport mechanisms can be used to supply explicit Layer 2 services to enterprise customers. These Layer 2 transport mechanisms are also used as internal infrastructure tools in service provider networks and as a means for service providers to offer specialist services to other service providers. Some examples are listed below:

- Layer 2 connections can be used to provide access circuits to other services. For example, if a service provider is offering a Layer 3 VPN service to a customer, rather than providing a traditional leased line connection from the CE to the Layer 3 VPN PE, they may use an Ethernet-based pseudowire between the CE and the Layer 3 VPN PE across an MPLS-enabled metro Ethernet infrastructure.

- Smaller service providers sometimes have fragmented networks, each based in a particular region or city. They can use Layer 2 transport services bought from larger service providers to provide interconnections between these isolated islands.

Some service providers also offer smaller service providers a connection to a public peering exchange by means of an Ethernet pseudowire. In this way, the customer of the service does not need to have a router at the peering exchange yet can still enter into peering agreements with other companies present at the peering exchange. This is illustrated in Figure 10.6. Service provider X has a router, PE1, at a peering exchange. Service provider X supplies an Ethernet pseudowire to service provider Y from PE1 to interface if1 on PE3 at the peering exchange. Service provider X also supplies an Ethernet pseudowire to service provider Z from PE2 to interface if2 on PE3 at the peering exchange. Once the pseudowires are set up, as far as peers A, B and C are concerned, service provider Y and

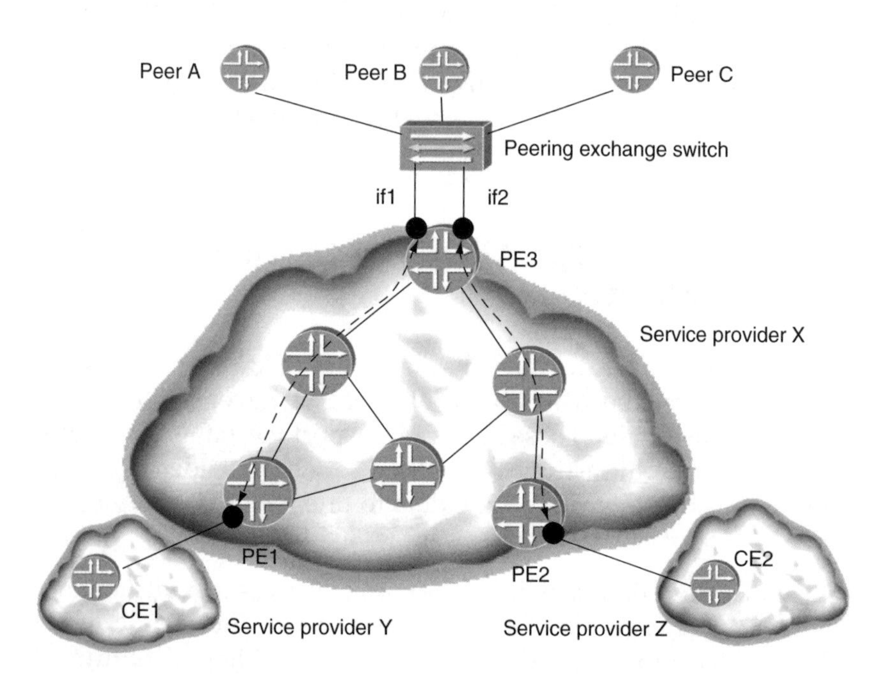

Figure 10.6 Use of pseudowires to connect small service providers to the peering exchange

Z are directly attached to the peering exchange. Service providers Y and Z can enter into peering arrangements with A, B and C and with each other without any involvement from service provider X, since the BGP sessions for the peerings involve peer A, peer B, peer C, CE1 and CE2 but not PE3.

A variation on this scheme is the use of pseudowires to interconnect smaller service providers for the purposes of private peering; e.g. service provider X could provide a pseudowire between CE1 and CE2 to enable service providers Y and Z to peer with each other without having to go through the peering exchange.

- Some service providers offer IP multicast as a service to their customers but do not run multicast protocols on their core routers. This might be as a matter of policy or because the core infrastructure is administered by a different department to the one running the edge routers and associated services. To solve this problem, multicast-enabled routers at the edge of the network are interconnected via pseudowires across the MPLS core. The consequence of doing this is a loss in bandwidth efficiency, since in effect the multicast routers are performing ingress replication rather than building a multicast tree across the core. However, if the volume of multicast traffic compared to unicast traffic is relatively low, this is often not regarded as a major issue. Note that an emerging alternative to such a scheme is one involving P2MP LSPs, in which the replication occurs within the MPLS network. This is discussed in more detail in the Point-to-Multipoint MPLS chapter of this book (Chapter 6).
- In mobile telephone networks, certain infrastructure interconnections are provided over ATM. As an alternative to using a native ATM core transport infrastructure to support these connections, it is sometimes advantageous to provide them using pseudowires over an MPLS network. This is especially the case if the service provider offers other services in addition to mobile services such as Internet connectivity or the Layer 3 VPN service as the MPLS network can also be used to support those services.

10.12 CONCLUSION

In this chapter, we have described the mechanisms underpinning Layer 2 transport over MPLS, and compared the two main control plane approaches that are in use. We have also discussed proposals for interdomain Layer 2 transport, which at the time of writing are under discussion in the IETF.

The ability to transport Layer 2 traffic over an MPLS network is a key ingredient of network convergence, allowing traffic to be migrated from ATM and Frame Relay networks and allowing new services based on Ethernet transport to be created. In the Ethernet case, the service is a natural extension of the technology already used within the enterprise. In the next chapter, we will see how the service provider can go one step further by offering an Ethernet multipoint service using a scheme called the Virtual Private LAN Service (VPLS).

10.13 REFERENCES

[CCC] K. Kompella, J. Ospina, S. Kamdar, J. Richmond and G. Miller, 'Circuit cross-connect', draft-kompella-ccc-01.txt (work in progress)

[GMPLS-TE] L. Berger (ed.), 'Generalized Multi-Protocol Label Switching (GMPLS) Signaling Resource ReserVation Protocol-Traffic Engineering (RSVP-TE) extensions', RFC3473 (work in progress)

[KOM-BGP] K. Kompella (ed.), 'Layer 2 VPNs over tunnels', draft-kompella-l2vpn-l2vpn-00.txt (work in progress)

[MRT-ENC] L. Martini *et al.*, 'Encapsulation methods for transport of Layer 2 frames over IP and MPLS networks', draft-martini-l2circuit-encap-mpls-09.txt (work in progress)

[MRT-TRS] L. Martini *et al.*, 'Transport of Layer 2 Frames Over MPLS', draft-martini-l2circuit-trans-mpls-16.txt (work in progress)

[PWE3-ATM] L. Martini, N. El-Aawar, J.Brayley, M. Bocci and G Koleyni, 'Encapsulation methods for transport of ATM over MPLS networks', draft-ietf-pwe3-atm-encap-09.txt (work in progress)

[PWE3-CON] L. Martini (ed.), E.Rosen, N.El-Aawar, T. Smith and G. Heron, 'Pseudowire setup and maintenance using LDP', draft-ietf-pwe3-control-protocol-17.txt (work in progress)

[PWE3-ETH] L. Martini, E. Rosen, N. El-Aawar and G. Heron, 'Encapsulation methods for transport of Ethernet over MPLS networks', draft-ietf-pwe3-ethernet-encap-10.txt (work in progress)

[PWE3-FR] L. Martini, C. Kawa and A.G. Malis (eds), Encapsulation methods for transport of Frame Relay over MPLS networks, draft-ietf-pwe3-frame-relay-05.txt (work in progress)

[RSVP-PW] R. Aggarwal, K. Kompella, A. Ayyangar, D. Papadimitriou, P. Busschbach and N. Sprecher, 'Setup and maintenance of pseudowires using RSVP-TE', draft-raggarwa-rsvpte-pw-02.txt (work in progress)

[TDM-PWE3] M. Riegel (ed.), 'Requirements for edge-to-edge emulation of TDM circuits over packet switching networks', draft-ietf-pwe3-tdm-requirements-08.txt (work in progress)

11

Virtual Private LAN Service

11.1 INTRODUCTION

In the previous chapter, we discussed point-to-point Layer 2 transport over an MPLS network. We discussed how the Ethernet case is especially attractive to enterprise customers as it is a natural extension of the technology already used on their own sites. In this chapter, we describe how to take this integration one step further, by enabling the service provider's network to appear as a LAN to the end-user. This scheme is called the Virtual Private LAN Service (VPLS).

11.2 THE BUSINESS DRIVERS

In previous chapters in this part of the book, we have discussed L3 VPN and L2 VPN services, and compared and contrasted the merits of the two schemes. Both schemes require some degree of networking knowledge on the part of the customer of the service. In the L3 VPN case, the customer may be required to configure a routing protocol to run between the CE and the PE, or at a minimum be required to configure a static route pointing to the PE. In the L2 VPN case, the customer builds an overlay network with point-to-point connections provisioned by the service provider and needs to run a routing protocol on that overlay network. Thus the degree of expertise required of the customer is somewhat greater than in the L3 VPN case. Both of these

MPLS-Enabled Applications: Emerging Developments and New Technologies Ina Minei and Julian Lucek

schemes may be fine for larger companies that have IT experts available to carry out the necessary designs and configurations.

However, with network-based applications becoming more prevalent in relatively small companies, there is also a need for such companies to have connectivity over the wide area. These companies might have a handful of sites and want to have connectivity between the LANs at those sites. For such companies, it is important to have an easy-to-use service as they may not have the luxury of IT experts that the larger companies have. VPLS achieves this by allowing them to interconnect their equipment over the wide area as if it were attached to the same LAN. Note that the customer plays no part in the emulation of the LAN service – the service provider's equipment does all the work. This is very attractive to the customer as deploying the service can be as simple as plugging an Ethernet switch at each site into an Ethernet port supplied by the service provider. In the case of the L2 VPN service described in the previous chapter, multiple VLANs are required between each customer site and the service provider PE (if a full mesh is required), as shown in Figure 11.1 (a). In the VPLS case, just one logical interface

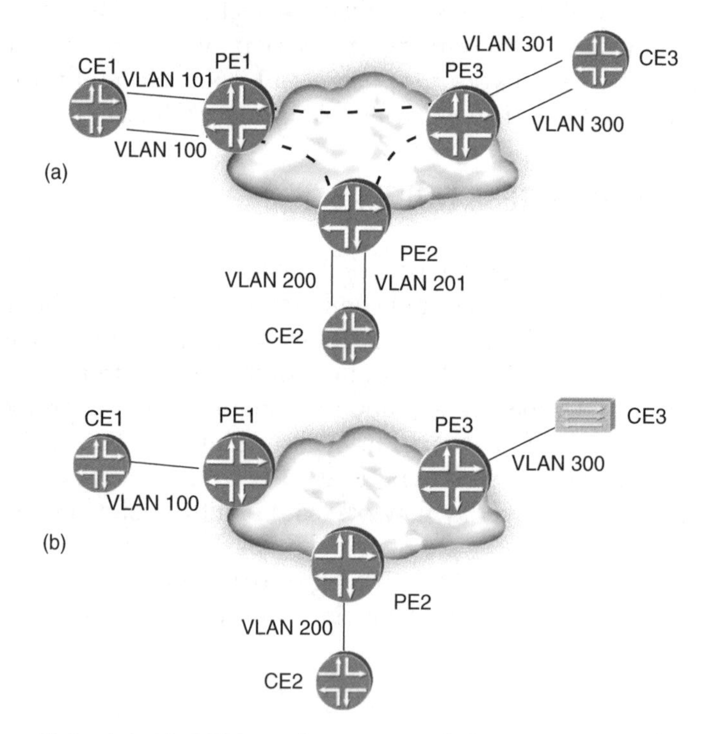

Figure 11.1 (a) L2 VPN service connectivity and (b) VPLS service connectivity

is required (e.g. a VLAN or an untagged Ethernet port), as illustrated in Figure 11.1(b). This is because the VPLS is a multipoint service, with the service provider's PE router taking care of which remote site, or sites, each frame needs to be delivered to. Like the L2 VPN services described in the previous chapter, any Layer 3 protocol can be carried over the VPLS, such as IPX and SNA.

Deploying VPLS allows the service provider to offer service to the small-to-medium enterprise sector that may have been difficult to address using L3 VPN or L2 VPN services. When the service is provided over native Ethernet (e.g. 100 Mbps or 1 Gbps Ethernet), it is easy for the service provider to offer a range of access rates and associated tariffs with the aid of a policer to enforce the access rate. As well as offering a native Ethernet access connection, other access media that are capable of encapsulating Ethernet frames could be used. This includes Frame Relay, using the method described in [RFC1490], and ATM, using the method described in [RFC2684]. Another possibility is to use a SONET/SDH circuit that supports the Generic Framing Procedure (GFP)[GFP] to encapsulate the Ethernet frame. A mixture of access media can be used within the same VPLS service instance, to cater for different types of site that the customer might have. For example, native Ethernet could be used to connect to city offices, but a DSL line with RFC2684 encapsulation could be used to connect to branch offices.

11.3 VPLS MECHANISM OVERVIEW

This section gives an overview of the VPLS mechanisms, using the service provider network shown in Figure 11.2 as a reference model. Shown in the diagram are the sites of two VPLS customers, X and Y. Customer X has sites attached to PE1, PE2 and PE3. Customer Y has sites attached to PE1, PE3 and PE4. From the point of view of each of the customers, the network appears to be a single LAN on to which that customer's, and only that customer's, CE devices are attached. That is to say, customer X belongs to one VPLS and customer Y belongs to another VPLS. This is illustrated in Figure 11.3, which shows the network from the point of view of customer Y. In Figure 11.2, each customer's device, whether a router or a switch, only requires a single Ethernet connection to the SP PE router (e.g. an untagged Ethernet interface or a VLAN), because the VPLS is a multipoint service, with the ingress PE taking responsibility of forwarding the frame according to its destination MAC

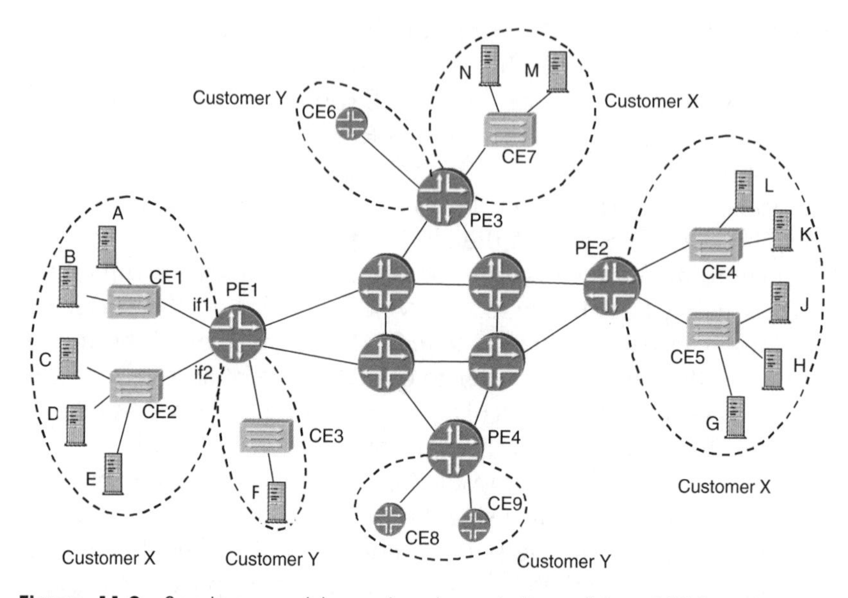

Figure 11.2 Service provider network and sites of two VPLS customers, customer X and customer Y

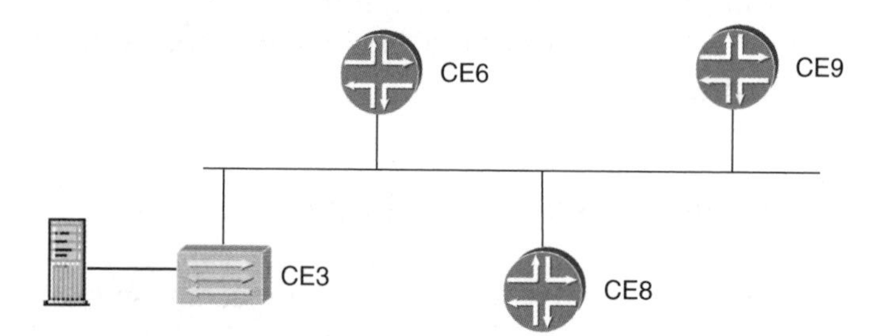

Figure 11.3 Network from the point of view of customer Y

address. For reasons of resilience, a CE can be attached to more than one PE. This is discussed in more detail in Section 11.5.2.2 of this chapter. Each site of customer X contains only a handful of PCs and the CE devices are all Ethernet switches. The sites of customer Y attached to PE3 and PE4 are offices containing a large number of PCs and the corresponding CEs are routers. Customer Y's site attached to PE1 is a small branch office and so a switch is used as a CE on that site. The repercussions of having a switch rather than a router as a CE are discussed later in this section.

For each VPLS, the PE routers are fully meshed with pseudowires. This is so that a PE receiving a frame from another PE can identify which VPLS the frame belongs to, on the basis of the pseudowire label. In addition, transport tunnels are required between the PEs to carry the pseudowire traffic. As with L3 VPN, LDP- or RSVP-signaled LSPs are typically used, but GRE or IPSec tunnels can be used as an alternative. For example, let us consider the connectivity between PE1 and PE3 in Figure 11.2. A pseudowire is required for VPLS traffic pertaining to the VPLS of customer X and another pseudowire for VPLS traffic pertaining to the VPLS of customer Y. In order to send traffic from customer X to PE3, PE1 identifies the corresponding pseudowire label, pushes it on to the Ethernet frame and then applies the tunnel encapsulation (e.g. another MPLS label if the tunnel is an LSP). This forwarding procedure is directly analogous to that for L3 VPN or Layer 2 transport, and indeed the same PE-to-PE tunnels can be used to carry traffic from all these services.

The question is, how does each PE discover which other PEs are members of a particular VPLS instance, so that it knows which PEs it needs to build pseudowires to? As with the point-to-point Layer 2 schemes discussed in the previous chapter, there are two main proposals for control plane implementation of the VPLS. The control plane of one scheme is based on LDP signaling and the control plane of the other is based on BGP signaling. The BGP version of VPLS has inherent autodiscovery mechanisms, which frees the user from having to configure the pseudowires manually. The LDP version of the VPLS has no inherent autodiscovery, so either the pseudowires must be manually configured or some external autodiscovery mechanism must be used. This is discussed in more detail later in Section 11.5.1 of this chapter.

As far as each customer is concerned, an Ethernet frame that is sent into the service provider network is delivered by the service provider to the correct site(s), on the basis of the destination MAC address. It is the task of each PE router to inspect the destination MAC address of each packet arriving from a locally attached site and to forward it to the appropriate destination site. This destination site may be attached to a remote PE or may be attached to another port on the same PE. If the destination site is attached to another PE, the ingress PE must forward the packet on the appropriate pseudowire to the remote PE. This means that the ingress PE needs to know which egress

PE to send the frame to.[1] In principle, two ways in which this can be achieved is to have a control plane signaling scheme to carry information about MAC addresses between PEs, or to have a scheme based on the MAC address learning. VPLS takes the latter approach, by having each PE take responsibility for learning which remote PE is associated with a given MAC address.

Each PE is functionally equivalent to a learning bridge, with a separate learning bridge instance for each VPLS. This is illustrated in Figure 11.4 for the case of customer X's VPLS on PE1. As can be seen, the bridge is regarded as having four logical ports, two of which are the local connections to CE1 and CE2. The other two ports are the pseudowires to PE2 and PE3. Note that PE1 is not aware of the detail of the connectivity behind each remote PE. For example, it does not need to know that customer X has two CEs attached to PE2 on separate ports. It simply needs to identify which frames need to be sent to PE2 and PE2 takes care of identifying which local port to forward the packet to. The essence of the learning function is as follows: by inspecting the *source* MAC address, say A, of a frame arriving on a port, whether an actual local port or a pseudowire from a remote PE, and by creating a corresponding entry in the forwarding table, PE1 learns where to send frames in the future having the *destination* MAC address, A. Thus with VPLS, neither centralized translation nor advertisement of MAC addresses are required.

It is important to highlight the effect of VPLS on the service provider's PE routers. In the case where Ethernet switches are

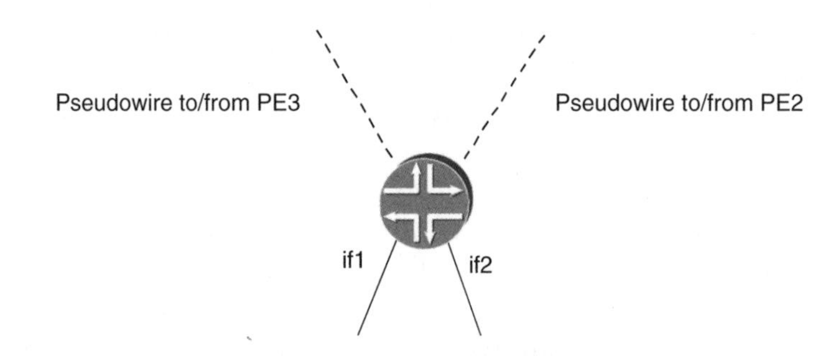

Figure 11.4 Illustration of a learning bridge on PE1 for customer X's VPLS

[1] An analogous situation exists in ATM LAN emulation (LANE). The LANE solution was to use central servers (LAN emulation servers) to provide a control plane function of translating destination MAC addresses to ATM addresses.

used as CE devices, the service provider's PEs need to learn the MAC addresses of individual hosts attached to the switches. As MAC addresses do not have a hierarchy, no summarization is possible, so PE routers need to have forwarding table entries for each individual MAC address. This means that if someone plugs their laptop into the office network served by CE2 in Figure 11.2, the effect will be felt by all the PEs that traffic to or from that laptop crosses. This is quite unlike the L3 VPN case, where PEs install forwarding entries for subnets, or Layer 2 point-to-point pseudowires, where PEs install forwarding table entries for attachment circuits. (In the L3 VPN case, a PE stores MAC addresses in its ARP cache, but only for hosts attached to a directly connected CE switch and not remote hosts.) As a consequence, the SP may decide to set a limit on the number of MAC addresses, and in practice it may be better for a large customer deployment to have routers as CEs rather than switches. Referring again to Figure 11.2, customer Y's site attached to PE3 is using a router as a CE, so the only MAC address from that site as far as the service provider is concerned is that of the Ethernet interface on CE6 facing PE3, and the service provider is not exposed to the network behind CE6.

Now that we have presented an overview of the VPLS model, let us examine the mechanisms in more detail. First, we discuss the forwarding plane mechanisms. Then we discuss the two control plane schemes for the VPLS in turn, the LDP-based signaling scheme and the BGP-based signaling and autodiscovery scheme. The LDP-based scheme is similar to the LDP-based scheme for point-to-point Layer 2 connections discussed in the previous chapter and the BGP-based scheme is similar to the BGP-based scheme for point-to-point Layer 2 connections. The forwarding plane mechanisms are very similar regardless of which signaling scheme is used.

11.4 FORWARDING PLANE MECHANISMS

In this section, we discuss the forwarding mechanisms associated with VPLS in more detail. The user needs to configure which local ports are members of each VPLS on each PE. Each PE maintains a separate forwarding table for each VPLS. For example, PE1 in Figure 11.2 maintains a forwarding table for the VPLS associated with customer X and another forwarding table associated with the VPLS in customer Y.

A requirement of the VPLS scheme is that, for each VPLS, the participating PE routers should be fully meshed with pseudowires. This means that if a PE needs to send a packet to a remote site, it can send it directly to the appropriate remote PE without the packet having to be processed by an intermediary PE. The advantage of full meshing is that the PE routers do not have to run a spanning tree algorithm to eliminate the possibility of loops (in the same way that Ethernet switches need to run such an algorithm if they are not fully meshed).

We will first discuss how unicast Ethernet frames are forwarded to their destination and then discuss the treatment of broadcast and multicast frames.

11.4.1 Forwarding of unicast frames

Let us examine how unicast Ethernet frames are forwarded from the source to the destination. Referring again to Figure 11.2, let us focus on host A and host J in customer X's network. The MAC address of host A is A and the MAC address of host J is J. Let us suppose that host A sends a frame with source MAC address A to host J with destination MAC address J. Suppose that PE1 does not know the location of MAC address J. As a learning bridge would do, PE1 floods the packet on all ports except the port on which it arrived (refer back to Figure 11.4 for an illustration of the ports). This means that the packet is flooded to the port to CE2, the pseudowire to PE2 and the pseudowire to PE3.

Let us consider what happens at PE2 and PE3. PE2 and PE3 know that the incoming frame belongs to customer X's VPLS, by virtue of the pseudowire on which the frame arrived. PE2 and PE3 each perform a lookup on the destination MAC address in their VPLS forwarding tables corresponding to customer X. If PE2 does not know the location of MAC address J, it floods the frame on its local ports facing CE4 and CE5. Note, however, that it does *not* flood the frame to any of the other PEs in the network – there is no need to do so, because all the PEs are fully meshed, so each receives a copy of the frame directly from the ingress PE. This split horizon scheme ensures that forwarding loops do not occur (otherwise PE3 might send the frame to PE2 which sends it to PE1 which sends it to PE3 again and so on). Similarly, PE3 sends the frame on to the port facing CE7 (but not CE6, since CE6 does not belong to customer X's VPLS).

Receiving frames with source MAC address A enables each PE to learn the location of A, in terms of the port on which the frame arrived. Thus PE1 puts an entry in its forwarding table creating an association with the port facing CE1 and PE2 and PE3 put an entry in their forwarding table creating an association between MAC address A and their respective pseudowires to PE1. At this stage, the forwarding table for customer X's VPLS on PE1 is as shown in Figure 11.5. As can be seen, PE1 has an entry for MAC address A pointing to interface if1, which is the interface facing CE1.

Let us now suppose that host J starts sending frames to host A. This is quite likely if host A has been sending frames to host J, since many applications are bidirectional. PE2 has a forwarding table entry for destination MAC address A pointing to the pseudowire to PE1, so it sends the frame on the pseudowire to PE1 (and does not need to flood it). When PE1 receives the frame it learns the fact that frames to host J should be sent on the pseudowire to PE2 and updates its forwarding table accordingly. PE1 already has an entry in its forwarding table for MAC address A so it forwards the frame on if1 to host A.

PE1 has now learnt that frames to J must be forwarded on the pseudowire to PE2, so it no longer needs to flood frames to all the other PEs. The forwarding table entry for MAC address J contains the pseudowire label expected by PE2 for frames belonging to customer X's VPLS arriving from PE1 and the transport tunnel required to reach PE2. The choice of tunnels is the same as in the L3 VPN or L2 point-to-point case. In the typical case, the transport tunnel would be an LDP- or RSVP-signaled LSP, but they could also be GRE or IPSec tunnels. If the same PEs are being used to offer all of these services, then the same transport tunnels between a pair of PEs can be shared among all those services.

Let us assume that some time later PE1 has learnt all the MAC addresses in customer X's VPLS. The forwarding table corresponding to customer X's VPLS is as shown in Figure 11.6. As can be seen, some of the entries correspond to hosts reachable via a local

MAC address	Next-hop
A	if1

Figure 11.5 Forwarding table on PE1 for customer X's VPLS, after learning MAC address A

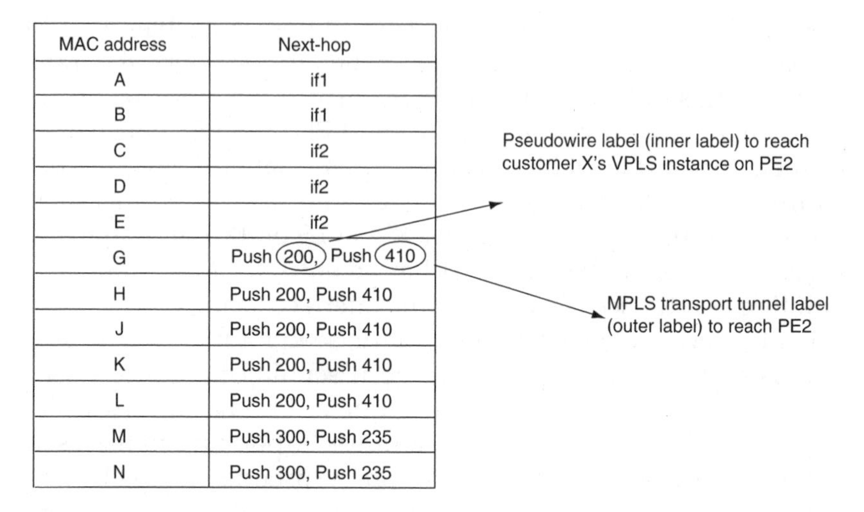

MAC address	Next-hop
A	if1
B	if1
C	if2
D	if2
E	if2
G	Push 200, Push 410
H	Push 200, Push 410
J	Push 200, Push 410
K	Push 200, Push 410
L	Push 200, Push 410
M	Push 300, Push 235
N	Push 300, Push 235

Figure 11.6 Forwarding table on PE1 for customer X's VPLS, after learning all the MAC addresses in the VPLS

interface and some of the entries point to hosts reachable via a remote PE. In the latter case, the forwarding table shows the pseudowire label and the transport tunnel label that must be pushed on to the Ethernet frame in order to forward it to the correct VPLS instance on the correct PE. For example, let us suppose that the pseudowire label expected by PE2 for frames arriving from PE1 belonging to VPLS X is 200 and the MPLS tunnel label required to reach PE2 is 410. As can be seen from the forwarding table, any frames that PE1 needs to send to hosts G, H, J, K or L have label 200 pushed on to the frame followed by label 410. Note that PE1 does not need to know the detail of the layout of the customer domain 'behind' the remote PE routers. For example, as can be seen from the forwarding table in the figure, PE1 knows that J and K are reachable via the pseudowire to PE2, but does not know (or care) that J and K are attached to different switches.

It should be noted that in the process described above, there was no advertising of MAC addresses using the control plane; the MAC addresses are always learnt. VPLS implementations have mechanisms for MAC aging, so that stale MAC addresses can be removed from the forwarding table. For example, an implementation may choose to remove a MAC address that has not been used for a certain number of minutes. Also, if the size of the table reaches its limit, the implementation may choose to remove the entries that have remained unused for the longest period of time.

11.4.2 Broadcast and multicast frames

Having discussed the treatment of unicast frames, let us discuss the treatment of broadcast and multicast frames. Let us suppose PE1 receives a broadcast frame sent by host B. The frame must be forwarded to all sites of customer X's VPLS. To achieve this, PE1 floods the frame on the pseudowires to PE2 and PE3 and on the port to CE2. In turn, PE2 and PE3 flood the frame to the attached CEs belonging to customer X, but, as a consequence of implementing the split horizon, do not send the frame to any PE.

The implementations deployed at the time of this writing treat multicast traffic in exactly the same way as broadcast; i.e. the frame is flooded throughout the VPLS. As a consequence, each PE that has a member of that VPLS attached receives a copy of the packet, even though it may not have any interested receivers attached. This may be fine if the amount of multicast traffic is relatively low; otherwise the bandwidth wastage may be of concern to the service provider.

Interestingly, it has been pointed out [WC2005] that the current scheme for dealing with multicast traffic in the VPLS works at the opposite limit to the current scheme for dealing with multicast traffic in L3 VPNs, in that the former has no multicast state in the core but wastes bandwidth as a consequence of using ingress replication for multicast traffic, whereas the latter is more bandwidth efficient but has a potentially large amount of multicast state in the core. However, there is no fundamental reason why the two types of VPN should be handled in a different way. Indeed, aspects of the next-generation solution set already described in the L3 VPN chapter for L3 VPN multicast are also applicable to the VPLS [MCAST], thus giving to the service provider a common set of tools to cater for the two cases. Let us look at how these tools are used in the VPLS case.

Instead of performing ingress replication, multicast trees can be used within the service provider network to distribute multicast VPLS traffic. The way in which these trees can be used is similar to the L3 VPN case: a single tree could be shared between all VPLS instances or multiple trees could be used with either a one-to-one or many-to-one mapping between VPLS instances and multicast trees. These schemes are equally applicable to IP multicast traffic and non-IP multicast traffic.

In cases where the customer is sending IP multicast traffic, certain multicast groups within a VPLS instance could be mapped to one multicast tree, other multicast groups to another multicast tree and so on, each of the trees involved also potentially carrying traffic from other VPLS instances. Each tree would only distribute packets to those PEs having sites that are members of the corresponding multicast groups. Note that in order to support such a scheme, each PE router needs to be aware of which sites contain members of each multicast group. Bear in mind that, unlike the L3 VPN case, with the VPLS there is no multicast protocol interaction between the PEs and the CEs so the PEs need to snoop PIM/IGMP protocol traffic passing between customer sites in order to ascertain multicast group membership.

As in the L3 VPN case, in all the VPLS cases described above, the multicast trees can be P2MP LSPs rather than PIM trees. This has the advantage that the same signaling mechanism can be used for the P2MP LSPs as for the tunnels for point-to-point traffic (i.e. RSVP or LDP). Also, if RSVP is used to signal the P2MP LSPs, the user can reserve bandwidth and has precise control over the path followed by each of the sub-LSPs within the P2MP LSP (e.g. to create a minimum cost tree rather than a shortest path tree in order to have the greatest bandwidth efficiency) and can take advantage of MPLS fast reroute for traffic protection.

11.5 CONTROL PLANE MECHANISMS

In the previous section, we discussed the forwarding plane mechanisms for the VPLS. Let us now turn our attention to the control plane mechanisms. There are two aspects to be considered:

- The discovery aspect. How does a PE know which other PEs have members of a particular VPLS attached?
- The signaling aspect. How is a full mesh of pseudowires set up between those PEs?

For example, in Figure 11.2, PE1 has members of VPLS X and VPLS Y attached. It needs to know that PE2 and PE3 have members of VPLS X attached and that PE3 and PE4 have members of VPLS Y attached. It then needs a means to signal a pseudowire to PE2 and a pseudowire to PE3 pertaining to VPLS X and a pseudowire to PE3 and a pseudowire to PE4 pertaining to VPLS Y.

As with the Layer 2 point-to-point transport discussed in the previous chapter, there are two alternative schemes for the signaling aspect. One of the schemes is based on LDP and the other on BGP. The LDP scheme is very similar to the LDP scheme for point-to-point transport and the BGP scheme is very similar to the BGP scheme for point-to-point transport discussed in the previous chapter. With regard to the discovery aspect, the BGP scheme has a built-in automated mechanisms for this discovery process (as is the case with L2 point-to-point transport signaled using BGP), so the process is known as autodiscovery. In contrast, the LDP scheme does not support autodiscovery. Therefore in the LDP case, either one must manually configure the pseudowires, in terms of which PE is the destination of each, or introduce some external discovery mechanism.

11.5.1 LDP-based signaling

The LDP signaling scheme for VPLS [LDP-VPLS] is very similar to the LDP scheme for point-to-point Layer 2 connections described in the previous chapter. LDP is used for the signaling of the pseudowires that are used to interconnect the VPLS instances of a given customer on the PEs. In order to signal the full mesh of pseudowires required, a full mesh of targeted LDP sessions is required between the PEs, or at least each pair of PEs that have VPLSs in common (unless H-VPLS, or hierarchical VPLS, is being used; see Section 11.5.1.2). In the absence of an autodiscovery mechanism, these sessions must be manually configured on each PE router. This LDP session is used to communicate the value of the 'inner label' or 'VPN label' that must be used for each pseudowire. The FEC element defined to carry the necessary parameters is the same as that described in the previous chapter, so this section repeats some of this information as a recap. The VC ID, which in the point-to-point case was used to identify a particular pseudowire, is configured to be the same for a particular VPLS instance on all PEs. Hence the VC ID allows a PE to identify which VPLS instance the LDP message refers to.

For example, referring to Figure 11.2, PE1 has an LDP session with PE2, PE3 and PE4. Let us suppose the network operator assigns VC ID 100 to the VPLS of customer X and VC ID 101 to the VPLS of customer Y. Over the LDP session with PE2, PE1 and PE3 exchange pseudowire labels for each of the two VPLSs that they have in common. In the absence of any autodiscovery mechanism, the VC IDs must be listed manually on each PE, and each VC ID must be associated with a list of remote PE addresses.

Let us look at the information that is communicated over the LDP session, in addition to the label value itself. The FEC element includes the following fields:

- VC ID.
- Control Word bit. This indicates whether a control word will be used.
- VC type. This indicates the encapsulation type. In the case of VPLS, this would be Ethernet or VLAN-tagged Ethernet.
- Interface parameters field. This contains information such as the media MTU.

It should be noted that just because LDP is being used as the signaling mechanism for the pseudowires this does not mean that LDP must be used as the signaling mechanism for the underlying transport tunnels used to carry the packets from the ingress PE to the egress PE. As stated in Section 11.4.1 of this chapter, the transport tunnels could be RSVP-signaled or LDP-signaled LSPs or could be GRE or IPSec tunnels.

11.5.1.1 Autodiscovery mechanisms

The LDP scheme does not have any in-built autodiscovery mechanisms. As a consequence, each LDP session and each pseudowire must be manually configured or some external autodiscovery mechanism must be used. In recognition of the fact that manual configuration is not an attractive option for service providers, work is being carried out in the IETF into potential external discovery mechanisms. In contrast, the BGP scheme for the VPLS discussed later in this chapter has inherent autodiscovery, so no external mechanism is required. Mechanisms that have been proposed for use with the LDP signaling scheme are as follows:

- *BGP* [BGP-AUTO]. Here BGP is only being used as an autodiscovery mechanism; LDP is still being used to signal the pseudowires.
- *RADIUS* [RAD-AUTO]. In this scheme, PE sends a request containing an identifier specific to a VPLS instance. The Radius

server returns a list of addresses of PEs belonging to that VPLS instance.

- *LDP*. Extending LDP to provide autodiscovery has also been proposed in the past, however at the time of writing there is no work on this topic in the IETF.

11.5.1.2 Hierarchical VPLS

Hierarchical VPLS (H-VPLS) is a scheme that was devised in order to address a significant limitation of the LDP-based signaling scheme, the fact that a full-mesh of LDP sessions is required between PE routers. In all but the very smallest of deployments, the requirement for a full mesh is a burdensome administrative overhead. The H-VPLS scheme removes this restriction, although, as we shall see, at the expense of introducing other issues. As discussed at the end of this section, H-VPLS also gives some bandwidth efficiency improvement when dealing with multicast or broadcast traffic.

In the H-VPLS scheme, instead of a PE being fully meshed with LDP sessions, a two-level hierarchy is created involving 'hub PEs' and 'spoke PEs'. The hub PEs are fully meshed with LDP sessions. Attached to each hub PE are multiple spoke PEs. The spoke PEs are connected to the hub PEs via pseudowires, one per VPLS instance. From the point of view of a spoke PE, it has local ports and a pseudowire 'uplink' port, leading to the parent hub PE. The spoke PE performs flooding and learning operations in the same way as a normal VPLS PE. However, the spoke PEs are not required to be fully meshed with LDP sessions.

Let us look at Figure 11.7 in order to examine this scheme. Let us suppose that all the PEs in the diagram provide a VPLS service, but the service provider wants to avoid the operational overhead of configuring a full mesh of LDP sessions and of adding to that mesh as more PEs are deployed in the future. The service provider could instead choose to implement an H-VPLS scheme by designating P1, P2, P3 and P4 as hub PEs and the PEs PE1 to PE13 as spoke PEs. Rather than having to fully mesh PE1 to PE13, as would be the case without H-VPLS, only P1, P2, P3 and P4 are fully meshed with LDP sessions in order to exchange pseudowire labels. In addition, there is an LDP session between each hub PE and each of its satellite spoke PEs. A pseudowire is created between each hub PE and each spoke PE for each VPLS instance (i.e. for each VPLS customer attached to the spoke PE).

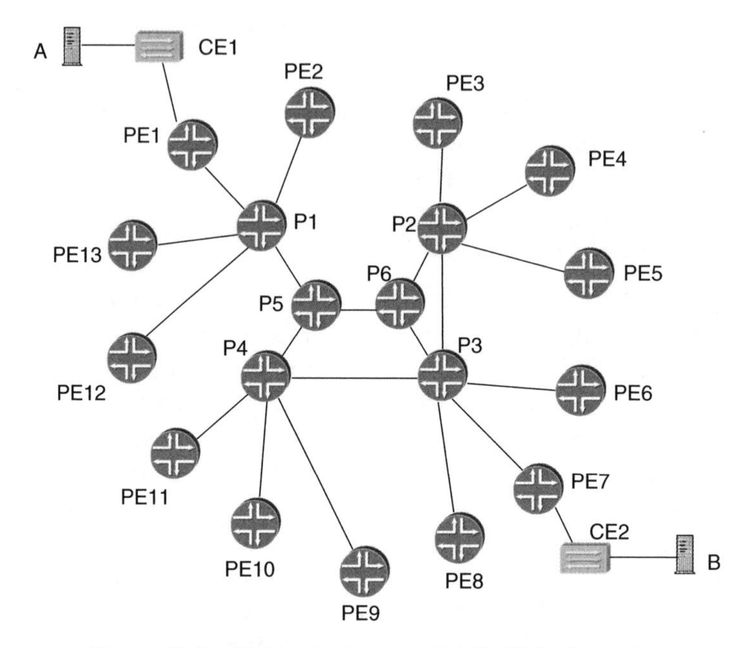

Figure 11.7 Network diagram for H-VPLS discussion

For example, in the figure, if P1 is the hub PE for spoke PEs PE1, PE2, PE12 and PE13, a pseudowire is created between P1 and each of those spoke PEs per VPLS that each spoke PE supports. The spoke PEs maintain a separate forwarding table for each VPLS and populate it by learning MAC addresses. Let us suppose that PE1 has two customers attached, X and Y. PE1 maintains a separate forwarding table for each of the two VPLS instances it is involved in. From the point of view of the instance pertaining to customer X, it has a logical port which is the pseudowire for that instance to P1 and a local port. It learns MAC addresses over those ports in the usual way. Let us consider the logical ports on the hub PE, P1, for that same VPLS. It has a pseudowire to each of the other hub routers in the LDP mesh and a pseudowire to each of the spoke PEs. It learns and floods across these. If host A needs to send to host B and the location of B is unknown to PE1, PE1 floods on all ports, including the uplink (pseudowire) to P1. P1, if it does not know the location of B, floods on all pseudowires, i.e. the pseudowires to its spoke PEs, PE2, PE12 and PE13, and the pseudowires to the other hub PEs. The other hub PEs in turn flood the frame to their spoke PEs. In this way, the packet is flooded to all locations on that customer's VPLS. If a new spoke PE is added to the network, or an additional customer is attached to an existing spoke PE, a new configuration is only required on that spoke PE and the parent hub PE.

Let us compare the properties of the H-VPLS scheme to a normal one-level VPLS in which the routers PE1 to PE13 are fully meshed with LDP sessions. In the H-VPLS scheme, the hub PEs, P1, P2, P3 and P4, have to learn MAC addresses. In contrast, if the service provider had deployed a one-level VPLS scheme with routers PE 1 to PE13 as the PEs then P1, P2, P3 and P4 are simply P routers and so do not carry any of this information. The MAC addresses needed to be stored in a hub PE in the H-VPLS scheme is roughly equal to the sum of the MAC addresses in its satellite spoke PEs. This is regarded as a significant limitation of the H-VPLS scheme, and represents a departure from the usual VPN model in which P routers do not carry any VPN-specific forwarding state.

Another property of the H-VPLS scheme to examine is the handling of broadcast/multicast traffic. If ingress replication is used to deal with this type of traffic, then the H-VPLS scheme is more bandwidth efficient than the one-level VPLS scheme. For example, PE1 only needs to send one copy of each broadcast/multicast packet to P1 which then floods to the other hub PEs and to its satellite spoke PEs. However, if the bandwidth inefficiency of ingress replication in a one-level VPLS scheme is a concern, a much better method of curing it is through the use of multicast replication trees, as already discussed in Section 11.4.2 of this chapter, rather than the use of H-VPLS. In contrast to H-VPLS, which only saves the bandwidth on the first hop, a multicast replication tree can give optimum bandwidth utilization end to end.

11.5.2 BGP signaling and autodiscovery

The BGP signaling and autodiscovery scheme for VPLS [BGP-VPLS] is very similar to that for L2 VPN and L3 VPN. Fundamentally it has the following components:

- A means for a PE to know which remote PEs are members of a given VPLS. This process is known as autodiscovery.
- A means for a PE to know the pseudowire label expected by a given remote PE for a given VPLS. This process is known as signaling.

The BGP NLRI takes care of the two components above at the same time, the NLRI generated by a given PE containing the necessary information required by any other PE. These components enable the automatic setting up of a full mesh of pseudowires for each VPLS without having to manually configure those pseudowires on each PE.

Like the BGP scheme for L3 VPN and L2 VPN, on each PE a Route Distinguisher and a Route Target is configured for each VPLS. The Route Target is the same for a particular VPLS across all PEs, and is used to identify which VPLS an incoming BGP message pertains to. This is exactly analogous to the L3 VPN and L2 VPN cases, in which the RT identifies which VRF or L2 VPN instance a BGP advertisement pertains to. As in the L3 VPN and L2 VPN cases, the Route Distinguisher is used to disambiguate 'routes'.

On each PE, for each VPLS an identifier is configured, known as a VPLS Edge Identifier (VE ID). Each PE involved in a particular VPLS must be configured with a different VE ID.[2] BGP is used to advertise the VE ID to other PEs in the network. This, along with other information in the NLRI, provides the means for remote PEs to calculate the value of the pseudowire label required to reach the advertising PE. The key advantage of the scheme is that each PE discovers the identities of all the other PEs in each VPLS without requiring any manual configuration of that information.

The VE ID is somewhat analogous to the CE ID in the BGP signaling scheme for L2 VPNs discussed in the previous chapter. One difference is that a single VE ID covers all the CEs that belong to a given VPLS instance on a PE, whereas a different CE ID is needed for each CE in a given L2 VPN instance on a PE. For example, in Figure 11.2, although PE2 has two CEs attached that are members of customer X's VPLS, it advertises one VE ID that encompasses both (and any other CEs that customer X might attach in the future to PE2). This difference is because, in the L2 VPN case, the pseudowire maps to a particular local Layer 2 tail circuit (VLAN, VC, etc.), whereas in the VPLS case, the pseudowire maps to the VPLS instance pertaining to a particular customer.

Note that, for a given VPLS, a given PE requires that each remote PE uses a different pseudowire label to send traffic to that PE. This is to facilitate the MAC learning process, as described in Section 11.4.1 of this chapter. Knowing which PE sent a frame

[2] An exception is the scheme for multihoming discussed later in this chapter.

means that the receiving PE can learn which PE is associated with the source MAC address of the frame. A PE in principle could simply send a list of pseudowire labels required to reach it, one per remote VPLS Edge in that VPLS. However, this could potentially mean having to send a long list of labels if there are a large number of PEs in the network that are involved in that VPLS instance.

Instead, as with BGP L2 VPNs described in the previous chapter, the necessary information is communicated in the BGP NLRI to enable each remote PE to *calculate* the pseudowire label expected by the advertising PE. The beauty of this scheme is that each PE in a given VPLS only needs to generate a small 'nugget' of information (as little as one NLRI per VPLS) to enable any remote PE to know that the PE in question is a member of that VPLS (by virtue of the route target and the BGP next-hop) and the pseudowire label expected by that PE.

Let us look in more detail at the information communicated by BGP to see how this scheme works. As can be seen, the scheme is very similar to the scheme for L2 VPNs described in the previous chapter. A BGP update message contains the following items:

- Extended community (route target). As with L3 VPNs, this allows the receiving PE to identify which particular VPN the advertisement pertains to.
- L2-Info extended community. This community is automatically generated by the sending PE. Encoded into the community are the following pieces of information:
  - Control flags, e.g. a flag to indicate whether a Control Word is required or not.
  - Encapsulation type (i.e. Ethernet with VLAN tagging or untagged Ethernet). This allows the receiving PE to check that the local and remote ports are configured in a consistent manner.
  - MTU (so that PE can check that remote ports are configured with the same MTU as the local ports).
- Other BGP attributes such as the AS path, etc.
- The NLRI. This contains the following items:
  - Route Distinguisher. As with L3 VPNs, this allows 'routes' pertaining to different VPNs to be disambiguated.
  - VE ID.

- o Label base.
- o VE block offset.
- o VE block size.

The label base, VE block offset and VE block size are the information required for a remote PE to calculate the pseudowire label to use when sending traffic to the VPLS in question on the advertising PE. A PE allocates 'blocks' of labels. Each block is a contiguous set of label values. The PE does not explicitly advertise each label within the block. It simply advertises the value of the first label in the block (the label base) and the number of labels in the block (the VE block size).

In simple cases, there is only one label block whose size is sufficient that each remote PE has a label to use within the block. In such cases, the label value that a remote PE, having a VE ID of value X, must use to reach the advertising PE is computed as follows:

$$\text{Label value} = \text{label base} + X - 1$$

For example, let us refer again to Figure 11.2. Suppose that PE1 advertises a label base of 100 000. PE2 and PE3 receive the advertisement, either directly or via a route reflector. Assume that the VE ID of customer X's VPLS instance on PE 1 is 1, that on PE2 is 2 and that on PE3 is 3. When PE3 needs to forward a frame to PE1, it calculates the pseudowire label expected by PE1 by adding its own VE ID (value 3) to the label base of 100 000 and by subtracting 1, yielding a label value of 100 002 to be used as the inner label. The BGP next-hop of the route is PE1, so PE3 knows that the frame should have an outer label pertaining to the tunnel (RSVP- or LDP-signaled LSP) that leads to PE1.

Sometimes, there may be more than one label block, e.g. if the original label block was exhausted as more sites were added to the VPLS. In this case, each block is advertised in a separate NLRI. Note that in the BGP NLRI, there is a 'VE block offset' parameter. This is equal to the VE ID that maps on to the first label in the block. For example, let us suppose that there are two label blocks for a particular VPLS, each with a range of 8. The first would have a label block offset of 1, so VE ID 1 would map to the label base of that block. The second would have a label block offset of 9, so VE ID 9 would map on to the label base of that block. Let us suppose that the label base of the second block is 100 020. A PE forwarding traffic from VE ID 12 would use the fourth label in the

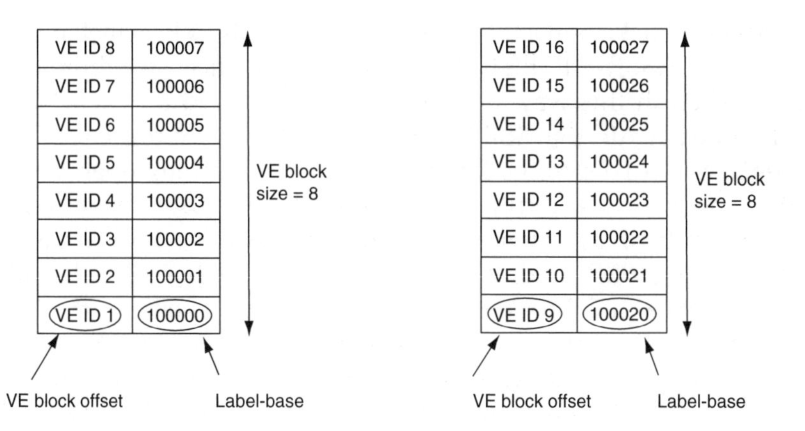

Figure 11.8 Illustration of label blocks and their mapping to VE IDs

second block, namely 100 023. These label blocks are illustrated in Figure 11.8.

11.5.2.1 Interprovider mechanisms for BGP-signaled VPLS

It is likely that some customers requiring VPLS service are based in multiple geographies. Often it is the case that no single service provider can provide service in all those geographies. Hence there is a need for interprovider capability for the VPLS, in such a way that a seamless end-to-end service can be offered to the customer across multiple ASs.[3] In the Hierarchical and Inter-AS VPNs chapter of this book (Chapter 9), we discussed three alternative schemes by which interprovider capability can be achieved for L3 VPNs. These schemes are known as option A, option B and option C after their section numbers in the IETF draft. The draft for BGP signaling for VPLS describes an analogous version of each of these for the VPLS case. Many of the mechanisms are the same for the L3 VPN and VPLS cases, which saves having to reinvent the wheel. Let us examine each of these in turn.

Option A: VPLS to VPLS connections between ASBRs

By analogy with the L3 VPN case, in which a back-to-back connection is made between VRFs on the two ASBRs, in the VPLS case

[3] Also, in some cases a service provider network may contain multiple ASs, hence requiring the use of interprovider mechanisms.

a back-to-back connection is made between the VPLS instances on the two ASBRs. If there are multiple instances needing to be connected in this way, a separate VLAN is used for each. To the PEs in each AS, the ASBR in that AS acts as any normal PE and hence is involved in the flooding and MAC learning operations in the same way as a normal PE. The fact that the ASBR needs to hold MAC address information is analogous to the L3 VPN case, in which the ASBR holds L3 VPN routes. To each ASBR, the other ASBR acts as a CE router. One issue with the scheme in the VPLS case is that if for reasons of redundancy multiple inter-AS connections are required between the two ASs between different ASBRs, then a scheme would be required to prevent forwarding loops occurring.

Option B: distribution of VPLS information between ASBRs by BGP

In this scheme, an EBGP session between ASBRs is used to advertise label blocks from one AS into another. For each label block advertised by a PE in the same AS as the ASBR, the ASBR creates an analogous label block which is advertised to the peer ASBR over the EBGP session. The label block created by the ASBR contains the same number of labels and maps on to the same VE IDs as the original label block, but the label base can be different. Why is this 'translation' of the label block necessary? If the ASBR simply relayed the label block with the same label base as that chosen by the originating PE, the label values could clash with labels that the ASBR had allocated for other purposes. For the same reason, in the L3 VPN case, the VPN label may be translated by the ASBR.

In the forwarding table, the ASBR installs a label swap operation between each label value in the original label block and the corresponding label value in the new label block. Note that the analogous scheme for the L3 VPN potentially may involve the ASBR holding a large number of L3VPN routes. However, in the VPLS scheme the number of 'routes' is small (as few as one per VPLS instance) so the scheme is very scalable in the VPLS case. Note that the ASBR is not involved in any MAC learning and does not need to hold any MAC address information.

Option C: Multihop EBGP redistribution of VPLS NLRI between source and destination AS, with EBGP redistribution of labeled IR4 routes between ASBRs

This is directly analogous to option C for L3 VPN, except that the EBGP multihop connection between PEs (or route reflectors) in the two ASs conveys VPLS NLRI rather than L3 VPN NLRI. As far as the ASBRs are concerned, the operations are the same as for the L3 VPN case – PE loopback addresses are advertised as labeled IPv4 routes to enable an MPLS forwarding path between PEs in the two ASs. Indeed, if the ASBRs are already exchanging labeled IPv4 routes for the purposes of interprovider L3 VPN, no change in configuration is required to accommodate the interprovider VPLS service since the same PE-to-PE MPLS path can be used by both services. The ASBRs are not involved in any of the VPLS forwarding or control plane operations, in the same way that they are not involved in the L3 VPN operations in the L3 VPN case. At the time of writing, this inter-AS scheme had already been deployed by a pair of service providers.

Experience with interprovider deployments for L3 VPN has shown that option B is a popular one, despite potential scaling issues, because the peering arrangements are analogous to those for Internet peering, involving EBGP connections between directly connected border routers. It is likely to become popular for the VPLS case for the same reasons. Option C requires EBGP multihop connections between PE routers or between route reflectors (in addition to EBGP sessions between border routers), which makes it a less popular choice. In summary, the BGP-based scheme for VPLS is based on tried and tested techniques developed for L3 VPN interprovider schemes. This emphasizes a key advantage of using the same signaling protocol for VPLS as is used for L3 VPN.

11.5.2.2 Multihoming

In the diagram shown in Figure 11.2, each customer site was attached to one PE only. In practice, customers may wish to have some or all sites attached to more than one PE for reasons of resilience. In such scenarios, it is important to avoid Layer 2 forwarding loops occurring, bearing in mind that, unlike the case of IP forwarding loops, there is no time-to-live (TTL) mechanism to limit the number of circulations a packet can make. If the customer's CE devices are routers, then loops will not occur. If, however, the customer's CE devices are Ethernet switches, then there is the danger of loops occurring unless countermeasures are taken.

Let us look at Figure 11.9 to see how such loops could occur in the absence of any countermeasures. The Ethernet switch CE1 is homed to PE1 and PE2. Host A sends an Ethernet frame addressed to host B. Let us assume that none of the PEs in the network know the location of host B. PE3 floods the frame to all the PEs in the network. PE1 and PE2 each send a copy of the frame to CE1. Let us assume that CE1 does not know the location of host B. If CE1 is not running a spanning tree, then on receiving the frames from the PEs, it floods them on all ports, except the incoming port. As a result, PE2 receives a copy of the frame that had been sent to CE1 from PE1. PE2 floods it on all ports (excluding the incoming one), so PE1 receives a copy of the same frame and floods it on all ports and so on.

One countermeasure is for the customer to run the spanning-tree algorithm on its switches, so that the switches can create a loop-free topology by selectively blocking ports. However, this involves the service provider relying on the customer to implement this correctly. This runs counter to the model used by service providers for other services, in which the mechanisms of the service and the practices adopted by the service provider protect the service provider from mistakes made by the customer. Mistakes made by one customer cannot be allowed to affect other customers or the service provider.

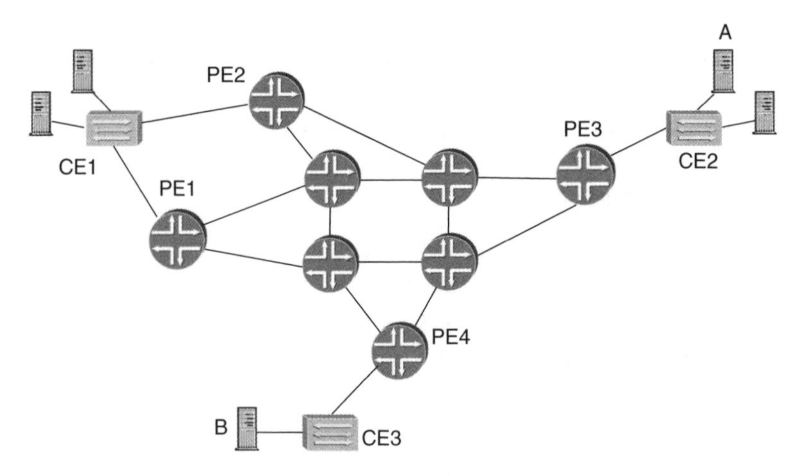

Figure 11.9 Multihoming of the customer's CE

The service provider can protect itself from such errors by only allowing one port to be active at a time. The question is how the PEs to which the customer site is homed know which should be the active port and how other PEs know which port is the active one. This can be achieved in the BGP case in quite a natural and straightforward way. Let us consider the dual-homed case shown in Figure 11.9, although the scheme operates in an analogous fashion for homing to more than two points.

The service provider configures the VPLS instance on PE1 and PE2 with the same VE ID, even though they are attached to different PEs. Each PE creates a BGP NLRI for that VE ID in the usual way. Thus each PE in the network receives two advertisements for the same VE ID. In the same way as, for example, IP path selection, BGP applies selection rules to determine which of the two NLRIs it installs in its routing table.

The goal is that traffic from other PEs only exits the network on one of the two ports facing CE1. Let us suppose that by default we wish traffic to exit from PE1. This can be achieved by having PE1 apply a higher (more favorable) local preference than PE2 when advertising the NLRI. Hence PE3, and any other PEs involved in that VPLS, choose to install the version of route advertised by PE1. PE2 sees the advertisement from PE1 and, seeing that it has a higher local preference than its own version of the route, knows not to accept packets on the port from CE1. If PE1 goes down, or the port or link facing CE1 on PE1 goes down, then the BGP session between PE1 and its peers (other PEs or route reflectors) times out or PE1 sends a BGP withdrawal message for the NLRI. This triggers the other PEs to install the version of the route sent by PE2 and triggers PE2 to start using its port to CE1. An additional advantage of the BGP path selection scheme (compared to simply relying on correct use of the Spanning Tree Protocol, or STP, by the customer) is that a remote PE (e.g. PE3) only sends broadcast traffic to the PE with the active port, rather than to both PEs, resulting in less bandwidth wastage in the core of the network.

11.5.3 Comparison of LDP and BGP for VPLS control plane implementation

Let us now compare the LDP and BGP schemes for the VPLS. Many of the differences between the two schemes are analogous to the differences between the LDP and BGP schemes for point-to-point Layer 2 transport discussed in the previous chapter of this book.

11.5.3.1 Control plane signaling sessions

A key difference between the LDP and BGP schemes is the fact that a full mesh of LDP sessions is required between the PE routers, whereas a full mesh of BGP sessions is not required between PE routers. This has two consequences:

1. Potentially a PE router is involved in a large number of targeted LDP sessions. This number grows in proportion to the number of PEs in the network and could be the factor that limits how much the network can grow. This is in contrast to the case in which LDP is used as a label distribution protocol for MPLS transport tunnels. In that case, the number of LDP sessions is typically low and fairly constant (one session between each directly connected pair of routers in the network).
2. The configuration burden of having a full mesh of LDP sessions is large. When a new PE is added to the network, a new LDP session has to be configured on every existing PE in the network. This is in contrast to the case in which LDP is used as a label distribution protocol for MPLS transport tunnels. In that case, LDP is very attractive precisely because of the ease of configuration, the configuration typically involving activating the LDP protocol on each core interface in the network. In addition, if md5 authentication is in use, there is the overhead of managing the md5 authentication keys associated with each LDP session in the mesh.

Note that a full mesh of BGP sessions would give rise to similar issues. In order to address this, BGP Route Reflection was developed. It is in widespread use today and enables service providers to run the Layer 3 VPN and Internet service without having to maintain a full mesh of BGP sessions between PE routers. The same route reflectors can also be used for the VPLS service.

H-VPLS, on the face of it, solves the two problems described above, but at the expense of introducing other problems, as discussed in Section 11.5.1.2 of this chapter. This is because H-VPLS is not a mechanism analogous to BGP route reflection. BGP route reflectors, by relaying signaling information, remove the need for direct sessions between PEs without affecting any forwarding plane operations. H-VPLS removes the need for LDP sessions between outer PEs, but in an indirect way, by having spoke PE routers forward frames to a hub PE. Because a spoke PE does not forward frames directly to another spoke PE, it does not need a pseudowire to it and hence does not need an associated LDP session. As described in Section 11.5.1.2 of this chapter, the consequence of this is an increased amount of learning and storage of MAC addresses on the hub PE, approximately equal to the sum of the MAC addresses stored on its spoke PEs.

Let us return to Figure 11.7 to examine how the BGP version of VPLS deals with the problem of a full mesh of control plane sessions. Instead of having a full mesh of BGP sessions between PE1 and PE13, route reflectors can be used. For example, P1, P2, P3 and P4 can be designated route reflectors and PE1 to PE13 can be router reflector clients. In so doing, the P routers do not get involved in having to learn and store MAC address information. They are simply holding and relaying the BGP reachability information received from the PE routers. As a consequence, there is no need to employ H-VPLS in the BGP version of VPLS. In order to deploy a new PE, one simply configures BGP sessions between that PE and its route reflectors.

11.5.3.2 Discovery of remote PEs

Another difference between the BGP and LDP versions of VPLS is the way in which a remote PE discovers which other PEs are involved in a particular VPLS. The LDP version of VPLS, as it stands today, does not have any in-built discovery mechanism, so the identities of the remote PEs must be manually configured on the routers. This means that if a new site is added to a customer's existing VPLS on to a PE that does not already have sites of that customer attached, then all the other PEs that have a site of that VPLS attached must be configured with a pseudowire to that PE. This runs counter to the operational model for L3 VPNs, where if a new CE is added to a customer's VPN, only the PE to which it is attached requires any configuration. The BGP scheme for VPLS in contrast, described in Section 11.5.2 of this chapter, has in-built autodiscovery. Therefore the operational model is very similar to that for L3 VPN: if a new site is added to an existing VPLS, only the PE to which the site is attached requires any new configuration.

As already discussed in Section 11.5.1.1 of this chapter, there are investigations in progress into external autodiscovery mechanisms that could be used in conjunction with LDP. One of the proposed schemes is to combine BGP autodiscovery with LDP signaling. However, it is difficult to see the advantage of this scheme compared to using BGP for autodiscovery *and* signaling and not using LDP at all. Once BGP is used for autodiscovery, the amount of extra information required to convey pseudowire MPLS label information is very small, with one NLRI taking care of both aspects.

As well reducing the operational burden, an autodiscovery scheme reduces the probability of configuration errors being introduced. As already discussed, a premise of the VPLS schemes is that the PE routers are fully meshed with pseudowires for each VPLS instance. If by accident one or more pseudowires are omitted, unexpected behaviour can occur within the customer domain, the cause of which can be difficult to pinpoint. For example, in Figure 11.2, let us suppose the pseudowire between PE1 and PE3 corresponding to customer X's VPLS service is missing. If host A sends an ARP corresponding to host M, then it receives no reply. If, however, it sends ARPs for host L, then it does receive a reply. If all the hosts involved were attached to a traditional LAN, then that would lead one to conclude that host M is turned off or the port to it is down, whereas in fact the problem is in the SP part of the network. Let us look at another example. Let us suppose that the CE routers in customer Y's VPLS are running OSPF and the pseudowire between PE3 and PE4 corresponding to customer Y's VPLS service is missing. CE6 is the designated router for the LAN and CE8 is the backup designated router. CE8 does not hear OSPF hellos from CE6 as a consequence of the missing pseudowire. This causes CE8 to take over as the designated router, confusing other CEs in the network, which are still receiving OSPF hellos from CE6.

11.5.3.3 Multihoming capability

As described in Section 11.5.2.2, the BGP version of VPLS has a multihoming capability, based on standard BGP path selection techniques. LDP has nothing analogous to BGP attributes such as local preference, so such a scheme is not possible in the LDP case. Currently many service providers offer multihoming options for their L3 VPN services at a premium to single-homing options. The customer benefits from greater resilience and the service provider benefits from greater revenue. It is reasonable to expect that multihoming will also be a requirement of many VPLS customers. Currently, the only way of multihoming in the LDP case is to employ H-VPLS and deploy a spoke PE on the customer's premises, which is homed to multiple-hub PEs. However, this introduces the disadvantages previously discussed for H-VPLS in Section 11.5.1.2.

11.5.3.4 Interprovider operations

In Section 11.5.2.1, we discussed the interprovider capability offered by the BGP scheme for VPLS and showed how the schemes are analogous to those for L3 VPN. In the LDP case, the only methods proposed at the time of writing are:

1. *'Brute force' meshing.* If interprovider capability is offered by ASs 1 and 2, a full mesh of LDP sessions is created between all the PEs in AS 1 and AS 2. In addition, all the PEs in the two ASs providing the VPLS service to a particular customer need to be fully meshed with pseudowires. This compounds the operational difficulties with adding a new PE or adding an additional site to an existing customer's VPLS, as extra configuration is required on all the PEs involved in the two ASs.
2. *Using a spoke pseudowire.* A spoke pseudowire (per VPLS) is provisioned between border routers, in order to interconnect the VPLS instances in the two domains. Methods for providing multiple connections for redundancy between different border routers without causing loops are under investigation. This problem is the same as in the option A BGP scheme described in Section 11.5.2.1. However, the option B and option C BGP schemes described in that section avoid the problem by the use of the AS path attribute. LDP does not have the capability to create schemes equivalent to these BGP schemes.

In summary, interprovider VPLS services are more readily created using the BGP version of VPLS, because BGP (by definition) was designed with inter-AS operations in mind. Also the BGP version of VPLS has the advantage that the machinery developed for L3 VPN interprovider operations can largely be reused.

Table 11.1 Comparison of LDP and BGP control plane schemes for VPLS

	LDP	BGP
Control plane sessions	Fully meshed, unless H-VPLS is used	Can use BGP route reflection or confederations to avoid full mesh
SP-controlled multihoming capability	Limited, requires H-VPLS	Yes
Commonality with operational model used for L3 VPN	None	High
In-built autodiscovery	No	Yes
Configuration burden of setting up mesh of N pseudowires	$O(N^2)$	$O(N)$
Interdomain capability	Difficult to achieve	Yes, using schemes analogous to those for L3 VPN

11.5.3.5 Summary of differences between LDP and BGP schemes for VPLS

The differences between the LDP and BGP signaling schemes for VPLS are summarized in Table 11.1. Although in principle it might be possible to modify LDP to accommodate all these issues, in effect one would be reinventing BGP, so the advantage of doing so is not clear.

11.5.4 Operational considerations

Let us consider some of the operational issues that occur when running VPLS services. Because of the nature of the VPLS service and the way it is more entwined with the customer's network than other VPN services, some of the operational issues discussed below do not have analogies in the L3 VPN or L2 VPN services. A consideration to bear in mind is that having a VPLS service does not allow the enterprise customer to exceed the best common practices with regard to the scope of a LAN, e.g. in terms of the number of attached hosts. The same scaling issues would occur as with a traditional LAN, where the amount of broadcast traffic becomes excessive if too many hosts are attached.

11.5.4.1 Number of MAC addresses per customer

A consideration for the service provider is the number of MAC addresses to be stored by each PE, bearing in mind that a PE might be providing a VPLS service to a large number of customers. Although implementations exist in which the number of MAC addresses stored can be large, there is always an upper limit. The service provider may need to protect themselves against an exhaustion of MAC address capacity by limiting the number of MAC addresses that are stored for each VPLS customer. VPLS implementations exist that allow the service provider to limit the number of MAC addresses on a per-VPLS basis or on a per-interface basis. This is by analogy with some L3 VPN implementations that allow the service provider to limit the number of L3VPN routes in a VRF. Being able to control the number of MAC addresses also opens up interesting billing opportunities for the service provider where the customer is billed according to the MAC address limit that they choose to purchase.

11.5.4.2 Limiting broadcast and multicast traffic

Another operational consideration for the service provider is to consider limiting the volume of broadcast and multicast traffic, bearing in mind that the cost to the service provider of sending such traffic could be high, especially in cases where ingress replication is used and there are a large number of PE members in a VPLS. As a consequence, some VPLS implementations allow the service provider to rate-limit this type of traffic.

11.5.4.3 Policing of VPLS traffic

If the VPLS service is delivered over 100 Mbps or 1 Gbps native Ethernet ports, service providers may need to police the amount of traffic that the customer sends into their network on each access port. The service provider can offer tiered services in terms of the amount of traffic that the customer is allowed to send, by analogy with many existing L3 VPN services.

11.6 CONCLUSION

In this chapter we have explored the Virtual Private LAN Service (VPLS). At the time of writing, the VPLS is gaining in popularity and is a valuable addition to a service provider's product portfolio. Because the service is simple for the customer to deploy, the service provider can address a wider range of customers than if they only offered L3 VPN and point-to-point Layer 2 services.

This chapter concludes the exploration of services enabled by MPLS. Because of the mission-critical nature of some of the traffic carried by these services, the ability to manage and troubleshoot the underlying network infrastructure is important. This is the subject of the next chapter in this book.

11.7 REFERENCES

[BGP-AUTO] H. Ould-Brahim, E. Rosen and Y. Rekhter (eds), 'Using BGP as an auto-discovery mechanism for Layer-3 and Layer-2 VPNs', draft-ietf-l3vpn-bgpvpn-auto-06.txt (work in progress)

[BGP-VPLS] K. Kompella and Y. Rekhter (eds), 'Virtual Private LAN Service', draft-ietf-l2vpn-vpls-bgp-05.txt (work in progress)

[GFP] *Generic Framing Procedure*, ITU-T Recommendation G.7041, 2001

[LDP-VPLS] M. Lasserre and V. Kompella (eds), 'Virtual Private LAN Services over MPLS', draft-ietf-l2vpn-vpls-ldp-07.txt (work in progress)

[MCAST] R. Aggarwal, Y. Kamite and L. Fang, 'Multicast in VPLS', draft-raggarwa-l2vpn-vpls-mcast-00.txt (work in progress)

[RFC1490] T. Bradley, C. Brown and A. Malis, *Multiprotocol Interconnect over Frame Relay*, RFC 1490, July 1993

[RFC2684] D. Grossman and J. Heinanen, *Multiprotocol Encapsulation over ATM Adaptation Layer 5*, RFC 2684, September 1999

[RAD-AUTO] J. Heinanen, G. Weber, W. Townsley, S. Booth and W. Luo, 'Using RADIUS for PE-based VPN discovery', draft-ietf-l2vpn-radius-pe-discovery-01.txt (work in progress)

[WC2005] Y. Rekhter, Paper D1–02, in MPLS World Congress, Paris, February 2005

12

MPLS Management

12.1 INTRODUCTION

In the previous chapters we have seen how MPLS is used as a key component for converging multiple services on to the same physical infrastructure. As service providers obtain more of their revenue from MPLS-enabled applications and as the traffic carried requires stricter SLAs, the ability to manage both the MPLS infrastructure and the services running on top of it efficiently becomes more important.

In this chapter we will take a look at some of the unique aspects of managing MPLS networks and services. Some of the topics covered are fault detection and the emerging mechanisms for detecting data plane failures, provisioning challenges and tools for gaining visibility into the network.

12.2 MANAGEMENT – WHY AND WHAT

From the provider's point of view, management is a broad definition of all the aspects that allow him to offer, deploy and bill for a service. This includes provisioning the service, detecting and isolating failures, avoiding downtime and accounting for billing purposes. The availability of tools for accomplishing these tasks and the capabilities of such tools affect the costs incurred for deploying the service and the revenue that can be derived from it. Good tools can

MPLS-Enabled Applications: Emerging Developments and New Technologies Ina Minei and Julian Lucek

ease the provisioning process, reduce the troubleshooting time when a fault occurs and provide granular accounting.

Management is a broad topic that could easily be the subject of an entire book. In this chapter, we focus on the router functionality that provides the necessary information and the basic tools for managing both the MPLS infrastructure and the services enabled by it. Some of this functionality, such as many accounting features, was added by vendors following customer demand and is specific to a given implementation. Other functionality, such as LSP ping, was defined in the IETF as new protocol machinery that routers must implement.

Because management is such an important piece of any proposed solution, work is done in this area in each and every one of the IETF working groups. In fact, a document cannot advance in the standards track in the IETF without having the appropriate management support, for example, in the form of an SNMP Management Information Base (MIB). In addition to the work in the IETF, the ITU has also produced a large number of standards for MPLS operations and management.

When discussing MPLS management, two questions need to be answered:

1. What are the management functions that must be provided?
2. At which layer must these functions be applied?

The answer to the first question is straightforward. The functions required from any management solution apply to MPLS management as well. In this chapter, we group them in the following categories:

(a) Fault detection and troubleshooting functions, such as the ability to detect misrouting of a packet in the network.
(b) Configuration functions, such as the ability to avoid misconfigurations or to automate the configuration process.
(c) Visibility functions, such as the ability to accurately account traffic or obtain information about a deployed service.

The answer to the second question is more complex. Because MPLS is used at different layers, the management functions must be provided at each of these layers. The different layers are:

(a) Device layer, e.g. the individual links and nodes in the network.
(b) Transport layer, e.g. the inter-PE tunnels in the core, set up with LDP or RSVP, used to transport MPLS-labeled traffic.

(c) Virtual connection layer, e.g. the per-VPN virtual tunnels created by the VPN labels.
(d) Service layer, e.g. the VPN service itself.

Because the same tools and methods are often used for managing different layers, we will discuss MPLS management from the perspective of the functions provided, rather than discussing it from the perspective of how each individual layer can be managed. Let us start with what is perhaps the most important aspect of MPLS management, detecting and troubleshooting failures.

12.3 DETECTING AND TROUBLESHOOTING FAILURES

Detecting the failure is the first step towards fixing it. As MPLS networks start carrying more and more services with strict SLAs, fast failure detection becomes a must. However, given the MPLS fast-reroute mechanisms discussed in the Protection and Restoration chapter (Chapter 3), is this entire discussion irrelevant? The answer is 'no'. The local protection mechanisms of MPLS protect against a physical link or a node failure, but other events, such as corruption of a forwarding table entry or a configuration error, can also cause traffic forwarding problems.

When talking about forwarding failures, there are two goals. The first, and most important, is to detect the problem quickly. For a provider, the worst possible scenario is to find out about the existence of a problem from the customer asking why the service is down. The second goal is to automatically recover from the failure. This may mean switching the traffic to a different LSP or even bringing down a service, with the correct indication, instead of blackholing traffic for a service that is reported to be up and running.

12.3.1 Reporting and handling nonsilent failures

From the point of view of failure detection, there are two types of forwarding errors: silent and nonsilent. We will discuss nonsilent failures first, and talk about both detection and fast recovery. Nonsilent failures are the ones that the control plane is aware of, such as the tear-down of a (nonprotected) LSP following a link-down event. If the control plane is aware of the failure, then the problem can be quickly reported. For example, most vendors support sending an SNMP trap when an LSP is torn down, not just for the primary path but also for protection paths. This error indication is important for the operator, who can correlate this information with other events in the network, such as VPN traffic being dropped for a particular customer or the potential failure of protection for an LSP.

The quick detection of the failure in the control plane does not guarantee that traffic will not be impacted. Here are two examples of how this can happen:

1. Delayed handling of the error. In a VPN setup, when a link-down event causes the PE–PE LSP to be torn down, the quick detection of the problem at the LSP head end does not necessarily guarantee that the VPN customers will not experience any traffic loss. Recall from the advanced L3 VPN chapter (Chapter 8) that if the reevaluation of the VPN routes is timer-based rather than event-driven, the VPN routes will not be immediately reevaluated, causing blackholing of customer traffic for a bounded amount of time until the reevaluation happens, and traffic is switched to an alternate LSP. Even if the reevaluation of the routes is event-driven, in cases where a large number of forwarding entries must be updated following the failure (to point to a different LSP), traffic loss will still happen until all entries are updated.
2. Insufficient propagation of the error. Let us look at a pseudowire service, where a failure of the transport tunnel results in a loss of connectivity between ingress and egress PEs. Assuming that this failure is detected, the PEs may send native indications over the related attachment circuits to notify the end-points of the fault condition. In such a case it is necessary to map the error correctly, using procedures such as those defined in [PW-OAM-MSG-MAP]. If the emulated service does not have well-defined error procedures, such as Ethernet, it is not possible to do so.[1] In such cases, although the error is known to the PEs, it cannot be correctly propagated over the attachment circuits.

[1] Work is underway in the ITU-T and IEEE to define in-band Ethernet OAM standards.

To summarize, nonsilent failures are reported in the control plane. The knowledge of the LSP failure can be used by the routers to update the forwarding state and by the operator to address the problem that caused the failure in the first place. However, the fact that the failure is reported in the control plane cannot guarantee that no traffic will be lost.

12.3.2 Detecting silent failures – MPLS OAM

Silent failures are the ones that the control plane is not aware of and are usually caused by a loss of synchronization between the control and data planes. The classic example of a silent failure is the corruption of a forwarding table entry. This is a popular example because some of the early implementations of MPLS suffered from this problem. As the implementations matured, the problem was resolved, but it remained one of the major concerns for providers because corrupted forwarding entries are particularly difficult to troubleshoot. They usually cause traffic blackholing, but may sometimes lead to traffic misrouting where traffic is incorrectly forwarded. Because the problem manifests itself in the data plane only, it is difficult to detect. For example, traffic to only a handful of destinations may be lost. Finally, this type of failure requires manual intervention to fix, usually rebooting the entire router.

The only way to detect a silent failure is by constantly monitoring the operation of the forwarding plane by sending test traffic. However, at what level should this be done? To answer this, let us take a look at a BGP/MPLS L3 VPN service. The options are:

- Transport layer. The inter-PE MPLS tunnels in the core, set up with LDP or RSVP, provide connectivity between the PEs and transport all VPN traffic across the core.
- Virtual connection layer. The per-VPN virtual tunnels created by the VPN labels are invisible in the core of the network. They provide the demultiplexing capability at the PE to steer traffic towards the correct customer site.
- Service layer. Rather than testing the individual building blocks providing the service, this approach tests the service itself, e.g. by sending test traffic between the customer sites in a VPN.

Regardless of which level the polling happens at, the next question is how often should the test probes be sent? There are two factors to take into consideration when answering this question:

1. The desired detection time. Clearly, the probes must be sent at intervals shorter than the desired detection time. Just how much shorter depends on how the detection mechanism works, e.g. the time it takes to receive feedback for a given probe and the number of failed probes that are necessary to declare a failure.
2. The resources spent on detection. Intuitively, it is easy to understand that a polling-based mechanism uses up forwarding-plane resources, because the probes themselves need to be forwarded along with the regular traffic. However, processing the probes may place a burden on the control plane as well, as will be seen in the following section. Thus, failure detection through polling comes at a cost and there is a tradeoff between quick detection and resources spent on doing so.

The data-plane mechanisms for detecting and pinpointing failures in MPLS networks are collectively referred to as MPLS OAM (operations, administration and management), implying data-plane OAM. When discussing MPLS management in general, it is important to distinguish between the OAM functions, which operate in the data plane, and other management functions, such as the misconfiguration avoidance schemes that will be discussed in Section 12.4.2, which operate in the control plane. Together, these mechanisms provide a complete set of tools for managing the network. In the next sections, we will take a look at some of the MPLS OAM mechanisms for failure detection, as defined in the MPLS, pwe3 and BFD Working Groups in the IETF.

12.3.2.1 LSP ping

Why define new methods for failure detection in MPLS, instead of just using IP ping for the traffic using the LSP? For example, to check the health of the LSP set up by the LDP from PE2 to PE1 in Figure 12.1, why not simply send periodic IP ping traffic to PE1's loopback address and ensure that this traffic is forwarded over the LSP? The answer is: because such an approach may not detect all failures.

Imagine a probe traveling on the LSP with label L4 on the hop between PE2 and C. Now assume that at node C, the forwarding

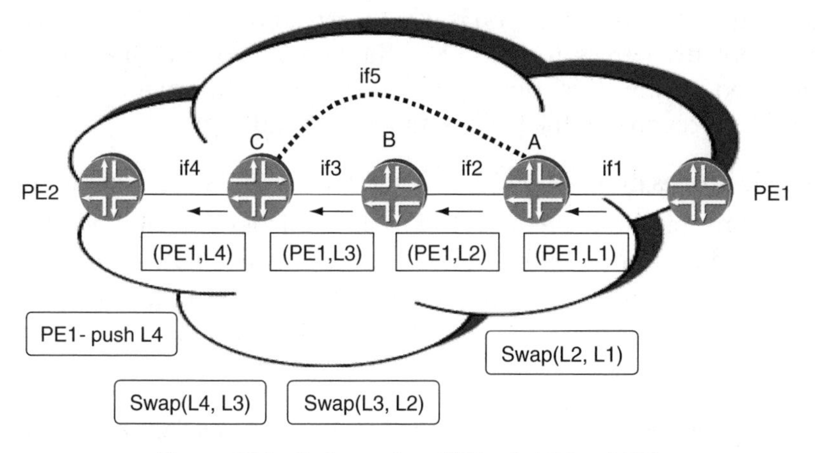

Figure 12.1 Failure of an LDP-established LSP

state is such that the label is popped instead of being swapped to L3. This could happen, for example, because of a corruption in the forwarding state entry, but may also be the result of a legal protocol operation, when LDP-independent control is used (as will be explained in Section 12.4.2.1). In any case, when the traffic arrives at C, the label is popped. The packet continues its journey to PE1 as a pure IP packet and the failure is not detected.

Therefore, what is needed is a way to do two things: (a) validate the forwarding of traffic in the data plane and (b) verify the data-plane state against the control-plane state. The basic mechanism for providing this functionality was defined in the MPLS Working Group and is documented in [LSPING]. The solution is modeled after the ping utility used in IP for failure detection, and is therefore referred to as LSP ping, or LSPing.[2] Similar to ping, it uses probe packets, called MPLS echo requests, and expects to receive back acknowledgments, called MPLS echo replies.

The idea behind LSPing is simple. Verify that a packet that belongs to a particular forwarding equivalence class (FEC) actually ends its MPLS path on the LSR that is the egress for that FEC. In the example shown at the beginning of the section, the FEC is PE1's loopback address. In the failure scenario, when the label is popped

[2] The IETF draft defining LSP ping also provides a mechanism for doing hop-by-hop failure localization, similar to the IP traceroute utility. This functionality will be discussed in Section 12.3.3.2.

at C, the MPLS path ends on router C instead of router PE1. Because C is not the egress for the FEC, the check fails and the error is detected. From this description, several requirements become apparent regarding the probes and their handling:

1. The probes must follow exactly the same path as the data packets.
2. The probes must be delivered to the control plane of the LSR on which they ended their MPLS path for verification.
3. The probes must contain enough information about the FEC to allow the receiving LSR to determine if it is indeed the correct egress.

Based on these requirements, the LSPing probe packets are defined as UDP packets as follows:

1. To ensure that the probes follow the same path as the data packets, they are forwarded using the same label stack as the data forwarded over the LSP they are testing.
2. To allow delivery of the probe to the control plane of the egress LSR, the router-alert option is set in the IP header. An interesting challenge arises with regards to the IP destination address. The problem is that the address of the LSR, which is the egress of the LSP, may not always be known. For example, LDP may advertise a label binding for an arbitrary FEC that is not necessarily associated with any address on the router. For this reason, the destination address is a random address in the 127/8 range (remember this is not a routable address).
3. To facilitate the check that the egress LSR must perform, the probe contains information about the FEC under test. This information is carried in the payload of the UDP packet and is encoded as a set of TLVs. The information is different based on the type of FEC, so TLVs are defined for each of the different types: LDP, RSVP, L3 VPN, pseudowire and so on. For example, for an LDP FEC, the prefix, its length and the fact that a binding was advertised for it using LDP is enough. For an L3 VPN FEC, the route distinguisher, the prefix and the length are needed. Note that the existence of the different FECs means that LSPing can be used to test different layers of the MPLS network. In a VPN setup, LSPing using the L3 VPN FEC tests the VPN tunnel, while LSPing using the LDP FEC can test the LDP LSP that carries the VPN traffic in the core.

4. To inform the originator of the probe of the result of the FEC test, the receiver must know its address. Therefore, the source address of the MPLS echo request probe is set to a routable address on the originator of the probe.

An LSR receiving an MPLS echo request validates it and sends a reply using an MPLS echo reply packet. The reply is also a UDP packet, sent to the source address of the MPLS echo request and forwarded in accordance with the route that is available for its destination address. Thus, the reply packet may travel as pure IP or it may be encapsulated in an LSP, if the path to the destination is through an LSP. The reply contains status information, such as success or failure and the reason for the failure. When the originator of the echo request receives this reply, it matches it against the outstanding requests and reports the appropriate status for the LSP under test. Two assumptions are made in this mode of operation: (a) the egress LSR can forward traffic back to the originator of the echo request and (b) the reply packets will arrive there. These assumptions may not always hold true. For example, if the incorrect source address is used, the egress LSR may not be able to send traffic back to the receiver. Furthermore, because the reply is a UDP packet, its delivery is not guaranteed. When this happens, the reply packets are not delivered to the originator of the LSPing, which will incorrectly infer that there is a problem with the LSP under test. This condition is referred to as a false negative.

In the discussion so far, the assumption was that a single echo reply is generated for each echo request and the two can easily be matched at the LSP head end. With the introduction of P2MP LSPs, as described in the Point-to-Multipoint LSPs chapter (Chapter 6), it is possible for an LSP to be set up from one head end to multiple tail ends. Can the OAM procedures for P2MP LSPs follow the LSPing model in this case? The IETF has not yet adopted a solution for liveness detection of P2MP LSP. However, a proposal exists [P2MP-LSPING] to extend LSPing to cover the P2MP case. If an MPLS echo request is sent on such an LSP, it will reach multiple egress routers; therefore multiple echo replies will be sent. This poses interesting questions, such as how to prevent the head end from being overwhelmed by a large number of replies, or how to report and detect a failure of just some of the destinations. Possible solutions to these problems are currently under discussion in the IETF. One option is to test only a subset of the receivers every time, another option is to ping each destination individually and yet other mechanisms are under development. Note that because of the nature of the data plane, all end-points will receive the probe packets, further complicating any solution relying on end-to-end testing of the P2MP LSP.

At this point, let us stop and note two properties of the LSPing solution:

1. The control plane of the router receiving the probe is involved in the validation of the echo request, with the following consequences:
   (a) If the control plane is busy and cannot process the echo request in a timely manner, the echo request may time out at the LSP head end, resulting in a false negative.
   (b) LSPing places a load on the control plane of the egress router. Because only a limited number of probes can be sent to the control plane and processed there, there is a limit on both the number of LSPs that can be monitored and on the frequency of the probes. For a point-to-point LSP, such as an RSVP-generated one, the number of probes arriving at the tail end of a particular LSP is directly proportional to the polling frequency at the LSP head end. For multipoint-to-point LSPs, such as LDP-generated ones, the load is more difficult to evaluate, because many routers in the network may be sending LSPing packets to the same egress LSR independently of each other.
2. The echo request tests just one of the possible paths to the destination. When several equal-cost paths can exist, such as the case of LDP, only one of them is tested by the echo request, so an error in a different path may not be detected. To test all the different paths, multiple probes must be sent, with different destination addresses. (This is one of the reasons why the echo request destination address is any address in the 127/8 range.)
3. The echo request tests the traffic flow in one direction only (from the source to the destination). The return path of the traffic is not tested.
4. The echo reply is sent as UDP; therefore its delivery is not guaranteed and false negatives are possible.
5. The receiver of the echo request may not support the LSPing procedures, causing a false negative.

Because of these reasons, using LSPing as a liveness detection mechanism should be handled with care. For example, there are proposals to use LSPing for constantly monitoring the health of the LSPs in the network in a way that ensures automatic recovery from failures. The idea is that when a failure is detected the LSP is torn down and the network 'heals itself'. The danger in such proposals is that they do not account for the possibility of false negatives, as explained above. A busy control plane on the egress LSR may cause a false negative. As a result, the LSP is torn down, which in turn generates more control-plane activity, causing failure of other LSPing probes and so on in a downward spiral. In this case, it is preferable to use LSPing for failure detection only and require operator intervention for actually fixing the problem. The operator can further troubleshoot the network and evaluate if tearing down the LSP is required. Note that the use of LSPing for troubleshooting is very powerful, since false positives are not possible.

However, is it possible to use LSPing as a liveness detection mechanism at all? The answer depends on the detection time desired. As explained previously, only a limited number of echo requests can be processed by the control plane in any given unit of time. This places a limit on the number of LSPs that can be tested and on the frequency of the probes. To overcome this scaling limitation, the BFD protocol, described in the Protection and Restoration chapter (Chapter 3), was extended for LSP connection verification.

12.3.2.2 BFD for MPLS LSPs

Based on the realization that verification of the data plane against the control plane is the main factor limiting the polling frequency for LSPing, the approach taken in BFD for MPLS LSPs is to validate the data plane only. Recall from the Protection and Restoration chapter (Chapter 3) that BFD is a simple hello protocol that can be used to test the forwarding path between two end-points of a BFD session at high frequency. How are the LSP end-points communicated to BFD? Is the mechanism useless if the data plane cannot be verified against the control plane?

The answer to these questions is to use a combination of LSPing and BFD. LSPing is used for bootstrapping the BFD session and for periodically (but infrequently) verifying the control plane against the data plane. BFD is used for doing fast failure detection by exchanging

BFD hello packets with high frequency, along the same data path as the LSP being verified.

It is outside the scope of this book to discuss the details of BFD or of the bootstrapping of the BFD session using LSPing. These are described in [MPLS-BFD]. Instead, the important thing to remember is that by combining the fast-failure detection of BFD with the extensive validation capabilities of LSPing, a scalable solution for LSP monitoring is achieved.

12.3.2.3 VCCV

Virtual circuit connection verification (VCCV) is the connection verification protocol for pseudowires set up using LDP (discussed in the Layer 2 Transport chapter, Chapter 10). VCCV was developed in the PWE Working Group in the IETF [VCCV]. The natural question is why was a different connection verification mechanism developed? Doesn't LSPing support a TLV for Layer 2 circuits? In fact, VCCV builds on the LSPing solution and actually reuses the Layer 2 circuit TLV defined for LSPing.

Let us take a look at some of the challenges of using LSPing in a pseudowire environment. Recall from the Layer 2 Transport chapter that a CE-facing interface is associated with a pseudowire. When traffic arrives over this interface, it is encapsulated in MPLS, labeled with the pseudowire label and sent to the remote PE. Because labeled traffic is forwarded between the PEs, a transport tunnel is necessary that can carry MPLS. Usually, this tunnel is set up with either LDP or RSVP. Sometimes, a Control Word is prepended to the L2 frame, as explained in the Layer 2 Transport chapter. Figure 12.2 shows how traffic is forwarded over the pseudowire.

Figure 12.3 shows what happens if an LSPing echo request is sent from PE1 to PE2 by applying the LSPing procedures described so far. The probe contains a TLV with information allowing PE2 to determine if it is the correct recipient, as described in [LSPING]. It is encapsulated with the same label stack as the data packets and is sent towards PE1. The inner label is the pseudowire label and the outer label is the transport tunnel label (in this example, RSVP-signaled). When the probe arrives at P2, the RSVP label is popped and the LSPing packet arrives at PE2 with a single label, the pseudowire label. At this point, if PE2 applies the normal forwarding procedures to this packet, the probe would simply be

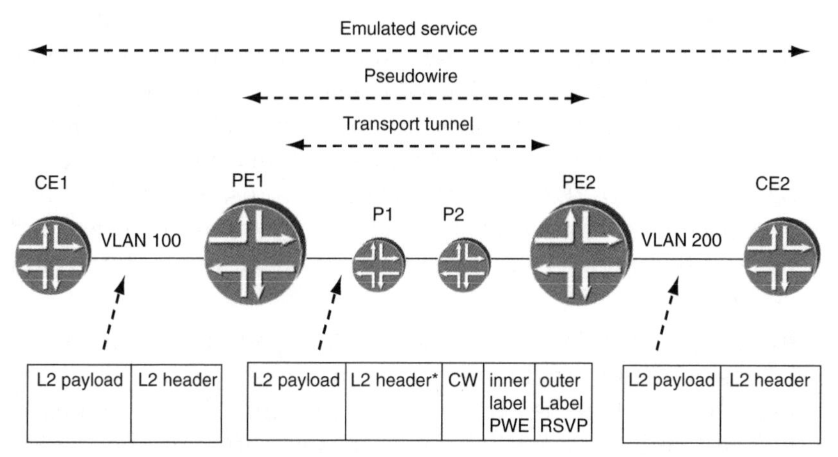

* Which parts of the Layer 2 header are transported over the MPLS core depends on the layer 2 protocol in question.

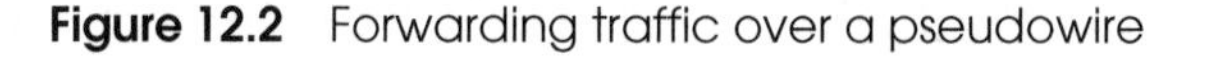

Figure 12.2 Forwarding traffic over a pseudowire

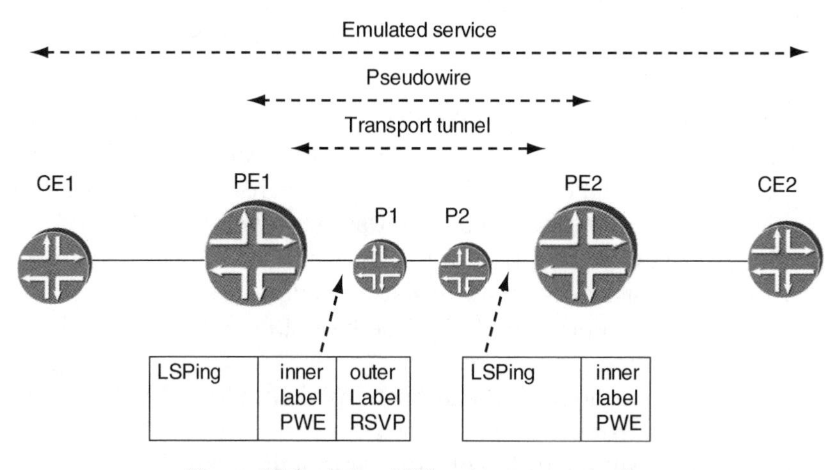

Figure 12.3 Using LSPing for a pseudowire

sent over the PE2–CE2 interface, instead of being processed by PE2. Note that this behavior is not unique to pseudowires and would be encountered in an L3 VPN setup as well.

Several options are available to fix this problem:

1. Set the TTL expiry to 1 on the inner label. Assuming the RSVP label (transport tunnel label) is popped at P2, when the probe arrives at PE2, the TTL on the inner label expires and the packet is delivered to the control plane at PE2.

2. Insert the router alert label between the pseudowire label and the transport tunnel label. The router alert label is a special label that causes the labeled packet to be delivered to the control plane. In this way, after the transport tunnel label is popped, the top of the stack is the router alert label, which will cause the packet to be delivered to PE2's control plane. The disadvantage of this approach is that data packets and LSPing packets are forwarded using different label stacks, so the data-plane verification is not as accurate.
3. Insert a Control Word after the LSPing packet and before the label stack. If forwarding for the data packets on the PE is done in such a way that the Control Word is examined and acted upon before sending the traffic to the CE, then bringing the LSPing packet to the control plane could be driven by evaluation of the Control Word. The idea is to set a bit in the Control Word to indicate that the packet should be delivered to the control plane rather than being forwarded, and use the Control Word when sending LSPing probes. (This approach is not shown in the figure.)

Thus, extra steps must be taken to ensure that the LSPing packet is delivered to the remote PE's control plane. Because several options are available, and because different platforms support different options, no one approach can be mandated. Therefore, it is necessary for the pseudowire end-points to negotiate which mechanism to use. VCCV defines how this negotiation is done and how the probe packets must be encapsulated, based on the negotiated values, reusing the LSPing procedures.

So why is there no such mechanism available for L3 VPNs? Will they not suffer from the same forwarding challenges? The reason is because the goals are different. For pseudowires, the goal is to build a control channel between the two PEs and do both failure detection and failure verification on this channel. This channel is either in-band, when the probes are taken out of the forwarding path based on the Control Word, or out-of-band, when they are taken out of the forwarding path based on the label. One of the goals is to use BFD over this control channel to monitor the liveness of the pseudowire and bring it down if a failure is detected. Therefore, it is required to determine at the time of the pseudowire setup whether such a channel can be built. The VCCV negotiation can provide this knowledge.

In other cases, such as L3 VPN, where LSPing is limited to troubleshooting a failure, the requirement for an indication on whether both ends support the same procedure is not required. This conclusion can be reached easily during the troubleshooting process.

To summarize, VCCV is used for connection verification of pseudowires. Although it is a new protocol, VCCV builds on both LSPing and BFD to provide failure detection and monitoring for pseudowires.

12.3.2.4 Pinging at the service level

The previous sections discussed tools for failure detection at the different layers used for building up a particular service, e.g. the PE–PE transport tunnel built with LDP or the virtual connection between the PEs providing a pseudowire. However, mechanisms such as ping can be used at the level of the service itself. For example, in an L3 VPN setup, two routers in customer sites can send ICMP pings to each other to verify connectivity. These ping packets will be forwarded just like any other VPN traffic and can therefore discover if there is any problem on the path between the two sites. This type of check is an example of an end-to-end check of the service itself.

The description above assumes that the ICMP pings are initiated by the CE routers, but they can also be initiated by PE routers, as long as the ping application is VRF aware and can send probes according to the forwarding information for the VRF under test. Note that in this case, only a segment of the service is being tested.

12.3.2.5 Failure detection summary

Fast-failure detection is required to avoid traffic loss following a failure. Nonsilent failures are reported in the control plane, which can act on the failure indication to prevent traffic loss.

Specialized mechanisms for failure detection are required for identifying silent failures, which are not reported in the control plane. The previous sections showed different layers at which failure detection can be applied. To summarize, failure detection can be run at the service layer end to end, at the service layer for just a segment of the service, at the virtual connection layer for the virtual tunnel created by the VPN label or at the transport layer for the PE–PE tunnel over which the VPN tunnel is transported, as shown in Figure 12.4 for an L3 VPN setup.

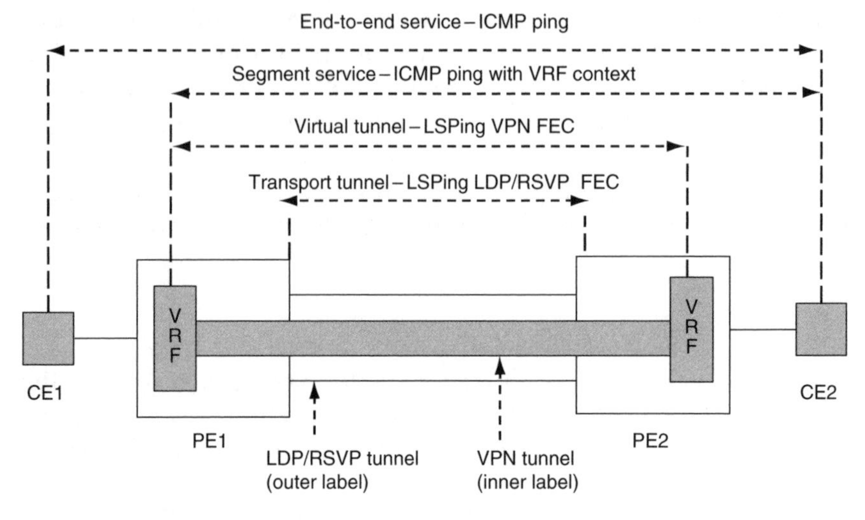

Figure 12.4 Applying failure detection at different layers

In the following section we will look at mechanisms for pinpointing the exact location of the failure.

12.3.3 Troubleshooting failures

In the previous sections we focused on the mechanisms used for detecting failures. The next step is to see how the location of the failure can be pinpointed. The most popular method for doing so is tracing the path of the traffic in the network. Path-trace can be run at the different layers, similar to how ping can be run at different layers.

12.3.3.1 ICMP tunneling

In Section 12.3.2.4 we saw that ping can be used at the service level within an L3 VPN for failure detection. Therefore, it is natural to want to use the traceroute mechanism for the localization of the failure. Traceroute sends a series of probe packets with increasing time-to-live (TTL) values. The hops where the TTL expires return an ICMP message indicating the TTL expiration to the originator of the probe, thus identifying the hops in the path. From this description, it should already be clear that traceroute operation in a VPN setup is not immediately applicable.

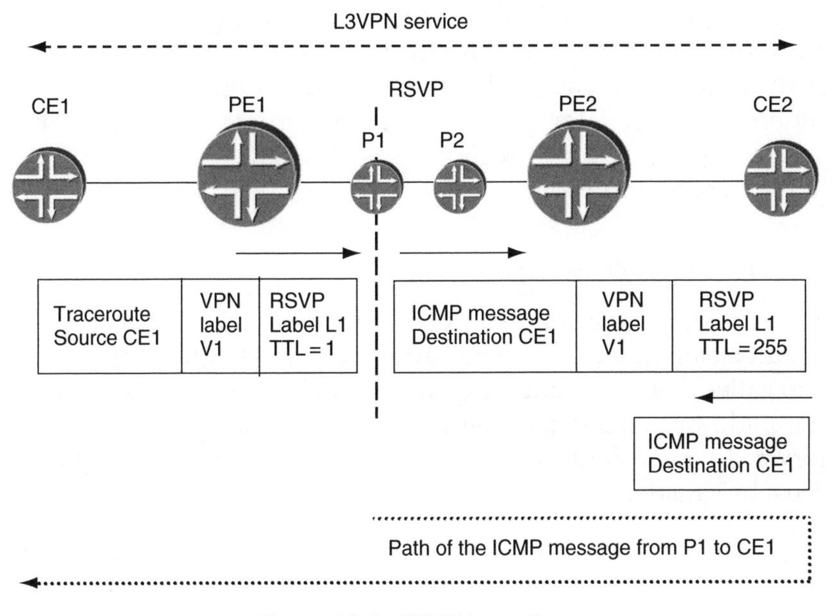

Figure 12.5 ICMP tunneling

Figure 12.5 shows an L3 VPN setup where the PE-to-PE tunnel is set up with RSVP. CE1 issues a traceroute for CE2's loopback address. At PE1, the probe is encapsulated with the VPN label and the RSVP label and sent towards PE2. When the TTL expires at P1, an ICMP message must be sent towards CE1. The problem is that CE1 is a private address in a VPN and therefore is not known at P1, so the message cannot be forwarded.

An elegant solution to this problem is described in [RFC3032]. The idea is that the destination of the probe is likely to have reachability to the source of the probe. Therefore, if the ICMP message reaches the probe's destination, it can be forwarded from there back to the source. To achieve this, P1 builds the ICMP message indicating TTL expiry at P1 and then copies the label stack from the original packet to this ICMP message (updating the TTL to a high value), as shown in Figure 12.5. This is called ICMP tunneling, because the ICMP message is tunneled all the way to the original probe's destination, CE2. At CE2, the message is looped back to CE1, because CE2 has forwarding information for CE1. The path of the ICMP message is shown at the bottom of Figure 12.5.

ICMP tunneling is essential for implementing traceroute in a VPN setup. However, it can also be used to report other errors such as the ones caused by the need for fragmentation. ICMP tunneling is used for providing the traceroute capabilities at a service level. In the next section we will see traceroute capabilities at the transport tunnel level.

12.3.3.2 LSP traceroute

The LSPing mechanism described in Section 12.3.2.1 also supports a fault isolation mode modeled after the IP traceroute functionality and called LSP traceroute. The idea is not just to report the hops in the path, but also determine if there is a mismatch between the forwarding and control planes. To accomplish this, two conditions must be satisfied:

1. The MPLS echo request probe must be processed by each hop in the path.
2. Enough information must be available in the probe to allow the transit LSR to determine that it is indeed a transit for the LSP under test.

The LSP trace uses the same packet formats, TLVs and echo request and reply mechanisms used by LSPing. For example, to trace the path of an LDP LSP, an MPLS echo request is sent from the head end, including the LDP FEC TLV, encapsulated with the correct label stack for the LSP, in this case the LDP label. To ensure that the echo request is received by each hop in the path, several such echo requests are sent, with increasing TTL values, just like a normal traceroute. To allow the transit LSR to check the control plane against the data plane, the LSR must know whether it is a correct recipient of the probe. Therefore, the MPLS echo request packet contains, in addition to the FEC TLV for the FEC under test, the list of acceptable recipients of the packet, from the point of view of the upstream LSR.

This list is encoded in the 'Downstream Mapping TLV' and contains identifying information (such as the router ID, label and interface) of all the possible downstream neighbors for the FEC, from the point of view of the upstream LSR. The LSR processing the echo request determines if it is a valid recipient of the traffic if it is listed in the Downstream Mapping TLV, with the correct label and interface information. If the check fails, it informs the head end of the failure in the echo reply, thus pinpointing the location of the problem. If the check succeeds, the LSR extracts all its valid downstream neighbors and labels for the FEC under test and sends this information in the echo reply, encoded in a Downstream Mapping TLV. This TLV is sent in the next echo request that will be sent with a TTL greater by 1, and therefore will reach its downstream neighbors.

This process is shown in Figure 12.6 for an LDP-signaled LSP between PE2 and PE1. The label distribution from PE1 to PE2 is shown along the links in the path. PE2 sends an MPLS echo request with a TTL value equal to 1. The request contains an LDP FEC TLV for FEC PE1 and a Downstream Mapping TLV listing neighbor C, as well as the label used by C, L4. When the TTL expires at C, the echo request is delivered to the control plane of router C for processing, along with the label stack with which the packet arrived. LSR C checks if it is one of the routers in the Downstream Mapping. Because the check is positive, C builds a new Downstream Mapping TLV, listing all its valid downstream neighbors for LDP FEC PE1

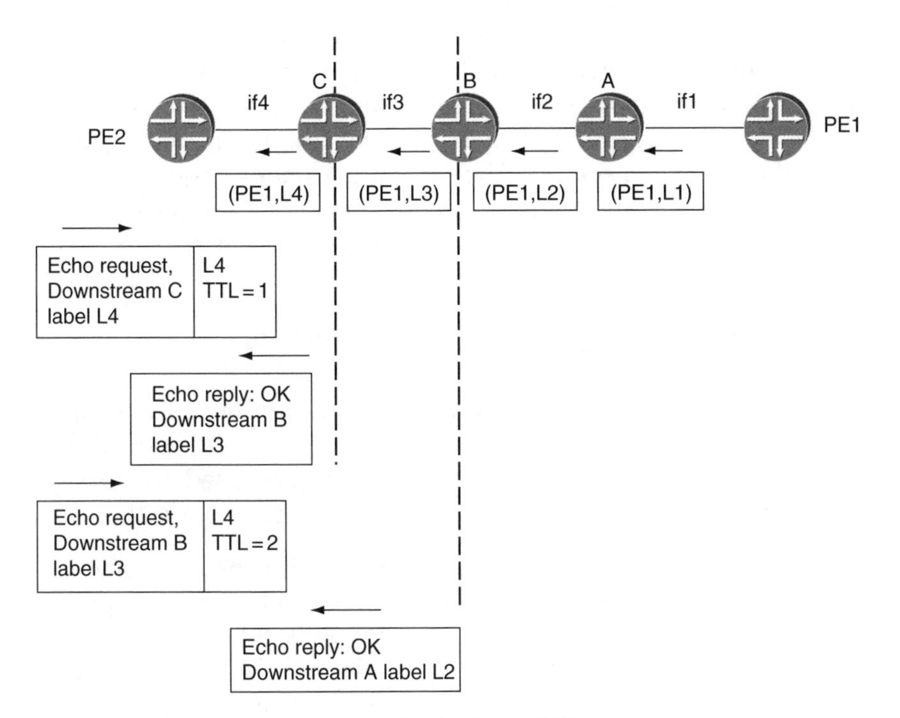

Figure 12.6 Trace LSP

(in this case, router B with label L3) and sends it back to the head end in the echo reply. This TLV is included in the next echo request probe sent by the head end, which is sent with TTL equal to 2 and will expire at B. When this probe is processed at B, the same process will be applied and so on (the rest of the steps are not shown in the figure). In this manner, the path of the LSP can be traced and at the same time the location of any failure can be reported.

To summarize, data-plane failures are difficult to detect and fix. For this reason, specialized tools such as the ones described throughout Section 12.3 were developed in order to discover their existence and pinpoint their location. However, data-plane failures are not the only source of problems in the network. In the next section we will look at configuration errors and their impact on the network.

12.4 CONFIGURATION ERRORS

Configuration errors are a common source of problems in network. Their impact can range from a service not coming up to traffic being routed to the wrong destination. There are two ways to deal with configuration errors. The first is to prevent the problem from happening in the first place, by improving the configuration process and by building mechanisms to reduce the amount of configuration needed to deploy a service. The second is to detect and report the misconfiguration, and try to protect the network from its ill effects. Let us discuss these two approaches separately below.

12.4.1 Preventing configuration errors

The basic idea behind preventing configuration errors is simple: fewer and simpler configuration statements means less probability for an error. Here are a few of the techniques used by various commercial implementations to put this idea into practice:

- Minimize the amount of configuration that must be applied to enable a feature. The more configuration statements required, the bigger the chance of an error, especially when the different statements are sprinkled in several places in the configuration file. For example, recall from the L3 VPN chapter (Chapter 7) that the definition of a VRF requires both an import and an export policy. For simple any-to-any connectivity, the two are often the same. Therefore, rather than requiring an explicit listing of each, both could be configured in one statement.

- Use intuitive configuration statements. If the configuration is not intuitive, the chance of errors is higher. For an example, refer to Section 2.4.3 discussing link colors in the Traffic Engineering chapter (Chapter 2).
- Avoid configuration when not necessary. For example, the bypass tunnels required for link protection can be dynamically computed and should not require manual configuration.
- Apply the same configuration in multiple places automatically. For example, to enable forwarding for MPLS packets on a large number of interfaces, it should not be necessary to use the same configuration statement multiple times. Instead, it would be better to have a way to apply the configuration to multiple interfaces in one statement.

However, requiring fewer configuration statements in one router's configuration file is not the end of the story. The choices that the operator makes regarding the label distribution protocols and the services deployed directly impact the amount of configuration necessary. Here are a few examples:

- The number of LSPs required. In the chapter discussing DiffServ Aware Traffic Engineering (Chapter 4), we saw that LSPs can be set up with reservations for a single class type or from multiple class types. When multiclass reservations are supported, the total number of LSPs that must be set up in the network decreases, and so does the amount of configuration necessary to set them up.
- Autodiscovery for the BGP-based Layer 2 VPN or VPLS solution. When using a BGP-based solution, no extra configuration is necessary to identify the other members of the service, as explained in the Layer 2 Transport chapter (Chapter 10). In contrast, when setting up a mesh of pseudowires using LDP, the end-points of each of the circuits and the targeted LDP sessions must be correctly configured on each box.
- RSVP as a label distribution protocol. When a full mesh of transport tunnels is required between all PEs, RSVP requires configuring tunnels on each of the PEs to all the other PEs. When adding a new PE to the mesh, tunnels must be set up from the new PE to all the existing PEs. However, the same tunnels must also be

configured towards the new PE from all the existing ones. The problem in this case is not just the fact that a large number of LSPs need to be configured, but also that this configuration is spread over a large number of PEs. Because of this property of RSVP, many deployments prefer to use LDP as the label distribution protocol, as explained in the Foundations chapter (Chapter 1).

In the last two examples listed above, the requirement for extra configuration work is due to the nature of the service or protocol used. Therefore, it makes sense to look for a solution at the same level. For instance, there are ongoing efforts in the IETF to provide autodiscovery capabilities to the LDP-based Layer 2 solutions, as discussed in the Layer 2 Transport chapter (Chapter 10). One thing to note is that many of the proposals rely on deploying another protocol for autodiscovery, which means that more configuration work is required to set up and maintain the protocol. When comparing two competing solutions, one must keep in mind not just the functionality provided but also the cost of deploying the solution, especially on a large scale. In this context, the configuration effort plays a big part.

An interesting solution at the protocol level for a configuration scaling problem has been proposed for the RSVP full-mesh problem in [AUTOMESH]. Providers want to use RSVP not just for traffic engineering but also for its fast-reroute capabilities. However, the burden of manually provisioning the full mesh of tunnels and of updating the mesh every time a new PE joins it constitutes a deterrent. The solution is to offload this burden on to the routing protocols themselves.

The idea is simple. Every PE has reachability to every other PE in the network because this information is distributed by the IGP. In principle, a PE could set up an RSVP LSP to any other PE in the network, if it knew that such an LSP was required. Thus, to build a full mesh of LSPs between PEs, all that is needed is for each PE to know who are the other members of the mesh. This can be easily accomplished by assigning an identifier to each group of PEs that must be fully meshed and have each PE advertise which groups it belongs to. Because the IGP distributes reachability information to all PEs and because it is already used for carrying TE information, it is an ideal candidate for distributing this mesh group membership information. The automesh proposal extends OSPF and IS-IS to carry a new TLV, the TE mesh group TLV, indicating what group(s) the PE belongs to and what is the address that should be used by the other PEs for setting up the LSPs towards it. Based on the advertisements received and based on the locally configured knowledge of mesh membership, each PE knows to which other PEs to set up LSPs.

When a new PE is added to the network, it is configured with the correct group membership. As soon as the IGP distributes the new PE's membership information in the network, all the other PEs can set up LSPs towards it, automatically (and the other way around). However, what are the properties of the LSPs that are set up this way? This is a matter of configuration of the mesh group properties. This configuration can be minimized if, for example, features like autobandwidth (discussed in the Traffic Engineering chapter, Chapter 2) are used.

To summarize, configuration errors can be avoided by minimizing the amount of configuration required. This can be done at the implementation level, by optimizing the configuration process, and at the protocol level, by building a mechanism to avoid the need for configuration. Examples of the latter are autodiscovery of VPN membership or automesh for RSVP-TE tunnels. However, as long as configuration is necessary, the possibility of errors in the configuration continues to exist. In the next section we will see how to detect and report misconfigurations.

12.4.2 Detecting and reporting misconfigurations

Because configuration always requires human intervention at one level or another, errors will continue to happen. Sometimes, the error causes an easily detectable problem, such as a routing peering not establishing. At other times, there is no immediate feedback on the problem and there is a need to define new mechanisms for detecting the failure and dealing with it. Let us take a look at a few examples of protocol extensions for detecting and reporting errors caused by misconfigurations.

12.4.2.1 Interface misconfiguration affecting LDP operation

Recall from the Foundations chapter (Chapter 1) that LDP label distribution follows the IGP. When a new interface is added to the network and LDP is not enabled over it, a failure will occur (assuming the new interface causes a change to the shortest path). This is shown in Figure 12.1. Assume the interface if5 does not yet exist in the network. The LSP for FEC PE1 (the loopback or router PE1) establishes along the router's path PE2–C–B–A–PE1. At this point, the operator decides to add the interface if5, and includes it in the IGP, but forgets to enable LDP on it. As a result, the IGP best path for PE1's loopback on router C will be C–A–PE1. Because the label advertisements arrive over a different interface (C–B) than the IGP best path, the forwarding state for the LSP will be removed. If independent control is used, the forwarding entry will be changed to pop the label, but C will continue to advertise its label towards PE2. If ordered control is used, the forwarding entry will be removed and the label advertisement will be withdrawn. (For a detailed discussion of this scenario, refer to the Foundations chapter, Chapter 1). In both cases, no labeled traffic can be forwarded between the two PEs, causing interruption of the service if the two PEs are providing MPLS/VPN services.

Note that the same problem of tearing down the LSP would happen even if LDP were enabled on the new interface. The condition would persist until such time as the LDP session establishes over the new link. To avoid this situation, a mechanism is defined in [LDP-IGP-SYNC]. The idea is simple: allow the user to specify the interfaces over which LDP is expected to run and advertise an infinite IGP metric for the link until the LDP session has come up and labels have been exchanged over it. In the example above, the new link A–C would be advertised with the infinite metric, and as a result the IGP best path would continue to go over the path C–B–A. Note that the cost of using this scheme is that the new link is avoided, because of its high metric. This affects not just the LDP traffic but the IP traffic as well.

What we have seen in this example is a mechanism for avoiding the ill effects of a configuration error for the LDP protocol. The undesirable consequences of the misconfiguration are avoided and an error can be reported when the condition is detected, thus allowing the operator to identify and rectify the problem.

12.4.2.2 Common misconfigurations for VPN scenarios

Assume a network with two VPN customers, A and B, using the same private address space. Two new sites are added on the same PE: CE1 in VPN A and CE2 in VPN B. Two common misconfigurations are possible:

1. Assigning the customer interface to the incorrect VPN. Recall from the introduction to the L3 VPN chapter (Chapter 7) that the decision to which VPN CE-originated traffic belongs is based on the interface over which the traffic arrives. If the link

from CE2, instead of the one from CE1, is connected to the port configured for VRF A, then CE2 becomes a member of the wrong VPN, as shown in Figure 12.7. Assuming the same address spaces in both VPNs, traffic originating in VPN B is forwarded to destinations in VPN A.

2. Configuring the wrong route target (RT). Recall from the introduction to the L3VPN chapter that correct access control between VPNs relies on accurate configuration of the route target. If a site in VPN A starts using the RT that was assigned for VPN B, then destinations in VPN B may become reachable from the site belonging to VPN A and vice versa (the exact outcome depends on whether the import RT, export RT or both are misconfigured).

In both of these cases, the problem is not just one of not providing the required connectivity to the new sites. Perhaps more importantly, the problem is one of violating the security guarantees offered to the two customers by allowing traffic to cross over from one VPN to the other.

For this reason, solutions are currently under discussion in the l3vpn Working Group in the IETF for handling such misconfigurations. Their goal is twofold: alert the operator of the problem and prevent misrouting of traffic. Let us take a look at some of the proposals for solving each of the problems mentioned above.

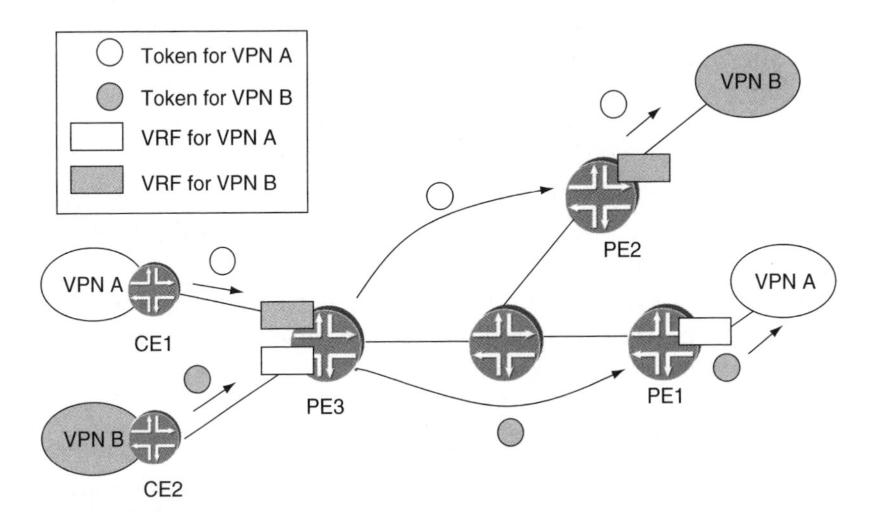

Figure 12.7 Using a token-based mechanism for VPN membership verification

Assigning the customer interface to the incorrect VPN

The most intuitive approach for handling this error is to enable authentication for the PE–CE routing protocol exchanges. Assuming different keys are used in each VPN, the routing protocol session does not establish and routes are not be propagated to/from the misconfigured CE to the rest of the VPN. Failure to establish the routing protocol session also triggers error messages that alert the operator to the problem. However, the approach of authenticating PE–CE routing protocol exchanges may not always be feasible, e.g. in setups where no routing protocol is running on the PE–CE link.

For this reason, a new mechanism is proposed in [VPN-CE-CE-AUTH] that provides a CE–based mechanism for VPN membership verification. The idea is to allow the customers to detect security breaches caused by a misconfiguration of the provider network. Here is how it works. To join a VPN, each site sends a token to the PE, which in turns relays it to all the members of the VPN, using similar mechanisms as the ones used for route distribution in a VPN. Customer devices use the token to verify VPN membership. The receipt of an unexpected token indicates that an unauthorized site joined the VPN and, as a result, an alarm is triggered to the operator. In addition, the VPN site receiving the unrecognized token may choose to protect itself from unauthorized access by withdrawing from the VPN, e.g. by discarding VPN traffic sent to it or by withdrawing its routes.

Figure 12.7 illustrates this mechanism. Note that at PE3, the two VPN sites are attached to incorrect VRFs. As the tokens propagate to other members of the VPN, the misconfiguration is detected. The actual details of the token implementation can be found in [VPN-CE-CE-AUTH].

Although this mechanism does nothing to prevent the misconfiguration itself, its deployment allows detection of the problem. One of its interesting properties is that it gives the customer control over detecting problems in the provider network and allows him or her to protect against security breaches caused by such problems, e.g. by withdrawing from the VPN.

Configuring the wrong route target

Controlling the access to a VPN is all about controlling access to forwarding information. Correct assignment of the interface to the VPN ensures that forwarding lookups are made in the table associated with the VPN. Correct configuration of the RT limits the information stored in this table. The problem is that these two mechanisms are disjoint.

When discussing incorrect assignment of interfaces to the VPN we saw that the simplest solution is to authenticate the PE–CE routing exchange. Could a similar approach be used to ensure that the correct RT is used? [VPN-RT-AUTH] attempts to take exactly this approach by validating PE–PE routing updates using routing update signatures. Note that the solution does not propose to authenticate the routing session but the individual route advertisements. When a VPN route is advertised between PEs, it carries a new BGP attribute, the BGP 'update authenticator'. This attribute contains a signature generated using an MD5 key. When receiving such an update, the signature is checked against the MD5 key of the remote site, as shown in Figure 12.8. If the check fails, the route is not added to the VRF and the operator is notified. In this way, routes that do not belong to the VPN are not added to the VRF, the information in the VRF is limited to only the routes that are certified as belonging to the VPN and misrouting is prevented. Note that for this approach to work, unique MD5 keys must be assigned to each VPN.

To summarize, preventing configuration errors is possible in many cases. Good operation practices and intuitive, easy-to-use configuration interfaces can prevent many misconfigurations. For others, it is necessary to extend the protocols with built-in machinery for detecting the problem and preventing its ill effects.

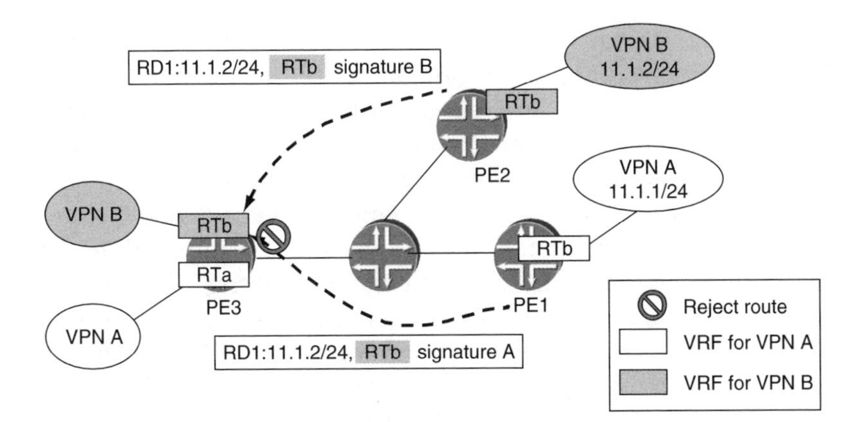

Figure 12.8 Detecting RT misconfigurations using routing update signatures

12.5 VISIBILITY

A big part of management is about gaining visibility into the network. What exactly is visibility? The answer is different for different people. For example, for a network operations engineer, it may be the ability to see the path that traffic is taking or to determine if a site in a VPN went down, while for an engineer doing capacity planning, it may be the ability to build the traffic matrix between two PEs.

Management information bases (MIBs) play a big role in affording visibility into the network. As explained at the beginning of this chapter, MIBs are available for all the protocols and applications developed in the IETF and can be used to manage the network as described in [MPLS-NM]. MIBs can be used not just for reporting the state of a service or protocol but also for finding network-wide information that is not readily available otherwise. For example, for RSVP, the path of an LSP is known from the RRO Object, but for LDP, this information is not available from the protocol. However, using the label information from the LDP MIB, the path of the LDP LSP can be traced without the need to issue an LSP trace.

SNMP traps can be sent to indicate errors, such as an LSP being torn down. The receipt of the traps at the network management station provides not just an alarm indication to the operator but also valuable information about what failures happened at the same time, allowing correlation of events in the network and identification of the root cause of a problem, such as an interface going down, causing an LSP to go down and causing a VPN site to become unreachable.

Traffic accounting is perhaps the most important visibility feature in the network. It is important not just for billing purposes but also for network planning and debugging. LSPs are interesting in this context because they are treated by many implementations as virtual tunnel interfaces and have the same accounting features as interfaces. For example, some implementations allow the user to see the amount of traffic forwarded over the LSP in real time. This is a useful debugging tool, because it can show if traffic is forwarded along a particular LSP. Another example is the ability of some implementations to apply firewall filters to the LSPs (as explained in the context of policers in the DiffServ Aware TE chapter, Chapter 4). Thus, granular accounting features, e.g. taking into account the class of service, can be made.

Because it is easy to count traffic entering an LSP, LSPs have become an increasingly popular way to measure the traffic demands between two points in the network. For example, to find out the traffic matrix, LSPs with zero bandwidth requirements can be set up and the traffic statistics can be monitored over these LSPs.

To summarize, the different MPLS MIBs give visibility into the different components of the MPLS network and LSP traffic statistics give visibility into the traffic patterns in the network. This information can be used for both billing and for capacity planning.

12.6 CONCLUSION

Entire books are dedicated to the subject of managing the MPLS infrastructure and the services running over it; attempting to cover this subject in one chapter cannot do it justice. In fact, the topic is much broader than just the functionality defined in the standards bodies. An entire industry is built around tools that can aid in tasks such as performance monitoring for a service or an LSP, VPN provisioning, network visualization, offline path computation, event correlation or capacity planning.

Rather than attempting to provide a view of all management functions and tools, this chapter focused on router functionality developed specifically for troubleshooting and managing in an MPLS environment, such as LSPing, VCCV, ICMP tunnelling and membership verification for VPNs. Although these topics constitute only a small part of the MPLS management story, they shed light on some of the unique issues that arise in MPLS environments.

12.7 REFERENCES

[AUTOMESH] J. P. Vasseur, J. L. Le Roux *et al.*, 'Routing extensions for discovery of Multiprotocol (MPLS) Label Switch Router (LSR) Traffic Engineering (TE) mesh membership', draft-vasseur-ccamp-automesh-00.txt (work in progress)

[LDP-IGP-SYNC] M. Jork, A. Atlas and L. Fang, 'LDP IGP synchronization', draft-jork-ldp-igp-sync-01.txt (work in progress)

[LSPING] K. Kompella and G. Swallow, 'Detecting MPLS data plane failures', draft-ietf-mpls-lsp-ping-09.txt (work in progress)

[MPLS-BFD] R. Aggarwal, K. Kompella, T. Nadeau and G. Swallow, 'BFD For MPLS LSPs', draft-ietf-bfd-mpls-02.txt (work in progress)

[MPLS-NM] T. Nadeau, *MPLS Network Management: MIBs, Tools, and Techniques*, Morgan Kaufmann, 2003

[P2MP-LSPING] S. Yasukawa, A. Farrel, Z. Ali and B. Fenner, 'Detecting data plane failures in point-to-multipoint MPLS Traffic Engineering – extensions to LSP Ping', draft-yasukawa-mpls-p2mp-lsp-ping-02.txt (work in progress)

[PW-OAM-MSG-MAP] T. Nadeau *et al.*, 'Pseudo wire (PW) OAM message mapping', draft-ietf-pwe3-oam-msg-map-02.txt (work in progress)

[RFC3032] E. Rosen *et al.*, *MPLS Label Stack Encoding*, RFC 3032, January 2001

[VCCV] T. Nadeau and R. Aggarwal, 'Pseudo wire virtual circuit connectivity verification (VCCV)', draft-ietf-pwe3-vccv-05.txt (work in progress)

[VPN-CE-CE-AUTH] R. Bonica *et al.*, 'CE-to-CE member verification for Layer 3 VPNs', draft-ietf-l3vpn-l3vpn-auth-01.txt (work in progress)

[VPN-RT-AUTH] M. Behringer, J. Guichard and P. Marques, 'Layer-3 VPN import/export verification', draft-behringer-mpls-vpn-auth-04.txt (work in progress)

12.8 FURTHER READING

[MPLS-OAM] D. Allen and T. Nadeau, 'A framework for MPLS operations and management (OAM)', draft-ietf-mpls-oam-frmwk-03.txt (work in progress)

[MPLS-REQS] T. Nadeau *et al.*, 'OAM requirements for MPLS networks', draft-ietf-mpls-oam-requirements-06.txt (work in progress)

[PWE-OAM] S. Delord *et al.*, 'PWE3 applications and OAM scenarios', draft-delord-pwe3-oam-applications-01.txt (work in progress)

[VPN-MGMT] Y. El Mghazli *et al.*, 'Framework for L3VPN operations and management', draft-ietf-l3vpn-mgt-fwk-08.txt (work in progress)

[VPN-MIB] T. Nadeau and B. Van Der Linde, 'MPLS/BGP Layer 3 virtual private network management information base', draft-ietf-l3vpn-mpls-vpn-mib-07.txt (work in progress)

[Y1710]ITU-T Recommendation Y.1710, in *Requirements for OAM Functionality for MPLS Networks*, 2002

13

Conclusions

13.1 INTRODUCTION

At the beginning of this book, we observed that in only a few years MPLS has evolved from an exotic technology to a mainstream technology used in a large proportion of service provider networks worldwide. One of the most successful MPLS-based services so far is Layer 3 VPN, now a lucrative revenue earner for many service providers. Also MPLS-based Layer 2 services are becoming available in many regions. However, in terms of revenue, MPLS-based services are still currently dwarfed by more traditional data services such as those based on native Frame Relay or ATM, although the growth rates for the MPLS-based services are higher.

A reason why MPLS has developed relatively rapidly is through the pragmatic way protocols have been developed, or existing protocols adapted, to support MPLS. In this book, we saw that an existing protocol, RSVP, was used as the basis for MPLS traffic engineering for its properties of session maintenance, resource reservation and admission control. Additional properties were added to cater for the requirements of MPLS, such as the ability to distribute labels and specify the path to be followed by the traffic. On the other hand, LDP was developed specifically for MPLS, because no existing protocol had the required properties. An existing protocol, BGP, was adapted to carry the routes (and associated labels) of L3 VPN customers, and the same scheme was then

MPLS-Enabled Applications: Emerging Developments and New Technologies Ina Minei and Julian Lucek

carried through to L2 VPN and VPLS. A lesson learnt from the development of these VPN services was that no single VPN service type suits all customers so, for example, L3 VPN is not inherently 'better' than L2 VPN or vice versa. Thus in order to address the widest possible range of customers, a service provider should consider offering the full range of MPLS-based VPN services. As more experience is gained of running MPLS-based services, additional features continue to be added to the underlying protocols, e.g. the addition of automated Route Target Filtering (RTF) to BGP to ensure that PE routers and route reflectors are not inundated with VPN routes that they are not interested in. Another example of the learning process was the realization that early schemes for supporting multicast over L3 VPN did not have good scaling properties, leading to the current work in the IETF to develop new schemes [VPN-MCAST].

In the rest of this concluding chapter, we take a look at some emerging trends in the field of MPLS, starting with an examination of converged networks.

13.2 NETWORK CONVERGENCE

A driving force for MPLS becoming the prevalent network technology in the future is network convergence, with many service providers beginning to consolidate disparate 'stovepipe' networks on to a single one based on MPLS. The end-point for this convergence is for an MPLS network to carry all of a service provider's traffic, including PSTN and mobile voice traffic, Layer 2 data, Layer 3 VPN, Internet and broadcast television. Although at the time of writing there are few, if any, fully converged networks carrying all of these services, several service providers have programs underway to build such networks.

Let us review the reasons why MPLS has made possible the deployment of critical services that router-based networks were previously regarded as not capable of supporting:

- *Flexibility with respect to connectivity.* An issue with native IP networks is that it is not possible to achieve end-to-end (PE-to-PE) bandwidth guarantees. MPLS achieves end-to-end bandwidth guarantees through traffic engineering and admission control, on a per class basis if required. This 'connection-oriented' approach is highly desirable to meet the QoS requirements of traffic such as PSTN voice, broadcast video and some Layer 2 services, e.g. those that emulate or replace ATM CBR services. On the other hand, other classes of traffic can be handled without any bandwidth guarantees (or can be allowed to oversubscribe their bandwidth reservation) in order to make use of statistical multiplexing. This 'mix and match' approach helps the service provider make good use of bandwidth resources without being too rigid (all traffic having to have an associated bandwidth reservation) or too loose (no bandwidth guarantees for any traffic).

- *Aggregation properties.* An LSP can carry all of the traffic of a particular class, or of multiple classes, between a pair of PEs. In general, many end-to-end microflows would be aggregated on to that LSP. As a consequence, the core of the network does not contain the state related to those individual flows. If the number of LSPs in the core of the network becomes an issue, because of the control-plane overhead associated with maintaining them, the hierarchical properties of MPLS allow LSPs to nest inside other LSPs. In this way, the growth of the network is not constrained by the number of microflows or the number of LSPs that the network is required to carry.
- *Single forwarding mechanism.* This is based on label swapping, regardless of the traffic type being carried. This makes it easier to carry new types of traffic, as only the PE routers need to understand the semantics of the native encapsulation of the traffic being carried. The core routers are shielded from this detail.
- *Failover mechanisms.* IP-based networks have a reputation for a slow response to events such as the failure of a transmission link. In contrast, MPLS fast reroute gives failover times comparable to SONET/SDH transmission networks.
- *Ability to support multiple services on common equipment.* An MPLS PE router can offer multiple services and can encapsulate multiple protocol types into MPLS, including IP packets, ATM cells and Ethernet frames. The variety of media types supported gives the service provider flexibility in the way customers are connected to the service provider, including over an ATM or Frame Relay access network, over Ethernet, over DSL or over SDH/SONET.

MPLS has helped change the reputation of router-based networks from being regarded as only suited for 'best-effort' service to being considered capable of carrying critical traffic. Other factors unrelated to MPLS have also contributed to this improved reputation. These include hardware-based forwarding, which allows better control over latency and jitter, as well as increasing the forwarding rate that the equipment is capable of handling. Also, vendors are introducing high availability features, including the following:

- The separation of control and forwarding planes on modern routers allows schemes such as graceful restart, where control processes can restart without interruption to traffic.
- Component redundancy, such as control processor modules and switch fabrics.
- In-service software upgrades, allowing software upgrades while the equipment is still running, thus reducing the reliance on maintenance windows.
- Bidirectional forwarding detection (BFD), allowing forwarding plane failures to be detected in a timely manner.

Within the book, we compared the properties of RSVP-signaled and LDP-signaled LSPs. We observed that at present there are more LDP deployments than RSVP ones, but that operators of converged networks are likely to deploy RSVP for its bandwidth guarantees and traffic protection advantages. We described new advances in RSVP-TE that allow point-to-multipoint LSPs to be created. As a result, a single control protocol, RSVP, can be used to create transport tunnels in the service provider network to cater for both unicast and multicast traffic.

A key advantage of building a converged network is capital expenditure (CAPEX) savings, as fewer pieces of equipment are required to support the range of services in the service provider's portfolio. As well as CAPEX savings, operational expenditure (OPEX) savings can be made because there are fewer networks to maintain and manage. However, even when having a single network to run, in order to fully realize OPEX savings, it is important to make the network simple to run, e.g. by having common signaling infrastructure and common operational procedures for each type of service being offered. Otherwise the convergence is incomplete, only taking place at the physical equipment level and packet forwarding level rather than also encompassing the control plane.

In this book, we have discussed the reasons why BGP was chosen as the signaling and autodiscovery mechanism for Layer 3 VPNs.

We also showed how the BGP-based mechanisms and associated operational procedures have been carried through to a BGP-based signaling and autodiscovery scheme for Layer 2 and VPLS services. Furthermore, we have discussed current proposals in the IETF to allow BGP to be used as a signaling mechanism within the core of the network for Layer 3 VPN multicast traffic. Thus BGP can be used to carry reachability information for all the MPLS-based services that exist today. Table 13.1 shows the services and the reachability information conveyed by BGP for each.

From the operational point of view, having a single protocol and a shared signaling infrastructure (comprising the BGP sessions and the route reflectors) to carry reachability information for all these services is a key advantage. The flexibility of the BGP protocol means that it can be used to support future services other than those listed in the table without having to compromise the fundamental protocol semantics. This is achieved by simply defining a new address family to carry the requisite reachability information. One example is a current proposal to use BGP to carry ATM addresses, e.g. Network Service Access Point (NSAP) addresses [ATM-BGP]. Another example is the use of BGP to carry CLNS reachability information.

BGP helps in the control plane scaling of networks as they grow, the use of route reflectors meaning that a full mesh of sessions is not required between PEs in order to convey reachability information. The autodiscovery properties of BGP mean that adding a new customer site to an existing service involves having only to configure the PE(s) directly attached to that site, rather than having to configure every PE in the network that serves the customer in question. Network operators take this property for granted when it comes to Layer 3 VPN and would find it unacceptable if they had to configure every PE in the network simply to accommodate a new customer

Table 13.1 Reachability information carried by BGP

Service type	Reachability information carried by BGP
Unicast L3 VPN	VPN-IP prefixes
Multicast L3 VPN	Multicast receivers
L2 VPN	CE IDs
VPLS	VE IDs

site. BGP gives the same ease of configuration to L2 VPN and VPLS services. This property will become increasingly necessary as networks grow and services begin to span multiple service providers.

The model for converged networks using the ingredients discussed in this chapter and throughout this book is summarized in Figure 13.1. The packet transport layer at the bottom of the diagram is likely to be based on DiffServ Aware TE, in order to provide transport tunnels between PE routers and to perform per-class admission control and give bandwidth guarantees to those classes of traffic that need it. These transport tunnels can be a mixture of point-to-point and point-to-multipoint LSPs, to cater for different traffic types. The signaling and autodiscovery function for all the MPLS-based applications is provided by BGP. The MPLS-based applications shown are Layer 3 VPN, Layer 2 VPN and VPLS. The box with the '?' symbol represents future MPLS-based applications, to convey the fact that these are also likely to be underpinned by the same BGP and DiffServ Aware TE infrastructure.

Shown in the diagram are various traffic types that map on to the MPLS-based applications. Layer 3 VPNs, as well as being used for explicit L3 VPN service to end customers, can be used as an infrastructure tool to carry PSTN traffic and Internet traffic. L2 VPNs can be used to carry emulated ATM and Frame Relay services and point-to-point Ethernet services, while multipoint Ethernet services are provided by VPLS.

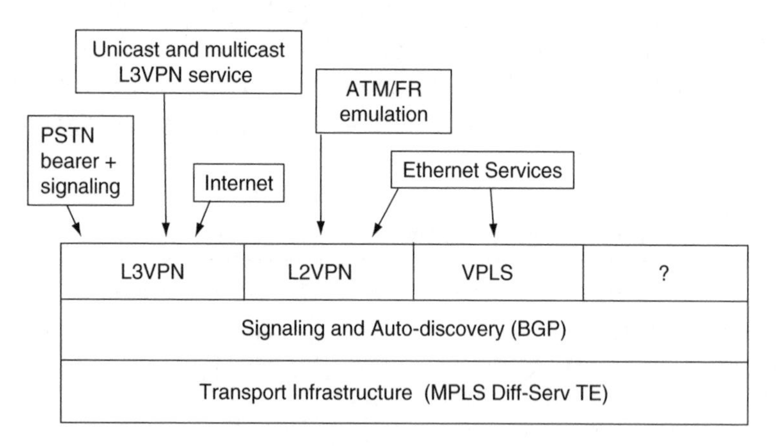

Figure 13.1 Model for converged network

13.3 INTERACTION WITH CLIENT EDGE EQUIPMENT

In earlier chapters, we discussed the admission control of LSPs into the network. It is anticipated that a further level of admission control will be required for some traffic types: admission control of client connections into LSPs by the ingress PE. As discussed in the DiffServ TE chapter of this book (Chapter 4), some implementations already have the ability to manually configure parameters such as the bandwidth requirement of a pseudowire on the ingress PE. The ingress PE then performs admission control of the pseudowire on to an LSP that goes to the appropriate egress PE. The natural next step is for the client edge equipment to signal its bandwidth and other connection requirements to the ingress PE, to avoid manual configuration on the PE. Such a scheme is currently under development in the MPLS and Frame Relay Alliance, in the context of signaling interworking between ATM switches and MPLS PE routers [MPLS ALL]. In this scheme, the ATM switch aggregates multiple VCs and VPs that need to pass to the same remote ATM switch into a single connection. The ATM switch signals to its local PE router the bandwidth requirements for that connection, which the PE translates into bandwidth requirements for a corresponding pseudowire to the egress PE. This is illustrated in Figure 13.2, in which an ATM switch, SW1, requires a trunk to SW3.

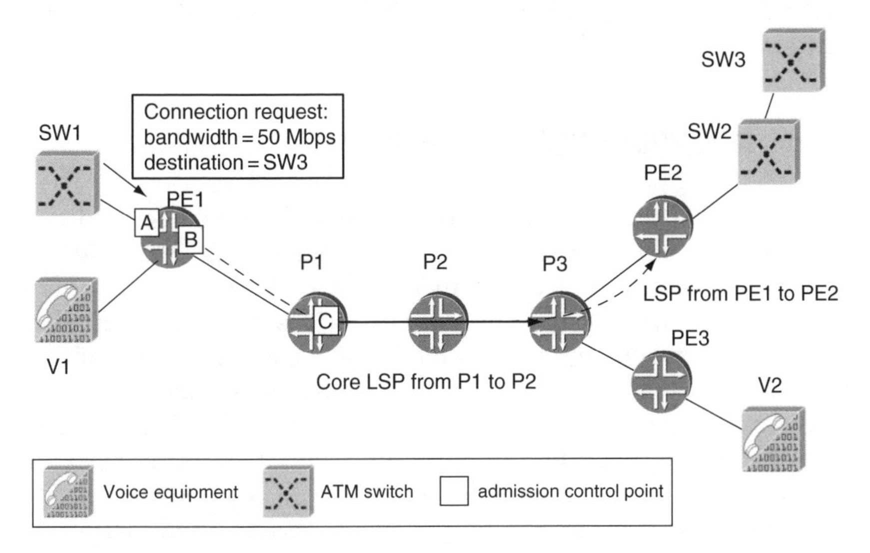

Figure 13.2 Resource reservation by client equipment

It signals the bandwidth requirements of the connection to the attached PE router, PE1. PE1 performs admission control of a pseudowire on to the LSP PE2, and the connection is admitted if sufficient resources exist.

In this way, in the network there is a hierarchy of admission control. The admission control points are shown in the diagram. At point A, admission control is performed of client circuits on to LSPs. At point B, admission control of edge-to-edge LSPs from PE1 into the network is performed. In turn, LSPs from the edge may be nested inside other LSPs, in which case additional admission control occurs, e.g. at point C. This hierarchy of admission control improves scalability. PE1 is aware of the connection from SW1, but is not aware of the individual VCs and VPs contained within. P1 is aware of the LSP from PE1 to PE2, but is not aware of the pseudowires or other services carried within. P2 is only aware of the core LSP from P1 to P3, and is not aware of any PE–PE LSPs carried within.

An interesting question is how does PE1 know that SW3 is in the ATM island 'behind' PE2? One proposal, already mentioned in passing in this chapter, is for BGP to carry ATM address reachability information between the PE routers [ATM-BGP].

The principle of client equipment signaling bandwidth requirements to the MPLS network could be applied to cases other than ATM. For example, a possibility is for voice equipment (e.g. V1 and V2 in the figure) to signal bandwidth requirements for voice bearer traffic to their local PE.

13.4 INTERPROVIDER CAPABILITY

Today the MPLS-based services described in this book are predominantly used in a single provider, single customer mode. That is to say, typically when data from a customer arrives on the service provider's network, it is carried across that network and is delivered to another site of that same customer. Usually the only common denominator that allows traffic to pass between different end customers and across multiple service providers is the Internet. This is a problem because the Internet does not have any guarantees with respect to bandwidth or treatment of packets at each hop. Also roaming corporate workers usually only have access to their corporate network via a VPN solution that runs over the Internet, which can result in poor performance for certain applications. Because many large corporations have presence in all corners of the world and increasingly need connectivity between sites to run their business applications, seamless global connectivity with quality guarantees is an important requirement.

However, no single service provider has the combination of global coverage and high penetration within each country required to offer such customers seamless global MPLS-based services. In some cases, certain service providers do have interconnection arrangements for MPLS-based services. To date, such arrangements are very fragmented, typically involving isolated pairs of service providers. As yet, there is no general interprovider connectivity where traffic can pass through a chain of service providers with a similar SLA to that experienced in the single provider case. Ideally, one would wish to arrive at a position analogous to the PSTN, where it is taken for granted that one can dial any number in the world regardless of which service provider the dialed number is attached to or which other service providers the call needs to pass through.

At the time of writing, work is in progress towards creating such a generic interprovider connection scheme [INFRA]. The scope of the work required extends beyond the MPLS layer to aspects such as:

- Negotiation of session parameters such as QoS, bandwidth and security requirements between the end customer and the carrier, or between carriers.
- Billing to the end customer and intercarrier settlement payments.

The work is still in the early stages, but aims to solve significant constraints in the way data networks are used today. As discussed earlier in this book, work is also being carried out in the IETF to enable better interprovider connectivity at the MPLS layer, e.g. the proposals for RSVP to signal interdomain pseudowires and the work on interdomain traffic engineering.

13.5 MPLS IN THE ACCESS NETWORK

To date, MPLS has been used in the core part of a service provider network, so the scope of the MPLS domain is inwards from the PE routers. There is also interest in using MPLS as an aggregation technology in access networks. The driving force behind this is network consolidation – having one type of access technology allows more efficient use of fiber resources and means that a smaller quantity and fewer different types of network devices are required. Typically today a service provider might operate several access networks in parallel. SONET/SDH networks are used to aggregate and transport customers' access circuits, such as E1/T1 or E3/DS3, to the service provider's PoP. At the same time, ATM aggregation is also used, either to provide access using ATM tails to enterprise customers or for aggregation of broadband traffic. Also, in some regions Ethernet-based access circuits are provided. In the MPLS access model, the service delivery would still be at the PE routers, but the access tails would all be MPLS-based, with a single transport technology underneath (e.g. Ethernet or SONET/SDH). IP packets and Layer 2 traffic would be encapsulated in MPLS to carry them from the customers' premises to the service end-points on the PE routers. Such a scheme gives more flexibility in the use of bandwidth compared to TDM circuits, which are more quantized, and gives the opportunity to take advantage of statistical multiplexing for certain classes of traffic.

13.6 MPLS IN THE ENTERPRISE

The focus of this book has been the use of MPLS by service providers to offer services to enterprise customers, as this is the main way in which MPLS is used today. However, an emerging trend is for larger companies to use MPLS as part of their own internal network infrastructure. Such an enterprise treats its network as a mini version of a service provider network, providing services to the various departments within the company. In some cases, there is a requirement for data separation between certain areas of the company. For example, some financial institutions have sensitive areas that should not be accessible from the company in general, or may need to maintain internal walls between different business units or subsidiaries for reasons of client confidentiality or avoiding conflicts of interest. A typical model is to use Layer 3 VPNs, to constrain the connectivity between different departments, while still having the ability to have shared resources accessible from all the departments that require it. This model is illustrated schematically in Figure 13.3.

While the actual routers used may be smaller in capacity than those used in service provider networks, the principles remain the

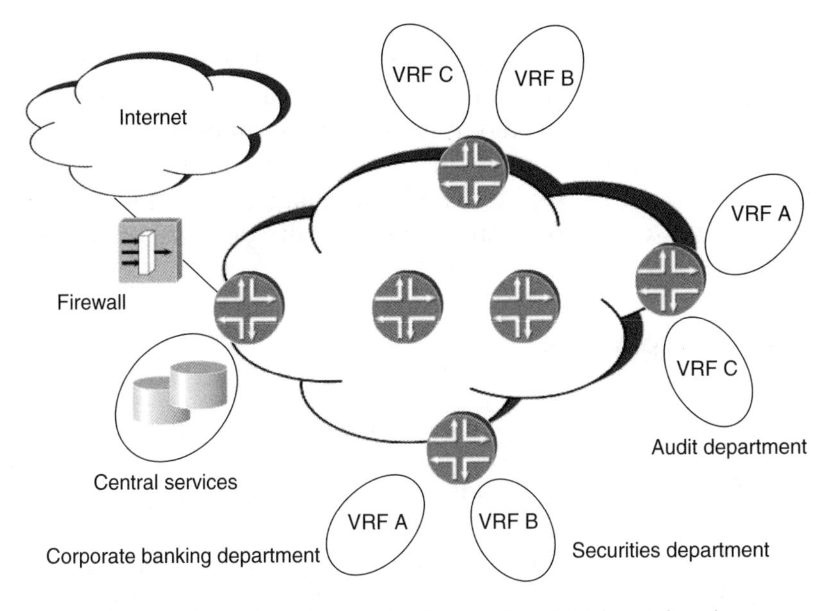

Figure 13.3 Use of Layer 3 VPN in an enterprise network

same, with P routers that perform MPLS forwarding without awareness of the routes carried within each VPN and PE routers with VRFs corresponding to the various departments that the network serves. For multinational companies having a presence in all continents, the geographical span of such networks can sometimes be greater than that of some service provider networks and attention is often paid to optimizing the path of intercontinental traffic to avoid excessive latency, either by juggling IGP metrics or through MPLS traffic engineering.

13.7 FINAL REMARKS

As can be seen from the examples given in this book, the scope of MPLS and the way it is being used has extended beyond what even the more optimistic proponents might have predicted when the work started only a few years ago. Indeed, the use of MPLS in the enterprise discussed in the previous section is an example of MPLS technology being used in an unexpected way. Other such examples include the use of MPLS to provide VPLS. This shows that the question is not whether the technology was originally intended for a particular purpose but whether it can fulfil a particular purpose efficiently. The scaling properties and extensibility of a solution is what determines its ultimate success and deployment. In this final chapter we have attempted to give a flavor of the directions MPLS may go in the future. Today it is mainly a technology used in the service provider core to provide a particular subset of data services. In future, it is likely to take center-stage as a technology that spans multiple providers and underpins the delivery of all voice and data services.

13.8 REFERENCES

[ATM-BGP] C. Kodeboyina, C. Metz and P. Busschbach, 'Carrying ATM reachability information in BGP', draft-ck-bgp-atm-nlri-01.txt (work in progress)

[INFRA] http://www.infranet.org/learn/white_papers/

[MPLS ALL] T. Walsh and R. Cherukuri, 'Two reference models for MPLS control plane interworking', MPLS/FR Alliance Technical Committee document mpls2005.050.00, March 2005

[VPN-MCAST] E. Rosen and R. Aggarwal, 'Multicast in MPLS/BGP IP VPNs', draft-ietf-l3vpn-2547bis-mcast-00.txt (work in progress)

Acronyms

AAL	ATM Adaptation Layer
ABR	Area Border Router. A router used to connect two OSPF areas
AC	Attachment Circuit. In the context of Layer 2 VPNs, the physical or logical circuit used to connect a CE to a PE
AF	Assured Forwarding DiffServ class
AFI	Address Family Identifier. In BGP, the identity of the network layer protocol associated with the network layer reachability information being advertised
AIS	Alarm Indication Signal. In SONET/SDH networks and ATM networks, a means of signaling in the downstream direction the existence of a fault
APS	Automatic Protection Switching. A method for providing protection at the SONET/SDH layer by moving the traffic to a standby link
ARP	Address Resolution Protocol
AS	Autonomous System. A collection of routers belonging to the same administrative entity and having a common external routing policy
ASBR	Autonomous System Border Router. A router used to connect two ASs
ATM	Asynchronous Transfer Mode
ATM PVC	ATM Permanent Virtual Channel
BC	Bandwidth Constraint. In Diff-Serv Aware Traffic Engineering, BCs determine the bandwidth availability on a link for a Class Type or group of Class Types

MPLS-Enabled Applications: Emerging Developments and New Technologies Ina Minei and Julian Lucek

BE	Best Effort DiffServ class
BECN	Backward Explicit Congestion Notification. In Frame Relay networks, a message sent towards the transmission source indicating the existence of congestion in the network
BFD	Bidirectional Forwarding Detection. A protocol to detect faults in the bidirectional path between two forwarding engines
BGP	Border Gateway Protocol. An interautonomous system routing protocol. The current version of BGP is BGP-4, described in RFC 1771
CAC	Call Admission Control
CAPEX	CAPital EXpenditure
CBR	Constant Bit Rate. An ATM service category having a constant maximum bandwidth allocation. Often used for real-time applications
CCC	Circuit Cross Connect. A scheme for the transport of Layer 2 frames over an MPLS network
CE	Customer Edge (usually designates equipment at the edge of the customer's network)
CIR	Committed Information Rate. In Frame Relay networks, the bandwidth associated with a logical connection.
CLI	Command Line Interface
CLNS	ConnectionLess Network Service. A service defined by the Open Systems Interconnect (OSI) that does not require the existence of a connection in order to send data
CLP	Cell Loss Priority. A bit in the ATM cell header that indicates whether the cell is a candidate for being dropped in the presence of congestion
CoC	Carrier of Carriers. In the context of BGP/MPLS L3 VPN, a carrier providing VPN transit to a customer who is himself a carrier
CoS	Class of Service
CPE	Customer Premise Equipment
CPU	Central Processing Unit
CR-LDP	Constrained-based Routing LDP
CsC	Carrier's Carrier – see CoC
CSPF	Constrained Shortest Path First. In traffic engineering, the algorithm used to compute the paths of MPLS LSPs

CSV	Circuit Status Vector. In BGP-signaled L2 VPNs, a means for a PE to communicate to remote PEs the state of its connectivity
CT	Class Type. In Differentiated Services Aware Traffic Engineering, a set of classes that have a common aggregate bandwidth requirement of the network
CV	Connection Verification
DE	Discard Eligible. A bit in the Frame Relay header that indicates whether the cell is a candidate for being dropped in the presence of congestion
DiffServ	Differentiated Services
DiffServ-TE	Differentiated Services Aware Traffic Engineering
DLCI	Data Link Connection Identifier. In Frame Relay networks, the means by which a logical circuit is identified.
DoS	Denial of Service
DSCP	DiffServ Code Point. A 6-bit field in the IP packet header that determines the class-of-service treatment received by the packet
DSL	Digital Subscriber Line
EBGP or eBGP	External Border Gateway Protocol
ECMP	Equal Cost Multi-Path
EF	Expedited Forwarding DiffServ class
EIGRP	Enhanced Interior Gateway Routing Protocol
E-LSP	EXP-inferred LSP (LSP for which the DiffServ behavior is inferred from the EXP bits in the MPLS header)
ERO	Explicit Route Object (used in RSVP-TE to encode path information)
EXP	Experimental bits in the MPLS header
FA	Forwarding Adjacency
FA LSP	Forwarding Adjacency LSP, used in LSP hierarchy as a container for other LSPs
FCS	Frame Check Sequence. A set of bits added to a frame in order to detect errors in the frame
FEC	Forwarding Equivalence Class. Packets that are to be forwarded to the same egress point in the network along the same path and with the same forwarding treatment along that path are said to belong to the same FEC

FECN	Forward Explicit Congestion Notification. In Frame Relay networks, a message sent towards the receiver indicating the existence of congestion in the network
FR	Frame Relay
FRR	Fast ReRoute. The process of quickly routing traffic around the point of failure
FTP	File Transfer Protocol
GFP	Generic Framing Procedure. A mechanism to encapsulate packets into SONET/SDH frames
GRE	Generic Routing Encapsulation. A protocol for encapsulation of an arbitrary network layer protocol over another arbitrary network layer protocol
HDLC	High-level Data Link Control
H-VPLS	Hierarchical VPLS
IBGP or iBGP	Internal Border Gateway Protocol
ICMP	Internet Control Message Protocol
IETF	Internet Engineering Task Force: www.ietf.org
IGMP	Internet Group Management Protocol. A protocol to enable the host to join or leave a multicast group. Described in RFC3376
IGP	Interior Gateway Protocol
IP	Internet Protocol
IPsec	IP security
IPX	Internetwork Packet eXchange. The network layer protocol in the NetWare operating system
IS-IS	Intermediate System-to-Intermediate System. A link-state IGP described in RFC1195
ISO	International Organization for Standardization
ISP	Internet Service Provider
ITU-T	International Telecommunications Union – Telecommunications
LAN	Local Area Network
LDP	Label Distribution Protocol. LDP is documented in RFC3036
LER	Label Edge Router
L-LSP	Label-inferred LSP. An LSP for which the DiffServ behavior is inferred from the label in the MPLS header
LMI	Local Management Interface. A set of enhancements to the basic Frame Relay specification

LOM	Local Overbooking Multiplier. In the context of DiffServ-TE, it is a factor by which the bandwidth for one particular CT is overbooked
LSA	Link State Advertisement. The advertisement sent by a link-state IGP such as OSPF or IS-IS, containing information about the state of the links
LSP	Label Switched Path
LSPing	LSP ping. A mechanism for detecting MPLS data plane failures, based on similar concepts as ping
LSR	Label-Switching Router. A router that can forward packets based on the value of a label attached to the packet
MAC address	Media Access Control address. A unique 48-bit identifier that represents the physical address of a device
MAM	Maximum Allocation Model. A bandwidth constraint model for DiffServ-TE. The model enforces strict separation between the bandwidth allocated to the different CTs
Mbps	Mega bits per second
MD5	Message digest 5. The MD5 algorithm is documented in RFC1321. Its purpose is to take as input a message of arbitrary length and produce as output a 128-bit fingerprint (signature)
MDT	Multicast Distribution Tree. In the context of VPN multicast, these are the multicast trees in the provider network that provide connectivity to all the PE servicing sites of a multicast-enabled VPN. Conceptually, the MDT creates the abstraction of a LAN to which all the PEs belonging to a particular VPN are attached. This property is very important for the C-instance PIM sessions between the PEs, which can consider each other as directly connected neighbors over this LAN
MIB	Management Information Base. A formal description of a set of objects that can be managed using SNMP
MP	Merge Point. In the context of MPLS FRR, it is the tail end of the backup tunnel and the point where traffic from the backup merges back into the protected LSP

MP2MP	MultiPoint to MultiPoint. An LSP is MP2MP if it has multiple ingress and egress points
MP-BGP	BGP with multi-protocol extensions, as described in RFC 2858, that allow BGP to carry routing information for multiple network layer protocols
MPLS	MultiProtocol Label Switching. A set of IETF standards to allow traffic to be forwarded based on labels rather than destination addresses
MPLS-TE	MPLS Traffic Engineering. The traffic engineering capabilities of MPLS, implemented through a combination of source-based routing and constrained-based routing.
MTU	Maximum Transmission Unit. The largest physical packet size (measured in octets) that can be sent in a packet or frame-based network
NLRI	Network Layer Reachability Information. In BGP terminology, a route prefix is referred to as NLRI. Different AFI/SAFI pairs are considered to be different NLRI types
NSAP	Network Service Access Point. Type of addressing used by ISO network layer protocols
OAM	Operations And Management, or Operations, Administration and Management. A set of network management functions covering fault detection, performance data and diagnosis capabilities
OPEX	OPerational EXpenditure
ORF	Outbound Route Filtering. A method for minimizing the number of BGP advertisements between two peers. The main difference between ORF and RTF is in the scope of the filtering: ORF operates between two peers while RTF can propagate filtering information across multiple hops
OSPF	Open Shortest Path First link-state IGP. OSPFv2 (version 2) is documented in RFC2328
P device	Provider device. Designates a router in the core of a provider's network
P2P	Point to Point. An LSP is P2P if it has exactly one ingress and one egress point
P2MP	Point to MultiPoint. An LSP is P2MP if it has one ingress and multiple egress points

PABX	Private Automatic Branch eXchange. A telephone switch used inside a corporation. It connects internal extensions with each other and provides access (by dialing an access number) to the public telephone network
PBX	Private Branch eXchange. Same as PABX.
PCC	Path Computation Client. A client of a PCE. The PCC may be either a router or another PCE
PCE	Path Computation Element. A network element that can compute TE LSPs for which it is not the head end. For example, an ABR or ASBR can play the role of a PCE. The PCE may also be an independent device in the network.
PDH	Plesiochronous Digital Hierarchy
PDU	Protocol Data Unit
PE device	Provider Edge device. Designates equipment at the edge of the provider's network, providing aggregation of the different CE devices
PHB	Per-Hop Behavior. In the context of DiffServ, defines the packet scheduling, queuing, policing or shaping behavior on a particular node
PHP	Penultimate Hop Popping. The act of removing the MPLS label one hop before the LSP egress
PIM	Protocol Independent Multicast. Defined in RFC2362, RFC3973 and in several documents in the pim Working Group in the IETF
PLR	Point of Local Repair. In the context of MPLS FRR, it is the head end of the backup tunnel and the point at which traffic from the protected LSP is locally rerouted around the failed resource using the backup tunnel
PoP	Point of Presence. Physical location at which a carrier establishes itself for obtaining local access and transport
PPP	Point-to-Point Protocol
PP VPN	Provider-provisioned VPN. VPNs for which the service provider (SP) participates in the management and provisioning of the VPNs
PSTN	Public Switched Telephone Network
PVC	Permanent Virtual Channel

PWE	Pseudowire. A method of emulating a Layer 2 service (such as FR or ATM) over an MPLS backbone by encapsulating the Layer 2 information and then transmitting it over the MPLS backbone
QoS	Quality of Service. A measure of performance that reflects both the quality of the service and its availability
RD	Route distinguisher. In the context of BGP/MPLS L3 VPNs, an 8-byte string that is concatenated to the VPN-IP prefixes, for the purpose of making them unique before advertising them over the common provider core
RDM	Russian Dolls Model. A bandwidth constraint model for DiffServ-TE. The model allows sharing of a bandwidth across different CTs
RFC	Request For Comments. A type of IETF document. An overview of the IETF process can be found in RFC1718
RIPv2	Routing information protocol version 2, described in RFC2453
RP	Rendezvous point. In the context of PIM-SM, a meeting point for multicast sources and receivers
RR	Route reflector. In the context of BGP, a route reflector acts as a focal point for iBGP sessions, eliminating the need for a full mesh of sessions. Instead of peering with each other in a full mesh, routers peer with just the reflector
RRO	Record Route Object. Object used in RSVP-TE to track the path along which traffic is forwarded
RSVP	Resource reSerVation Protocol. The base specification of the protocol is in RFC2205
RSVP-TE	RSVP with traffic engineering extensions. The RSVP extensions for setting up LSPs are defined in RFC3209
RT	Route Target. In the context of BGP/MPLS L3 VPN, the route target is an extended BGP community, which is attached to a VPN route. The RT is what accomplishes the constrained route distribution between PEs that ends up defining the connectivity available between the VPN sites
RTF	Route Target Filtering. A method for constraining VPN route distribution to only those PEs interested

	in the RT with which the route is tagged. The method relies on each PE advertising the RTs for which it is interested in receiving updates and can achieve significant savings in the number of advertisements sent and received
SAFI	Subsequent Address Family Identifier. In combination with an AFI it defines an NLRI type
SDH	Synchronous Digital Hierarchy
SE	Shared Explicit. A reservation style used by RSVP that allows an LSP to share resources with itself
SLA	Service-Level Agreement
SONET	Synchronous Optical NETwork
SNA	Systems Network Architecture. A set of network protocols originally designed to support mainframe computers
SNMP	Simple Network Management Protocol
SP	Service Provider
SPF	Shortest Path First. The shortest path computation performed by the IGPs
SRLG	Shared Risk Link Group. A group of links that is affected by the same single event
STP	Spanning Tree Protocol
TCP	Transmission Control Protocol. Reliable transport protocol used in IP
TDM	Time Division Multiplexing
TE	Traffic Engineering. The ability to steer traffic on to desired paths in the network
TED	Traffic Engineering Database. Database created from the traffic engineering information distributed by the IGPs
TE LSP segment	In the context of setting up an interdomain TE LSP using the stitching method, these are the smaller LSPs that get stitched together
TLV	Type-Length-Value. Type of encoding of information in protocol messages
ToS	Type of Service. A field in the IP header designed to carry information that would allow deployment of QoS
TTL	Time To Live

UDP	User Datagram Protocol. Unreliable transport protocol used in IP
VC	Virtual Circuit
VCI	Virtual Channel Identifier
VCCV	Virtual Circuit Connection Verification. The connection verification protocol for pseudowires set up using LDP
VE ID	VPLS Edge Identifier. In BGP-signaled VPLS, a means of uniquely identifying a site within a VPLS
VLAN	Virtual LAN
VoIP	Voice over IP
VP	Virtual Path
VPI	Virtual Path Identifier
VPLS	Virtual Private LAN Service. A scheme in which a service provider's customer site appear to be attached to the same LAN
VPN	Virtual Private Network. A private network realized over a shared infrastructure.
VRF	VPN Routing and Forwarding. The per-VPN routing and forwarding tables that ensure isolation between different VPNs
WAN	Wide Area Network

Index